Distillation Symposium 2016

Topical Conference at the 2016 AIChE Spring Meeting and 12th Global Congress on Process Safety

Houston, Texas, USA
10-14 April 2016

ISBN: 978-1-5108-2482-9

Distillation Symposium 2016

Topical Conference at the 2016 AIChE Spring Meeting and 12th Global Congress on Process Safety

Houston, Texas, USA
10-14 April 2016

AIChE
The Global Home of Chemical Engineers

Printed from e-media with permission by:

Curran Associates, Inc.
57 Morehouse Lane
Red Hook, NY 12571

Printed by Curran Associates, Inc. (2016)

For permission requests, please contact AIChE
at the address below.

AIChE
120 Wall Street, FL 23
New York, NY 10005-4020

Phone: (800) 242-4363
Fax: (203) 775-5177

www.aiche.org

Additional copies of this publication are available from:

Curran Associates, Inc.
57 Morehouse Lane
Red Hook, NY 12571 USA
Phone: 845-758-0400
Fax: 845-758-2633
Email: curran@proceedings.com
Web: www.proceedings.com

TABLE OF CONTENTS

443436 The Colwell Equation - More Than Just a Clear Liquid Height Model

Abstract:

The Colwell equation was proposed as a means of calculating the equivalent clear liquid height on trays. The Colwell equation can also be used to estimate other variables that are important in the design of trays. In this paper the methodology of using the Colwell model for these calculations will be discussed and it will be shown that this approach yields good agreement with experimental measurements.

443967 cMIST™: Novel, Compact Gas Dehydration System

Abstract:

The natural gas industry is increasingly pursuing compact and lower weight processing technology to meet the technological and economic demands of offshore, onshore, remote, and challenged gas processing. Increased modularity of process equipment and process intensification require innovative solutions. This paper describes ExxonMobil Upstream Research Company's (EMURC) new inline dehydration and separation technology meant to replace conventional TEG tower and associated separator vessels to meet pipeline dewpoint specifications. Compact Mass transfer and Inline Separation Technology (cMIST™) achieves this goal with significant reductions in weight and footprint.

EMURC is qualifying cMIST™ for upstream natural gas dehydration. The results of a field demonstration are discussed to establish the effectiveness and advantages of this novel technology. The technology was demonstrated using a skid-mounted test unit in the Woodford trend area of Oklahoma, USA at an XTO Energy production facility. cMIST™ was tested under a range of field conditions to examine the dehydration capacity, operability, and reliability of the technology.

cMIST™ relies on state-of-the-art technological advances in contacting and inline separation to achieve pipeline specification dry gas and/or prevent hydrate formation with compact equipment. The technology is highly modular to allow for simple transportation into remote, challenged, or offshore environments and provide installation configuration flexibility. One case study showed a 50% weight reduction and 80% footprint reduction for the absorption system on an offshore platform dehydrating 700 MMscfd of gas. During EMURC's field demonstration and qualification program, cMIST™ achieved pipeline specification over a range of process conditions and during sustained operation. cMIST™ was demonstrated to have a broad application in gas dehydration.

What Process Troubleshooters Should Know about Downcomer Gamma Scans

By

Henry Z. Kister, Fluor, Aliso Viejo, CA, USA

and

Chuck Winfield, Quantum Technical Services LLC, Pasadena, TX

Dedicated to the memory of Professor Ian D. Doig, University of NSW, an engineering maestro, teacher, and mentor, whose inspiration, warmth and legacy will remain with us and in chemical engineering forever

Presented at the Distillation Topical Conference, AIChE Spring Meeting, Houston, Texas, April 10-14, 2016

WHAT PROCESS TROUBLESHOOTERS SHOULD KNOW ABOUT DOWNCOMER GAMMA SCANS

Henry Z. Kister, Fluor Corp., Aliso Viejo, CA and Chuck Winfield, Quantum Technical Services LLC, Houston, TX

Gamma scanning distillation and absorption towers is one of the most powerful diagnostic techniques available. Most gamma scans are shot through the active areas. In many cases, invaluable insight to the nature of the problem or bottleneck can be gained by also scanning the downcomers.

Due to the narrow width of downcomers (compared to active areas), scanning downcomers can be trickier than scanning tray active areas. If attention is not paid to the pitfalls, one can end up with a misleading diagnosis. Misinterpretation of downcomer gamma scans is not uncommon.

This paper presents cases where downcomer scanning was central to obtaining a good diagnostic of a tower problem, and other cases where pitfalls veered an investigation in a wrong direction. The paper presents guides of how to avoid such pitfalls, properly interpret downcomer scans and promote correct diagnostics.

WHAT GOES WRONG IN DOWNCOMERS

Downcomers are simply drain pipes. In a tower, they drain liquid from one tray to the tray below. They are analogous to the drain pipe beneath the bathroom sink. Like their bathroom counterparts, downcomers work well when they are not overloaded, plugged, filled with foamy material, or have lots of gas flowing up through them. And just like in the bathroom, the liquid will build up if its drainage is restricted. In the tower, this buildup leads to flooding. Flooding and unsealed downcomers are the most common phenomena that bottleneck downcomers. Both are readily identifiable with judicious gamma scans.

FLOODING

Flooding is accumulation of liquid in the column. This accumulation propagates from the lowest flooded region upward, until the whole column fills with liquid or until an abrupt change in tray design or flow conditions (e.g., feed point) is reached. Flooding may or may not propagate above that point.

In existing towers, the onset of flooding limits the ability of the tower to process additional feed rates, which often bottlenecks the entire plant. New tower diameter is set to sustain operation at a comfortable margin (typically 20%) below the flood point. In tower revamps, the maximum throughput is restricted by the approach to (typically 10%) the flood point.

Flooding is by far the most common capacity limitation in distillation and absorption towers. When a tower floods, tray efficiency diminishes, separation deteriorates, and products go off-spec. Massive liquid entrainment may occur from the top of the tower and reach downstream units, causing contamination and sometimes equipment damage. Tower flooding often sets instability in the tower, with intermittent buildup and dumping of liquid. This instability may be transferred to downstream or thermally coupled units. Liquid accumulation in the tower reduces the bottom flow rate, sometimes starving the bottom pump and causing it to cavitate. To avoid the onset of flooding, operators cut throughput, and the plant loses capacity.

Troubleshooting for flood is mandatory when attempting to diagnose the root cause of poor performance in a tower. There are many other causes that may give poor performance and are not associated with flooding. Low tray efficiency and poor separation may also be caused by inefficient trays, poor distributor design, excessive weeping, or internals damage. Liquid entrainment may be caused by poor design of reflux entry piping or due to reflux flash. Instability may be caused by inadequate control, poor controller tuning, tray dumping, or fluctuating feed rates. Pump cavitation may be caused by undersized suction piping or by air leaking into the pump suction.

In order to reach a good solution to an operating problem or to debottleneck a tower, it is imperative to correctly determine the root cause of the problem or bottleneck. With flooding being such a common root cause, flood troubleshooting is of utmost importance. Further, it is equally important to correctly identify the location of the flood and its nature. We have seen many experiences where tower flooding was misdiagnosed, and as a result the solution or debottleneck fell short of achieving its expectation. When misdiagnosed, the problem persists, even worsens, and engineers lose their credibility, if not their jobs.

Once flood and its cause are adequately identified, corrective actions and a proper solution can be devised. Different types of flood respond to different solutions. The solution usually involves re-optimization (e.g., changing feed preheat, tower pressure, or other operating conditions) or hardware modifications to tower internals or auxiliary equipment. A solution that may alleviate one type of flood may aggravate another. Implementing the wrong solution due to a misdiagnosed flood type can aggravate a problem instead of fixing it. Therefore it is important not only to determine whether flooding is taking place, but also what is the flood mechanism and its root cause.

Flood Mechanisms All floods are characterized by liquid accumulation. There are four different mechanisms that lead to this liquid accumulation in trays [1-4]:

1. Entrainment (Jet) Flood. Froth or spray height on trays rises with vapor velocity. As the froth or spray approaches the tray above, some of the liquid is aspirated into the tray above as entrainment. Upon further increase in gas flow rate, massive entrainment of the froth or spray begins, causing liquid accumulation and flood on the tray above.

2. Downcomer Backup Flood. Aerated liquid backs up in the downcomer because of tray pressure drop, liquid height on the tray, and frictional losses in the downcomer apron. All of these increase with increasing liquid rate. Tray pressure drop also increases as the vapor rate rises. When the backup of aerated liquid exceeds the tray spacing, liquid accumulates on the tray above, causing downcomer backup flooding.

3. Downcomer Choke Flood (also called downcomer entrance flood or downcomer velocity flood). A downcomer must be sufficiently large to transport all of the liquid downflow. Excessive friction losses in the downcomer entrance, and/or excessive flow rate of vapor venting from the downcomer in counter-flow, will impede liquid downflow, initiating liquid accumulation (termed downcomer choke flooding) on the tray above.

4. *System Limit Flood (also called ultimate capacity flood).* This is an ultimate jet flood, and takes place when the vapor momentum force acting to lift the large liquid drops above the tray exceeds the gravity force. This flood is independent of tray geometry and tray spacing.

Judicious downcomer gamma scanning can give a good indication on whether the flood initiates in the active areas or in the downcomers. If in the active area, jet flood or system limit flood are implicated. If in the downcomers, downcomer flood is likely.

When to Look for Downcomer Flood Downcomer backup floods are common when there is plugging on the trays or in the downcomer. As tray holes plug up, the friction pressure drop rises parabolically (with the velocity squared) causing the downcomer backup to rapidly rise. Similarly, downcomer plugging will cause the liquid friction in the downcomer exit to rapidly escalate. Downcomer backup floods are also common at small tray spacing (<18 inches or so), when the tray open areas are small (<7% of the active area), in foaming systems (downcomer liquid highly aerated, causing tall backups), and when maldistribution is experienced in multi-pass trays.

Downcomer choke flood is common in high-pressure distillation (>100-150 psig), in foaming systems, and in high liquid load systems (downcomer inlet velocities > 0.4 ft/s for non-foaming low pressure systems).

Damage and blockage can often lead to either downcomer backup or downcomer choke flood.

DOWNCOMER SEAL LOSS

When a downcomer loses its liquid seal, gas rises up through it, and interferes with liquid descent. This interference has led to capacity bottlenecks, poor separation, instability, and inability to start towers up.

On conventional trays, at liquid flow rates above about 2-3 gpm/in of outlet weir length, the outlet weir generates a frothy pool (Fig.1a). As long as the downcomer clearance is lower than the liquid head in the pool, this liquid pool will seal the downcomer. In contrast, at low liquid flow rates (below about 2-3 gpm/in of outlet weir length), and especially at high gas velocities, the gas atomizes the tray liquid, forming a dispersion known as the spray regime[1-4] or drop regime. In this regime, there is no frothy pool and the liquid resides as atomized drops in the inter-tray space (Fig. 1b). The continuous gas phase reaches the downcomer outlet. There is no static seal preventing gas rise up the downcomer. In the spray regime the downcomers are prone to seal loss.

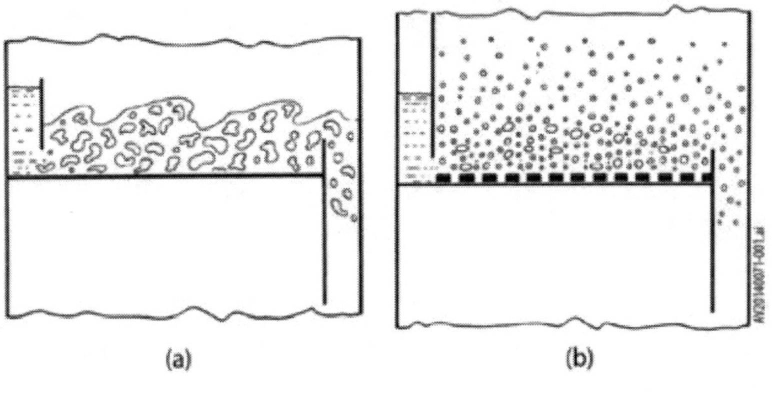

Figure 1: Flow Patterns on trays (a) Froth regime (liquid phase is continuous) (b) Spray regime (gas phase is continuous). (Henry Z. Kister, excerpted by special permission from Chemical Engineering, September 8, 1980; copyright by McGraw-Hill, Inc., New York, NY, 10020).

Fig. 2 illustrates how downcomer unsealing can bottleneck a tower in the spray regime. This is a common scenario at start-ups with low liquid rates. Consider a tray with very little liquid on it. The gas, looking for the easiest path, ascends through both the trays and downcomers (Fig. 2a). Upon liquid introduction, some reaches the downcomer across (Fig. 2b). If the gas velocity through the downcomer is too high, it will not allow the liquid to descend into the downcomer (Fig.2c). Unable to descend, the liquid will accumulate on the tray above. The tray will flood, and liquid will carry over the top of the tower (Fig. 2d).

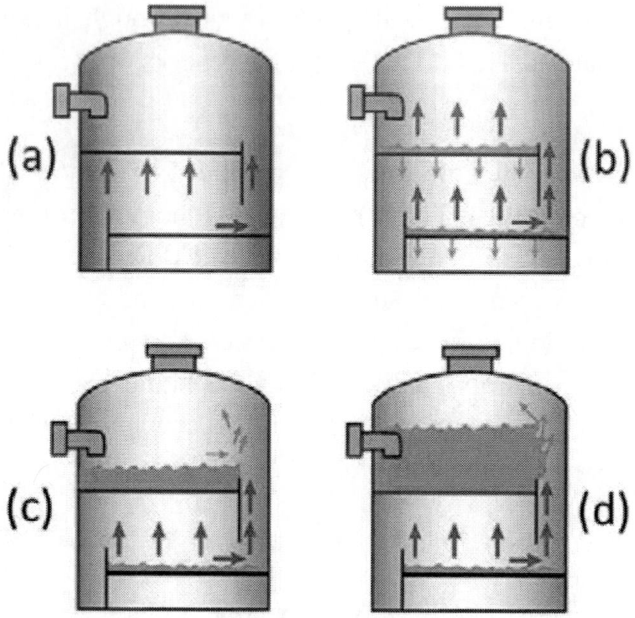

Figure 2: How Downcomer Unsealing Bottlenecks a Tower in the spray regime (a) Gas distributes between tray and downcomer; (b) Liquid introduced, reaches the downcomer, attempts to descend; (c) When gas velocity up the downcomer is too high, the gas does not permit liquid descent; (d) Unable to descend, liquid accumulates on the upper tray, generating flood. (from H.Z. Kister, "Practical Distillation Technology," Course Manual. Copyright © 2013; reprinted with permission.)

Downcomer seal loss is also encountered in many high-capacity trays, in which maximizing capacity leads to downcomers that terminate in the inter-tray vapor space, well above the floor of the tray below. Being above the tray below, there is no positive liquid seal, and vapor can easily break into these downcomers. Downcomer seal loss can also be caused by tray damage or by tray plugging. With a plugged tray, the pressure drop through the active area rises, incurring high resistance to vapor flow, and going up a downcomer becomes the path of lesser resistance. When the tray plugging is extensive, vapor can be induced up a downcomer, especially when there are multiple downcomers.

PERFORM DOWNCOMERS SCAN: YES OR NO?

Adding downcomer scans incurs additional costs. However, the main cost of gamma scanning is the travel time and transport of the scanners and their equipment to the sites, getting through the required safety and work permits, and getting their equipment set up. While the scanners are already on site, the cost of performing additional scans is relatively low. Still, it is money that should be saved if the scans are unnecessary.

Counteracting this, the cost of failure to cure a problem due to misdiagnosis can be tremendous. Typically, we see downcomer scans performed about 10% of the time. In contrast, our experience is that at least half the time downcomer scans can contribute enough to the diagnostics to justify their costs. Downcomer scans are not performed often enough in the industry.

In order to make an educated decision as to whether to proceed with a downcomer scan, consider the following:

1. Review the previous sections of this article and check if any of the above issues are on the troubleshooting list. Pay special attention to those issues that cannot be easily predicted by calculation: e.g. plugging, damage, foaming, seal breaks in truncated downcomers. If any of these are high-possibility items on the troubleshooting list, there is a strong incentive to perform the downcomer scans.

2. Perform hydraulic calculations. These calculations evaluate the proximity of the operating conditions to downcomer bottlenecks. These can be done using proprietary software. For members of Fractionation Research Inc, their software is excellent for this analysis. For non-members, good results can be obtained using the correlations in Perry's Handbook[3], other texts (e.g. 1, 4), vendor software, or the software available on major simulators with appropriate selection of the suitable correlations..

3. Perform judicious tests to check various theories. Reference 5 is a classic example showing how a combination of judicious tests and a hydraulic analysis led to a correct identification of a downcomer bottleneck and a successful fix.

4. If previous active area gamma scans are available, review them closely. Active area gamma scans showing low froth heights and clear vapor spaces indicate that the trays are not approaching a limit, so if there is a limit in the tower at slightly higher rates there is a good chance it will be in a downcomer (or downcomers).

CORRECT AND INCORRECT SCANNING OF DOWNCOMERS

Column gamma scans can be inconclusive, even misleading, unless scanning pitfalls are avoided. The guidelines below can help avoid common pitfalls that have, in the authors' experience, adversely affected the validity and quality of information supplied by downcomer gamma scans:

1. It is important to supply the scanners with a good set of drawings of the column which include accurate information on tray location, orientation and column shell thickness. It is also beneficial to give the scanners a summary of the column's history with an emphasis on the problem or problems it is experiencing. Include specifics such as rates or times at which the problem occurs. This information is essential for good planning and interpretation of the scans.

2. Interpretation of gamma scans may at times be difficult and requires good knowledge of both scanning technology and column operation. In many cases, a gamma scan contractor is brought into the plant, shoots a scan, and then writes a report. The authors have seen many such reports containing interpretations that were way off the mark. To avoid this, it is important to have a person or persons who are familiar with the tower design and operating history communicate closely with the contractor both when the column is being scanned and when the scans are being interpreted.

3. It is important to ensure stable plant operation while scanning. Otherwise variations due to instability may be misinterpreted as column abnormalities. Also, the active area scans and the downcomer scans are shot at different times, and a change in plant conditions may be misinterpreted as an issue with tray or downcomer hydraulics.

4. Avoid shooting gamma scans at extreme weather conditions. These often lead to column instability while slowing down the gamma scans. Both may lead to variations that can be misinterpreted. Due to the narrow width of downcomers, when scanning in windy conditions it is often necessary to use hard-lines or guide wires to make sure the scan path remains constant.

5. When troubleshooting for flood or other capacity limits, always perform a flooded and an unflooded scan. The flooded scan can give a good indication of the location of the flood, but does not give much information on what causes it. The unflooded scan should be shot at loads just below the flood point, and is invaluable in providing information on what is likely and what is unlikely to be the cause.

6. Some commercial scanners prefer to shoot downcomer scans perpendicular or angled to the outlet weir (e.g., Figure 3e) instead of parallel to the weir as shown in Figure 3b. The authors

have had very few satisfactory experiences with downcomer scanning perpendicular or angled to the weirs. Such scans are very difficult, often impossible to correctly interpret due to the coupling between the tray and downcomer dispersions, making them breeding ground for misleading diagnostics. The problem is aggravated when the downcomer is sloped. Further, the gamma rays used in scanning the top of every downcomer often travel through support beams, which may be quite thick; their presence shows up in the same manner as extra liquid. For these reasons, the authors recommend always performing downcomer scans parallel to the weir (Figure 3b or 3c) and not perpendicular or angled to it (Figure 3e).

7. Scanning downcomers parallel to the weir (Figure 3b) also has pitfalls. If the scan chord is shallow, scattering and reflection may lower its accuracy. When the column shell is thick (e.g., high-pressure services), the metal can absorb a larger portion of the radiation than the fluid. Consider a scan along a 12 inch long chord in a downcomer containing liquid of 0.5 specific gravity at 50 percent aeration. If the column shell is 1 inch thick, it will absorb approximately five times more of the radiation than the process fluid. For this reason, in high-pressure columns it is best to scan center rather than side downcomers (Figure 3c). When side downcomer scanning is needed in such situation, it is better to use neutron backscatter, a technique that does not suffer from these limitations (item 8 below).

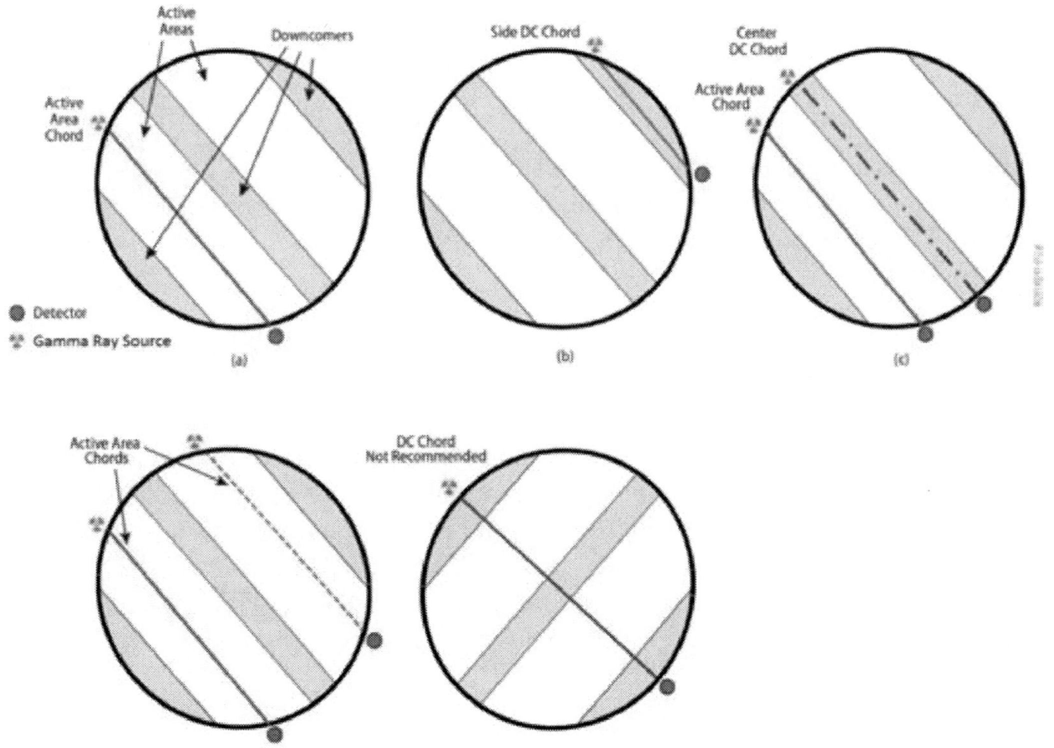

Figure 3: Gamma chords in tray towers, shown for a tower with 2-pass trays (a) Active area scan, (b) Side downcomer scan, (c) Active area and center downcomer scans, (d) Scans through active areas on both sides, looking for channeling, (e) Poor practice of a scan chord that passes both through active areas and downcomers. (from H.Z. Kister, "Practical Distillation Technology," Course Manual. Copyright © 2013; reprinted with permission.)

8. In order to minimize scattered radiation, side downcomer scans should be run at a high threshold setting on the radiation counting equipment. A high threshold setting electronically discriminates low energy or scattered radiation resulting in higher resolution data.

9. Neutron backscatter is invaluable for scanning side downcomers, especially those with shallow chords and thick walls as described in item 7 above.

   Neutrons are high-energy particles capable of penetrating a substantial thickness of metal, but travel only short distances, typically 6 inches, into the vessel. Upon collision with hydrogen nuclei, slow neutrons are reflected back toward the source. This is analogous to rebound of balls on a pool table. The intensity of the reflected neutrons is proportional to the concentration of hydrogen atoms in the fluid adjacent to the source and is measured by a detector. A more detailed description of this technique is provided elsewhere[6,7].

   Neutron backscatter techniques are suitable for measuring froth densities along a downcomer containing hydrocarbons, organics, or water, all of which contain hydrogen atoms. In a liquid, molecules are close together, giving a high neutron reflection rate, while in a vapor the molecules are far apart giving a low reflection rate.

   Neutron backscatter is difficult to apply through wall thicknesses exceeding 1-1/2 inches. Wet or icy insulation can lead to misleading measurement due to significant concentration of hydrogen atoms in the insulation. Neutron backscatter cannot be applied where hydrogen atoms are absent (e.g., carbon tetrachloride).

   Figure 4 shows a typical neutron backscatter device along with an illustrative neutron scan of a vessel. The source and detector are mounted on the same hand-held sweeper. The neutron source and detector are positioned near the wall of the vessel and are moved up or down along the vessel wall.

   The scan shows low reflected neutron intensity from the vapor space near the top of the vessel. As the device is lowered down the outer wall, higher intensity reflected neutrons are seen in the liquid rich region in the mid-section of the vessel. This is due to the higher concentration of hydrogen atoms. The change in hydrogen atom concentration makes this technique well suited for distinguishing a normal operating downcomer (which has mostly clear liquid in the bottom and progressively more vapor as one ascends it) from a flooded downcomer (which is near liquid full). Low reflections throughout the downcomer indicate an unsealed downcomer. Low reflections near the bottom, with high reflections above, would indicate possible plugging with low hydrogen-bearing solids.

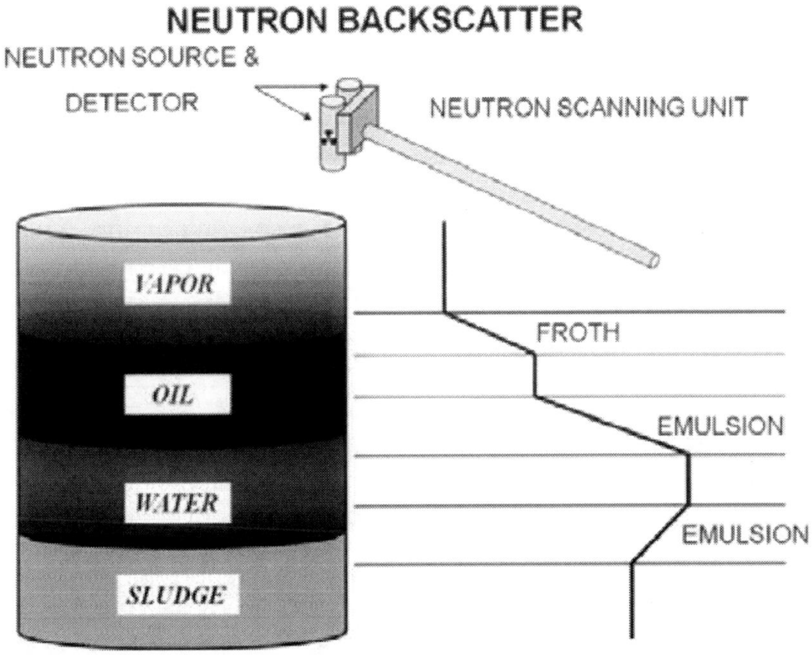

Figure 4: Neutron Backscatter Device

10. It is important to combine the information provided by the scans with information from other troubleshooting techniques such as flood tests, vapor liquid sensitivity tests, and hydraulic analysis. Questions such as "Do the gamma scans show flood where the hydraulic analysis predicts flood?" or "Has there been foaming experience in the service where the gamma scans interpreted the trays to foam?" can go a long way to eliminate "lying gamma scans".

11. It is a usually good idea to begin with a top-to-bottom active area scan (Figure 3a or the red chord in 3c). The scan chord should be placed strategically, at least 6 inches away from the downcomer walls to positively avoid reflections.

12. Some towers have practical constraints that may interfere with downcomer scans. One tower had wind deflectors, which are curved, thick protrusions on the outside wall, designed to reduce the wind bending moment on the tower. These made it difficult to get even a good scan of the active areas, and downcomer scanning was completely out of the question.

13. When there are multiple downcomers, especially when these are truncated (e.g., Figure 5) correct selection of the chords and trays to be scanned is critical. Incorrect choice of scan chords and scanned trays has been, in our experience, a common cause of reaching incorrect or inconclusive diagnostics.

14. The desired scan intervals need to be set. Scanners usually shoot gamma rays every two inches as they descend the tower. This is too large an interval when scanning short downcomers. The scanning can easily be done at smaller intervals (every 1 inch), but this raises the cost.

15. With multiple downcomers, sketch the chords to scale on tower elevation and /or plan diagrams clearly showing the dimensions of each chord and where they are relative to downcomers. Figure 5 shows a scan plan for high-capacity sieve trays with truncated multiple downcomers. On these trays, the downcomers do not extend all the way to the tray below, but terminate about half way down the vapor space, with liquid issuing through holes in the downcomer floors. Successive trays are rotated 90 degrees to each other. As these trays were at close 16 inch tray spacing, a 1 inch scan interval was the only satisfactory choice. To keep down the cost, the strategy was to focus on a selected region, with limited comparison to others. Six scan chords were selected, one of which passed through the downcomers. Reference 8 describes in detail the various active area scans. Since the hydraulics of the trays and downcomers was being examined, they were only scanned at unflooded conditions. Highly stable tower operation was sustained.

    In this case, only about one third of the total number of trays was scanned. Even though the number of chords selected may appear on the lean side, it was sufficient to provide the desired diagnostic that led to solving the operating problem.

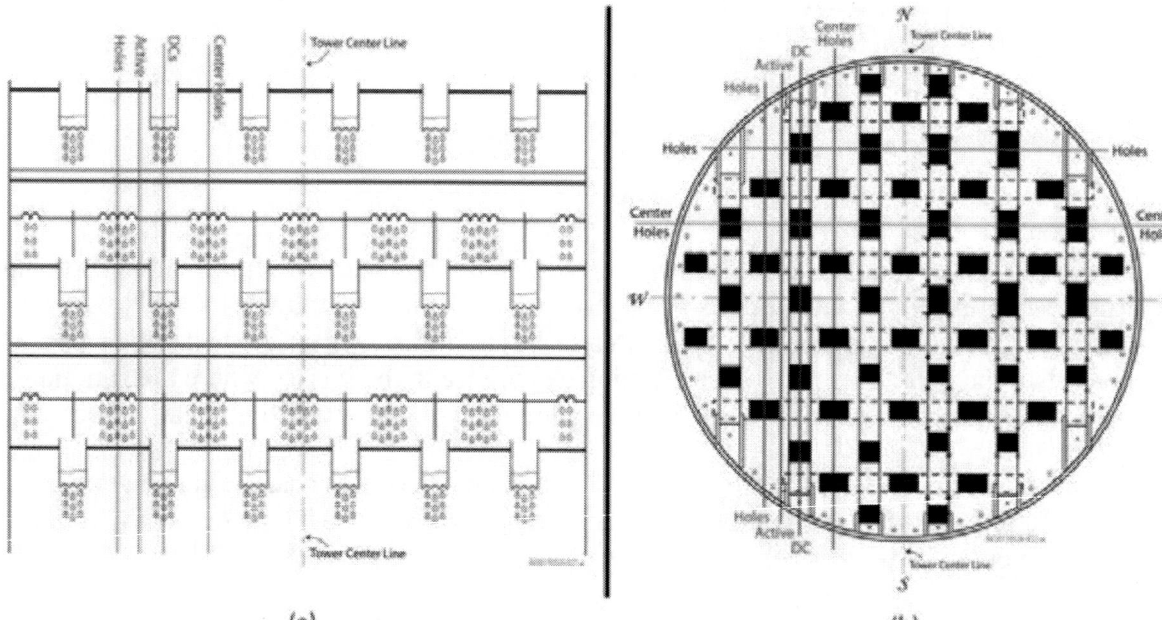

(a) (b)

Figure 5: Scan Lines for high-capacity sieve trays with truncated multiple downcomers. The trays are rotated 90 degrees to each other. (a) elevation, (b) plan. (from H. Z. Kister, Chem. Eng. Progr., April, 2013, reprinted courtesy of The AIChE)

16. Once the desired scan chords are laid out, the scanners should be consulted. Practical difficulties in performing the desired scans are frequent due to restricted access, nozzle or pipe interference, and other issues. The scanners can critically review the plans, foresee difficulties, and advise what it would take to resolve each difficulty. Work with them on a compromise. Sometimes the compromise is moving a chord only a few inches, or skipping a few trays, which will make very little difference to the results. At other times it may be better to switch from one side of the tower to the other. Additional scaffolding or other equipment may be needed, and should be set up ahead of the test. There are also times when the

practical difficulties simply make it impossible to shoot some desired chords, and an alternative plan may need to be devised.

There is much more that the scanners bring to the table. Gamma radiation does not propagate as a perfect collimated beam. It tends to scatter, producing "fuzzy" pictures of the downcomer froth heights. There is much that the scanners do to contribute by good choice of the sources used and the degree of collimation. In most of our investigations, the scanners' reviews made changes to our plans.

17. The investigation leader on the process/operation team should be available throughout the scanning period, ideally on the site. Practical issues always arise during the tests, forcing important on-the-spot decisions. Some chords yield unexpected results, weather or plant conditions may suddenly change, all raising the need for changing plans. Any such decisions should be made jointly between the process/operations team leader and the scan team leader. A poor decision may lead to a need for costly repetition of several measurements or to an impaired diagnosis.

CASE STUDY 1: ESTABLISHING THE FLOOD MECHANISM FOR DEBOTTLENECKING A HIGH PRESSURE STABILIZER

Downcomer gamma scans have been invaluable for identifying the flood mechanism that restricts the capacity of a tower. In the following case, a large gas plant stabilizer equipped with 4-pass valve trays was operating right at its maximum throughput limit, and bottlenecked the capacity of the entire gas plant. Any increase in rates would generate flooding in the bottom section of the stabilizer, as was seen by instability and high pressure drop. With a turnaround approaching, it was desired to debottleneck the stabilizer to raise the capacity of the plant by 25%. The economics was such that any modifications external to the tower (e.g., adding a preheater) were out of the question. Downtime was at a premium, so the revamp was a one shot deal: a failure would mean 5 years or more at the failed conditions, with a huge economic penalty.

There was debate as to what was the flood mechanism, with some calculation methods suggesting jet flood, others downcomer choke. Correct determination of the limiting flood mechanism was central to achieving an effective debottleneck of the tower. It therefore became necessary to positively identify the flood mechanism.

Results of our hydraulic calculation at the rates where flood was observed are shown in Table 1. These were our calculations, and were challenged by others.

Table 1: Hydraulic Calculations			
Tray Number	Bottom	4-6 Trays From Bottom	Just Below Feed
% Jet Flood	52	52	17
% Froth in Downcomer	61	60	41
% Downcomer Choke	98	99	38

To positively determine the flood mechanism, three scan chords were shot: one through the active areas, one through the center downcomers, and one through the off-center downcomers. All scans were shot parallel to the downcomer walls, as recommended earlier in the paper. Side downcomers were not scanned due to concerns about interference from thick tower walls, as described earlier. Each of the downcomer scans was performed twice: a "flooded" scan at the loads where flood was observed to initiate, and an "unflooded" scan, at loads about 3% below that point, with the tower operating at stable conditions. The active area scan was shot only at the unflooded condition.

Figure 6 shows an overlay of the unflooded scans of the center downcomer scan and of the active area scan near the bottom of the tower. The active area froth heights were 10-12" out of 24" tray spacing, and the vapor peaks approached the clear vapor line. This means that the active areas were nowhere near jet flood. When a tray approaches jet flood the spray heights get close to the tray spacing and one cannot see clear vapor. Another interesting observation is that the peaks above the downcomers from trays 2 and 4 are very similar to those above the active areas, which validates the scan quality. The downcomers from trays 2 and 4 were full of froth nearly to the top, and were getting close to a limit. This strongly supports a downcomer flood mechanism.

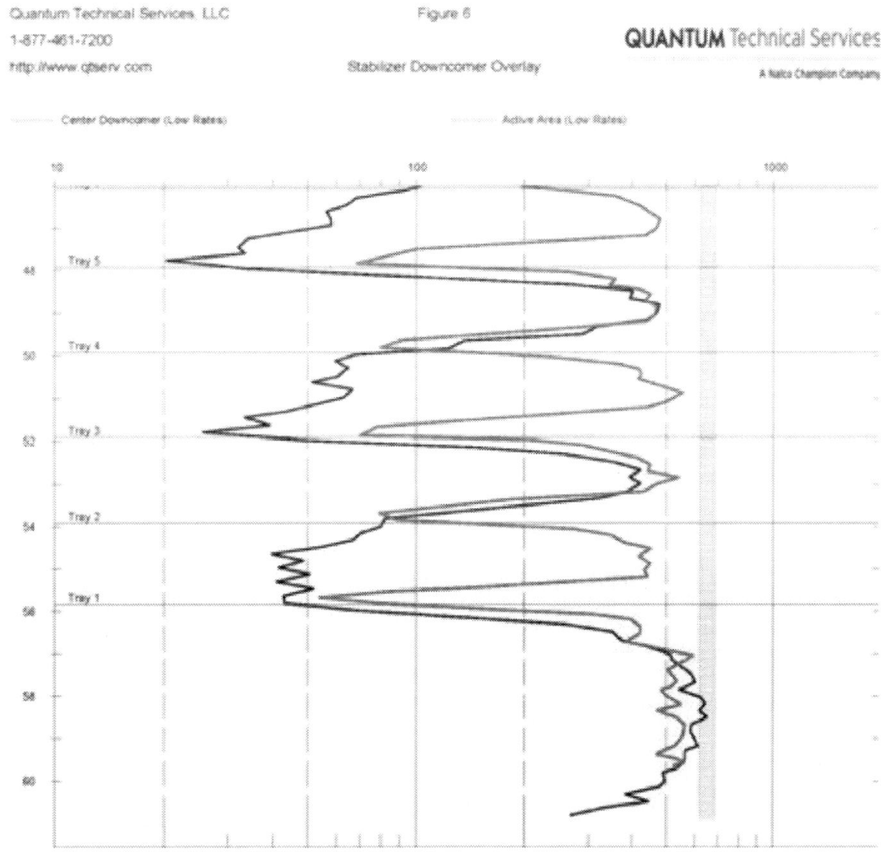

Figure 6: Stabilizer unflooded active area scan and center downcomer scan overlay, with the tower operating at stable conditions at rates 3% below flood (Reprinted courtesy of Quantum Technical Services, LLC)

Figure 7 shows an overlay of the center downcomer scans for the flooded and unflooded conditions. There is a clear change from the unflooded to the flooded scan: in the flooded scan, the downcomers from trays 6 and 8 were full of froth and have begun to flood. The downcomers from trays 12, 14, and 16 look normal, and were not flooded in either scan. It can be seen that in the flooded scan, the active areas above the downcomers from trays 6 and 8 were beginning to fill up with froth, and the peaks above these trays no longer approached the clear vapor line. This is all a clear indication of flood initiating in the downcomers and propagating into the active areas.

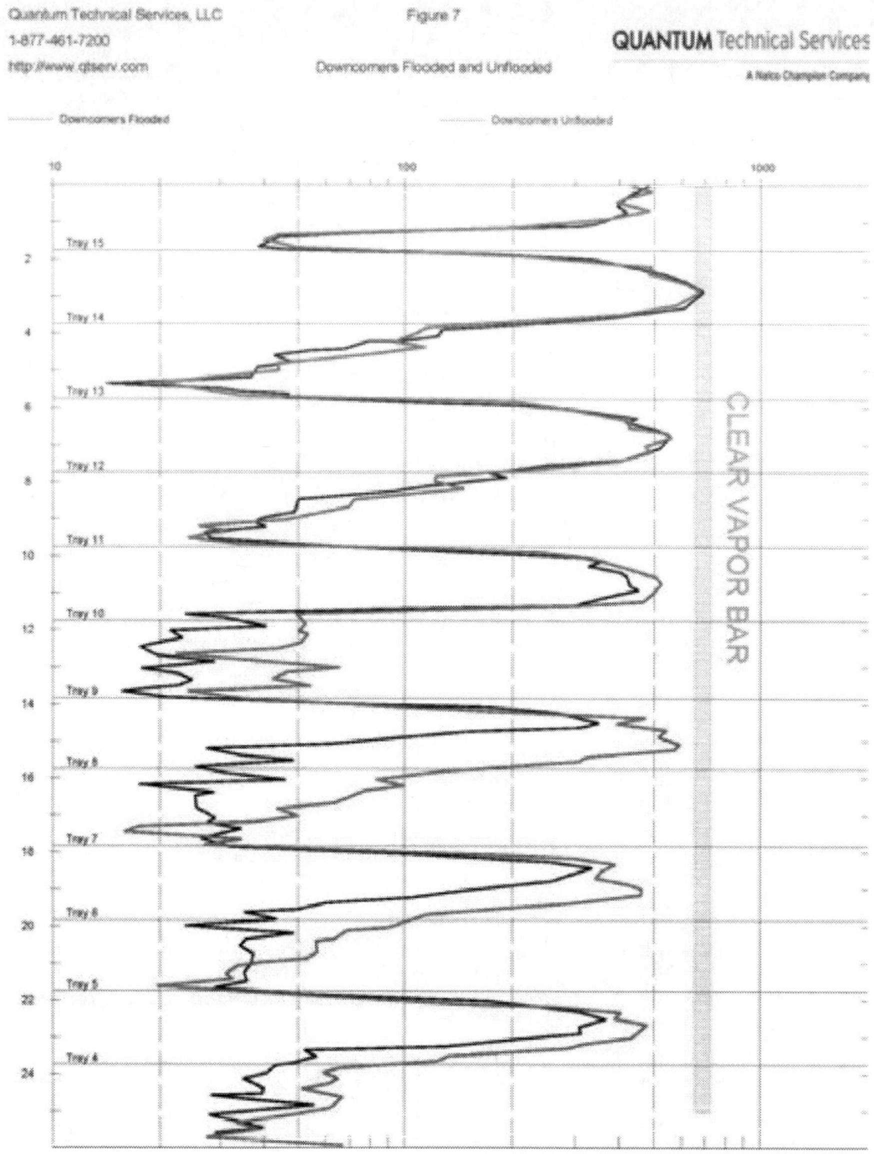

Figure 7: Overlay of center downcomer scans under flooded and unflooded conditions (Reprinted courtesy of Quantum Technical Services, LLC)

Overall, the gamma scans conclusively showed that the tower flooded by downcomer flooding in the lower 8 trays.

While downcomer scans can distinguish jet flood from downcomer flood, they cannot distinguish downcomer backup from downcomer choke flood. Both these downcomer mechanisms show up as full downcomers. The distinction can be drawn by calculation, which in the stabilizer case favored downcomer choke. Also, as stated earlier, downcomer choke is common at high pressure distillation, like the stabilizer. Further, in the absence of fouling and foaming, downcomer backup is uncommon in well-designed trays of 24" spacings and higher.

All these strongly supported downcomer choke.

Based on this diagnosis, the downcomer top areas were largely increased. To maximize throughput and minimize the loss in the active areas, the downcomers were heavily sloped, with the downcomers top areas being three times their bottom areas. The debottleneck solution exceeded expectations, with a capacity gain of 29%.

CASE STUDY 2: DOWNCOMER GAMMA SCAN CAN SHOW WHAT CALCULATIONS AND TESTS CANNOT

Downcomer gamma scans can reveal what calculations would not, but only if performed and interpreted correctly, as illustrated in the following case. A tower in a fouling service, equipped with 2-pass valve trays at 36" spacing, experienced flooding mid-way through the run. The flooding showed up as high pressure drop and loss of separation. To plan modifications for the next turnaround, and to devise an operation strategy for the rest of the run, it was necessary to understand the nature and location of the flood.

Both active area and downcomer scans were shot. The flooded active area scans showed large froth accumulation on the top tray. On the lower trays, froth heights were low, and clear vapor was reached less than 12" above the trays. Since the flooding occurred on the top tray, it was essential to scan the downcomers from that tray. The challenge was that these were side-downcomers, and practical difficulties precluded scanning them parallel to the weir.

Fortunately, this was one of the few cases in which one could get away with a chord that intersects the downcomer wall at a small angle (about 30 degrees). Usually, when a chord crosses the downcomer wall into the active area, it is impossible to decouple the downcomer dispersion from the tray dispersion. But here the tray dispersion terminated well below the tray above, so from 12 inches above the tray and up the scan saw only the downcomer dispersion.

Figure 8 shows the scans for the two side downcomers. The two downcomers from tray 1 showed widely different dispersions. The red curve shows a dense dispersion on the top half, with a gradually increasing froth density (ascending from the bottom of the downcomer upwards) below. At the bottom the froth density was about the same as the tray froth density above tray 3. The green curve showed a low-density dispersion throughout the downcomer, which looked very similar to the tray froth above tray 3 below. This suggests that this downcomer passed vapor, just like a tray does.

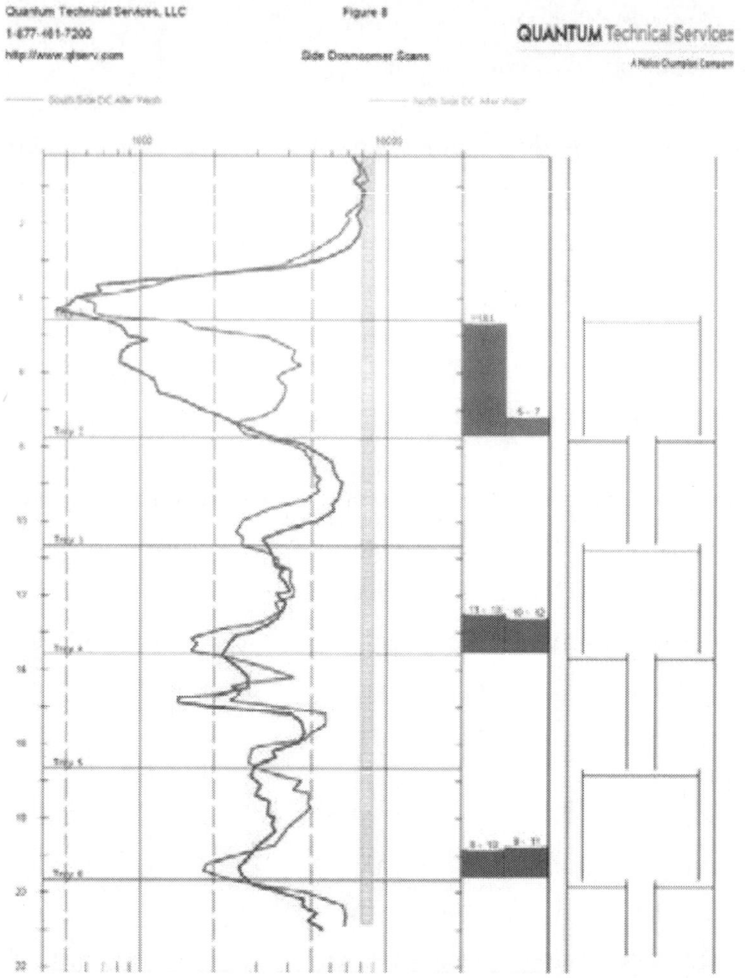

Figure 8: Side downcomer scans (Reprinted courtesy of Quantum Technical Services, LLC)

Returning to the red chord, the initial interpretation was of a seal loss in this downcomer holding up liquid from the tray above. A close check of the scan chord against the drawing revealed that this was a misinterpretation. The side downcomers were sloped (Figure 9), and below the top 18 inches, the scan chords passed through progressively shorter lengths of downcomer as one descended. Near the bottom of the downcomers the chords missed them altogether, catching only the tray dispersion (Figure 9). The switch from dense to highly aerated dispersion did not indicate loss of seal; rather, it indicated the downcomer chord becoming progressively shorter due to the slope. It is therefore likely that the downcomer through which the red chord passed was all full with the dense liquid seen in its upper part.

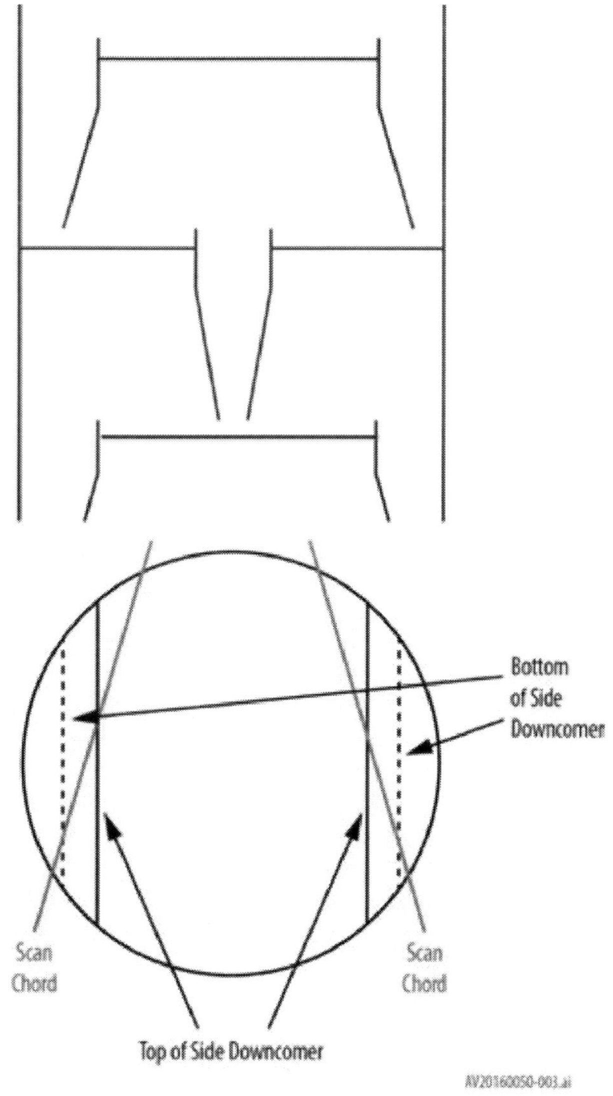

Figure 9: Side downcomer scan chords in Case Study 2

Putting the information together, tray 1 appears plugged, incurring a very high pressure drop which in turn incurred a high downcomer backup, leading to a downcomer backup flood. The plugging was non-uniform, with most of the liquid descending via the side downcomer that showed high dispersion density. Due to the high pressure drop across the top tray and the liquid maldistribution, it appears that the vapor managed to break the seal of the opposite downcomer, causing its dispersion to appear like part of the active area. This vapor breakthrough would channel more liquid to the downcomer that had the high dispersion density.

This interpretation highlighted the likelihood that reflux maldistribution in the tower contributed to the plugging problem. A turnaround inspection later confirmed the non-uniform nature of the plugging, and there were additional water marks pointing to the maldistribution.

CASE STUDY No. 3: UTILIZING NEUTRON BACKSCATTER DOWNCOMER SCANS TO IDENTIFY DRY/FULL DOWNCOMERS

Figure 10 represents a case where a US refiner wanted to extend their run length but was experiencing severe throughput limitations. A standard active area gamma scan was conducted that identified tray deck fouling resulting in isolated downcomer backup flood on trays.

It became critical to identify the regions where the plugging was most severe in order to develop a strategy to minimizing the throughput limitation. Downcomer scans are precious for providing this type of diagnostic. However, here there was a unique practical difficulty: this tower had abnormally large stiffening rings, which would have severely interfered with side downcomer gamma scans, producing too much scatter. So side downcomer gamma scans were not practical in this tower.

The only reliable way to scan the side downcomers was with neutron scans. However, for effective neutron scans, the side downcomers need to be adequately accessed by the scanning device (Figure 4), and the tower platforms did not afford anywhere near the required access. To provide the required access, an innovative technique was devised. The neutron scanning unit was taken off the pole and suspended on a standard scanning cable, allowing a large area to be scanned. It was also necessary to employ a crane with a man-basket to be lowered down the sides of the tower so a technician could guide the neutron backscatter device on the outer wall while the device was being lowered, ensuring correct readings. To the best of our knowledge, this is the first time such a technique was devised and implemented.

The innovative technique paid great dividends: neutron backscatter downcomer scans gave a concise identification of the plugged and dry downcomers (Figure 10). The neutron scans identified the following downcomers to have lost liquid seals:

- Zone B - Trays 13, 15, 17 north side downcomers
- Zone D - Trays 7 and 9 south side downcomers

The neutron backscatter scans also clearly show that liquid seal loss in downcomers can occur in distinct areas of a tower, while other areas would be operating normally. In this case, Zones B and D had fouled trays, while the downcomers in other zones appeared to operate normally. The dry downcomers suggest severe fouling on the side panels right above them. Vapor unable to ascend through the plugged active area would break the seal on the downcomer and ascend through it.

The neutron backscatter technique provided a clear picture of the side downcomers in a situation where a traditional gamma scan would not have been possible. As a result of the data obtained with the neutron backscatter device, the refiner was able to identify the severely plugged areas and perform online cleaning of the tower. This in turn eliminated the need for an immediate shutdown and allowed continued operation at acceptable throughput rates.

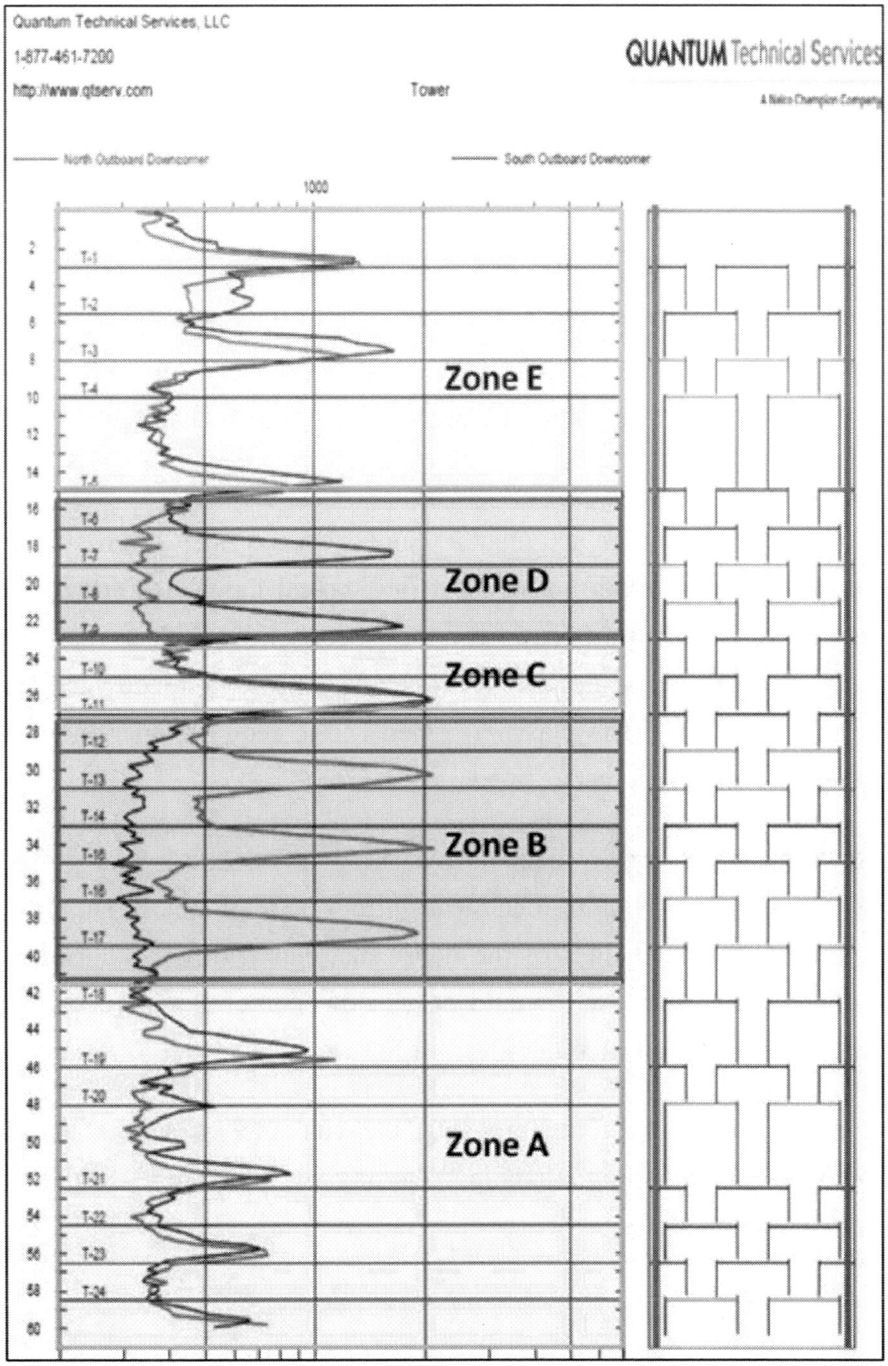

Figure 10: Neutron backscatter technique - An illustrative neutron backscatter scan of outboard downcomers. Zones A, C, and E show liquid in all outboard (side) downcomers and healthy disengagement to vapor with no signs of flooding or seal loss. Zones B and D showed no signs of liquid in alternating outboard downcomers. No liquid in a downcomer indicates a loss of liquid seal. Most interesting, the loss of liquid seal in the outboard downcomers for Zone B was limited to the South outboard downcomers and the loss of liquid seal in the outboard downcomers for Zone D was limited to the North outboard downcomers. The downcomers on the opposite sides appeared to operate normally. (Reprinted courtesy of Quantum Technical Services, LLC)

CASE STUDY No. 4: DOWNCOMER SCAN IDENTIFIES ROOT CAUSE OF FLOODING

Downcomer scans can often be the critical piece of the puzzle when looking for the root cause of flooding. In this case a Gulf Coast refiner had just come up from a turnaround and was experiencing operational issues with a tower. The tower was experiencing high pressure drop when approaching normal operating rates. An active area gamma scan was conducted and showed flooding throughout the bottom half of the tower originating at Tray 1 (Figure 11a). It is difficult to determine tray integrity when trays are in a flooded condition, so it was suggested that the customer reduce rates to unload the column and re-scan the active areas to rule out or verify tray damage.

Figure 11b shows the results of the active area scan at reduced rates. Here we see that all of the trays appeared to be in place with no signs of mechanical damage. Further, the froth heights appear low, almost all less than a third of the tray spacings, with clear vapor spaces above all trays, showing no proximity to jet flood. This directed attention to the downcomers. A third scan of the odd numbered downcomers was performed at the reduced operating rates to look for abnormal loadings in the downcomers.

Figure 11c shows the downcomer loadings measured in the third scan. Most downcomers showed a few inches of dense liquid, then clearing into a vapor space. The Tray 1 downcomer was different, it was completely full of the dense material with no vapor space above. Any additional rate increase to the tower would cause downcomer liquid to backup into the tray above, initiating flood in this downcomer.

Knowing the precise location of the problem, the customer was able to perform a quick shutdown, opening only the bottom manhole. The tray 1 downcomer seal pan was cleaned and the tower restarted with no issues.

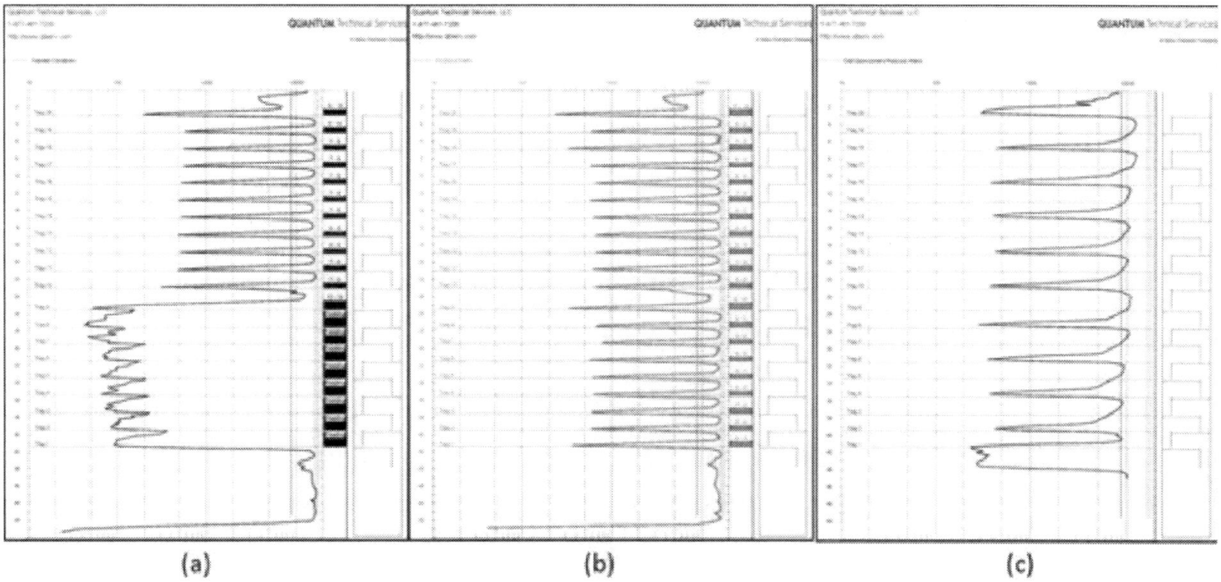

Figure 11: (a) – First gamma scan detected flooding in bottom half of tower. (b) – Reduced operating rates show no signs of active area damage. (c) – Downcomer scans show clearly flooding originates in Tray 1 downcomer.

CASE STUDY No. 5: DOWNCOMER SCAN DIAGNOSES FOAMING AND VERIFIES EFFECTIVENESS OF ANTIFOAM ADDITION

Combining downcomer scans with active area scans is a powerful method to evaluate foaming as well as the effectiveness of antifoam addition. Foams usually bottleneck downcomers rather than trays, so it is important to scan downcomers when troubleshooting for foam or evaluating the effectiveness of antifoam addition. This is shown in this case study.

A chemical tower was experiencing reduced throughput capacity with very poor separation. Gamma scans of the tray active areas and even numbered downcomers revealed that foam was present throughout the height of the column. The foam can be recognized by the uniform dispersion on the flooded trays as seen in Figure 12a; the absorption of gamma radiation near the tray floors was much the same as near the top of the vapor spaces, almost like shooting a scan through a bubble bath (which is foam). The scan of the even-numbered downcomers in Figure 12a shows much the same, with uniform dispersion through the scanned downcomers. The foamy uniform dispersion in each downcomer extended all the way to the tray above. Both the active area and downcomer scans provided conclusive evidence for foaming.

A chemical treatment company was contacted to help eliminate the foam detected by the scans. Antifoam does not always eliminate a foaming problem. The success of a treatment depends on the agent used and its concentration. Antifoams are surface-active agents, and too much antifoam can be worse than too little[9]. Several trials with different agents and concentrations are sometimes required. Gamma scans through the active areas and downcomers are powerful for evaluating the effectiveness of an antifoam injection.

Figure 12b shows the active areas and downcomer gamma scans used to evaluate the effectiveness of an antifoam injection into the tower. It shows that the foam was completely removed from the trays. The active area scan shows that once the foam was eliminated, the trays were not highly loaded, with froth heights well below half the tray spacing and clear vapor spaces above. The downcomer scans still show some foam in the downcomers, but the foam heights had subsided and there was good clearing near the very top of the downcomers. Although the foam was gone from the trays and no longer bottlenecked the tower, there wais still a good height of foam in the downcomers that could build up to the tray during changes or upsets. So the follow-up scans showed that the chemical treatment was effective in removing the foam seen on active areas and alleviating the problem of foam in the downcomers.

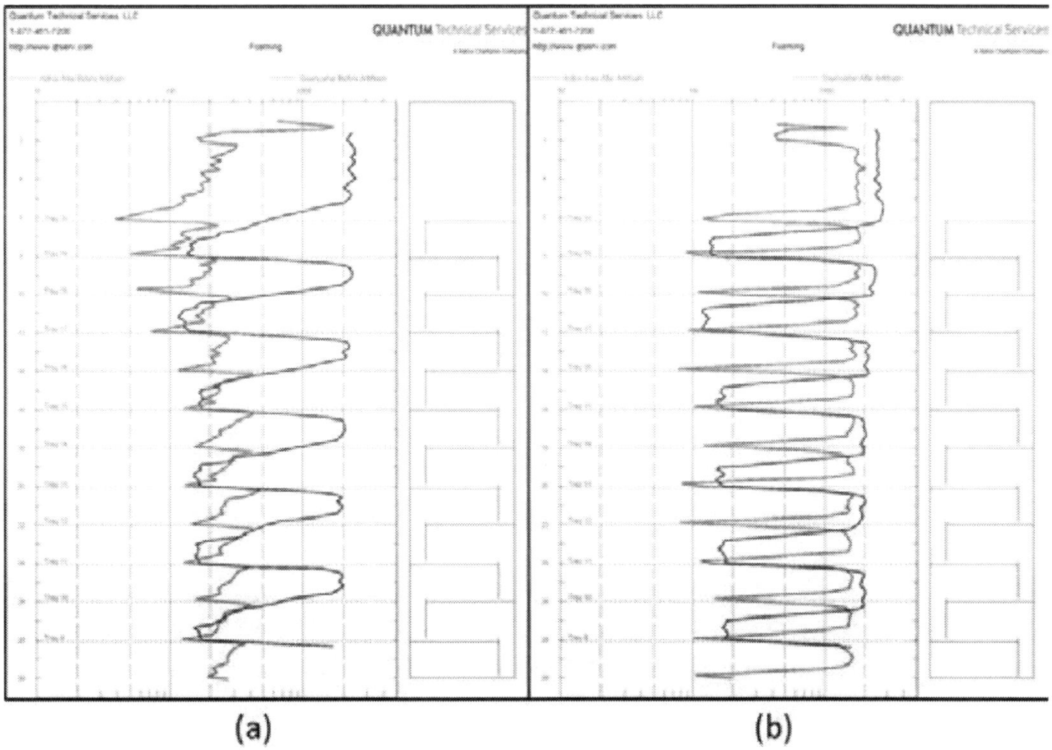

Figure 12: (a) – Gamma scans show foaming on active areas and in downcomers. (b) – Gamma scans show effectiveness of antifoam on active areas and in downcomers.

IN CONCLUSION

Typically, we see downcomer scans performed about 10% of the time. In contrast, our experience is that at least half the time downcomer scans can contribute enough to the diagnostics to more than justify their costs. Downcomer scans are not performed often enough in the industry.

Like any art, there are know-how and tricks involved in achieving an effective downcomer scan and correctly interpreting it. In this article, we went to great lengths to document and discuss the know-how, tricks and lessons that we learnt during late nights at the school of hard knocks and to place them into the hands of the reader.

This invaluable technique pulled many towers from the "unsolved mystery" files into a clear diagnostic with an effective solution. It is our hope that it will do the same for our readers.

ACKNOWLEDGEMENT

The authors acknowledge with appreciation the contribution to this article by Paul Chila from Quantum in sorting, collecting and compiling scan data into many of the diagrams presented in this paper

REFERENCES

1. Lockett, M.J., "Distillation Tray Fundamentals," Cambridge Univ. Press, Cambridge, UK, 1986.

2. Yanagi, T., Chem. Eng., November, p.120, 1990.

3. Kister, H. Z., P. Mathias, D. E. Steinmeyer, W. R. Penney, and J. R. Fair, "Equipment for Distillation, Gas Absorption, Phase Dispersion, and Phase Separation", Section 14, in R. H. Perry and D. Green "Chemical Engineers' Handbook", 8th Ed., 2008.

4. Kister, H.Z., "Distillation Design", McGraw-Hill, NY, 1992.

5. Kister, H. Z., and N. M. Mohamed, "Troubleshoot and Solve a Gas Treater Downcomer Unsealing Problem" Part 1, p.39, Gas Processing, January-February 2015, and Part 2, ibid, p.49, March-April 2015.

6. Charlton, J. S. (ed.), "Radioisotope Techniques for Problem Solving in Industrial Process Plants", Gulf Publishing, Houston, Texas, 1986.

7. White, R. L. "On-Line Troubleshooting of Chemical Plants" Chem. Eng. Progr. 83(5), p.33, 1987.

8. Kister, H. Z. , "Apply quantitative Gamma Scanning to High-Capacity trays", p.45, Chem. Eng. Progr., April, 2013

9. Kister, H. Z. "Distillation Troubleshooting", Wiley-Interscience, 2006.

10. Shakur, S. M., N. F. Urbanski, M. R. Resitarits, D. R. Monkelbaan, and D. A. Bucior, "Gamma Scanning a Column Containing Closely Spaced Trays", Paper presented at the AIChE Annual Meeting, Dallas, Texas, November 4, 1999.

445112 Charlie Colwell's Achievements and Correlation

Abstract:

This presentation is comprised of two parts. Part 1 will review the technical achievements of Dr. Charles Colwell as outlined below:

- Public domain papers pre-Exxon based on his PhD work at Northwestern University
  - Experimental and theoretical work on mass transfer adsorption on ion exchange resins
-
- Public domain papers while at ExxonMobil Research and Engineering Company
- Correlations for predicting clear liquid height and froth density on sieve trays
- Estimating sieve tray efficiency in the weeping region
- His extensive work in distillation and fractionation in the upstream, downstream and chemicals sectors
  - Sieve tray mass transfer efficiency model
  - Packing mass transfer efficiency model
  - Packing capacity correlations
  - Direct contact heat transfer correlations for packings, trays and shed decks
  - Tray downcomer capacity correlations
  - Fractionator troubleshooting
  - First time applications of packings (random and structured) and high capacity tray technology
  - Computer design/rating programs for trays and packings
  - Reactive mass transfer efficiency estimation method for caustic towers

In the second part of the presentation, the Colwell correlations for predicting clear liquid height on sieve trays will be investigated further when compared against publically available experimental data.

Vacuum Distillation – Process Configurations

Andrew W. Sloley

Advisian
Energy Center II
575 N Dairy Ashford St
Houston, TX 77079

Prepared for Presentation at the
American Institute of Chemical Engineers
2016 Spring Meeting
Houston, TX April 10-14, 2016

UNPUBLISHED

1. Summary

Vacuum tower configuration choices control product yields and energy recovery. Conventional fuels vacuum towers produce products for downstream cracking and have reasonable energy efficiency. However, less common configurations may allow for either diesel recovery, more effective energy recovery, or both.

Dieselization of the European auto fleet has shifted world-wide refining patterns. In all regions, diesel now often makes up the most valuable product from the refinery. The most profitable diesel is the diesel already in the crude oil. Adept molecule management to recover that diesel maximizes refinery profits.

Current energy prices do not favor intensive investment for direct energy savings. Nevertheless, energy savings are becoming more important. This is mainly driven by regulatory pressure from:

- Energy intensity regulations.
- Avoiding emissions (nitrous oxides and others).

Dieselization and energy recovery combine to focus attention on the crude vacuum unit. Significant diesel remains in most vacuum tower feeds. Many refiners find recovering diesel as a vacuum tower product attractive. Diesel recovery also changes temperatures in the vacuum unit, this may change heat recovery. Understanding how diesel recovery and temperature levels interact is critical to evaluating both new units and revamping existing units. Multiple vacuum tower configurations can allow the top vacuum tower product to meet a diesel boiling range specification.

Results are shown for running a typical crude oil for the most common configuration changes. These include:

- Additional pumparounds for heat recovery improvement.
- Diesel recovery with internal fractionation with net top-bottom reflux.
- Diesel recovery with internal fractionation with zero top-bottom reflux.
- Diesel recovery with no internal fractionation.

The results show that:

- Increased fractionation will increase diesel yields.
- Increased fractionation reduces pumparound temperatures and makes heat recovery more expensive (more capital) or less efficient (less heat recovered).

The more diesel the unit produces, the more difficult the heat integration. Trade-offs for adding diesel recovery to both existing plants and to new units must balance the value of diesel recovery against capital costs, yields, and energy recovery.

2. Process Configurations – Fuels and Lubricants

Refinery vacuum units fall into two distinct types: fuels units and lubricants units. Fuels units directly produce fuel streams (diesel) and feeds for downstream conversion units that make fuels. Modern downstream units typically include Fluid Catalytic Cracker (FCC) units and Hydrocracker (HC) units. Products going to FCC units may be hydrotreated before reaching the FCC. Lubricants units produce lubricant base oils for further processing into lubricants. Lubricants units often produce some fuel streams as well.

In concept, the conventional fuels unit vacuum tower operates as a series of drums that combine de-entrainment and absorption heat removal in one physical structure. No intentional reflux is present. Fuels units that recover diesel as a product often add limited internal fractionation capability.

In contrast, a lubricants unit has significant internal reflux and fractionation capability. The fractionation is required to meet viscosity or other lubricant specifications.

The focus here is to examine operation of fuels vacuum units. Table 1 shows the cases evaluated. These cases cover the range of the most economically important options.

Unit *configuration* is the mechanical layout of the unit. Six configurations will be discussed. Results will be presented for four configurations. These four have been selected to illustrate the advantages and disadvantages of the major options examined. Configuration options covered here include:

- Number of pumparounds (heat removal levels)
- Number of fractionation sections
- Location of fractionation sections
- Disposition of internal streams

Unit *operation* is the process objective in making a stream. The options examined here focus on the highest side-draw product from the vacuum column. Plants may use specific names for the top product. The results of three different types of top products are discussed. These are:

- Top product without a fixed specification.
- Top product to a diesel specification.
- Top product to a hydrocracker feed specification.

The appendix lists standard terms and acronyms used.

Table 1 Case Summary

Case	1	2	3	4	5	6	7	8	9
Pumparounds	2	3	3	3	2	2	2	2	3
Products	2	3	3	3	2	2	3	2	3
Diesel Prod					Yes	Yes	Yes	*	Yes
Diesel at specification					Too light	Yes	Yes	*	Yes
MPA duty from		TPA	BPA	Both PA					
Diesel fractionation bed					Yes	Yes	Yes	No	No
Internal flow to HVGO					Yes	Yes	No	No	No
Notes	* infeasible – zero diesel production								

2.1. Case 1 – Conventional Vacuum Unit – 2 Pumparound, 2 Product

Figure 1 shows the conventional refinery vacuum tower. This remains the most common unit configuration. Two pumparounds remove duty from the vacuum tower. Even though built in one shell, this tower operates as three independent sections:

- De-entrainment and washing above the feed entry.
- Bottom pumparound for heat removal.
- Top pumparound for heat removal.

No fractionation or reflux is present between the TPA and BPA sections. The wash oil return from the BPA and the wash bed create some fractionation between the feed and BPA. However, the composition gradient created is small. The wash bed is mainly a de-entrainment operation.

Case 1 is used as the base case for comparison of the different options.

2.2. Cases 2, 3, 4 – Addition of a Pumparound – 3 Pumparound, 3 Product

Figure 2 shows a vacuum unit with three pumparounds and three products. An additional pumparound is added. The purpose of the third pumparound is to increase the heat recovery possible. The extra MPA section adds an additional temperature level for heat integration. It also shifts duty from the TPA and the BPA into the MPA.

The individual cases vary how the heat shifts:

- Case 2 shifts duty from the TPA into the new MPA.
- Case 3 shifts duty from the BPA into the new MPA.
- Case 4 splits the duty in the MPA. Some comes from the TPA and some from the BPA.

The new pumparound operates at a temperature in-between the BPA temperature and the TPA temperature.

2.3. Cases 5, 6 – Top Product as Diesel, Internal Fractionation

Figure 3 shows a vacuum unit with two pumparounds and two products. The top product meets diesel distillation specifications. The tower includes a fractionation section between the TPA (diesel) and the BPA. The reflux from the bottom of the fractionating beds falls into the BPA section. This case is internal fractionation with net TPA-BPA reflux.

The cases vary the diesel product rate:

- Case 5 has a product rate matched to Case 7. This gives a lower diesel endpoint distillation.
- Case 6 has a product rate maximized for the configuration.

The intent of this option is to produce more diesel product and less gas oil product.

2.4. Case 7 – Top Product as Diesel, Internal Fractionation – 3 Product

Figure 4 shows a vacuum unit with two pumparounds and three products. The top product meets diesel distillation specifications. The tower includes a fractionation section between the TPA (diesel) and BPA. The reflux from the bottom of the fractionating bed is drawn as a LVGO product. The intent of this operation is to produce more diesel product and less gas oil product. Case 7 has the same number of theoretical stages as Case 5/6. This case is internal fractionation with zero TPA-BPA reflux.

2.5. Case 8 – Top Product as Diesel, No Fractionation

Figure 5 shows a vacuum unit with two pumparounds and two products. The configuration is the same as Case 1. However, in this case the attempt is made to meet diesel product specifications on the top product. No fractionating bed has been added in this case.

2.6. Case 9 – 3 Pumparound, 3 Product, Top Product as Diesel

Figure 6 shows a vacuum unit with three pumparounds and three products. This is the same configuration as Cases 2/3/4. In this case, the top product meets a diesel product specification. No fractionating bed has been added. This case is diesel production with no internal reflux.

2.7. Other Fuels Configurations

Many other unit configurations are possible and have been built. Figure 7 shows another example. This configuration adds a fractionating bed between the top pumparounds of a unit that is similar to Cases 2/3/4.

In fuels units with fractionation, the distillation specification of the top product may be either set by diesel product requirements or downstream processing requirements. Many units that have gas oil feeding a hydrocracker may have a distillation requirement to protect the reactor catalyst. Figure 8 shows a vacuum unit producing hydrocracker feed. In concept, the hydrocracker feed units are the same as the diesel production units. The major difference is in the distillation specification used.

2.8. Lubricants Vacuum Units

For comparison, Figure 9 shows a vacuum lubricants unit. Lubricants units typically have multiple fractionating beds. Product specifications are set by a combination of performance criteria (viscosity, viscosity index) and product properties (density, distillation). Compared to fuels units, vacuum units often include a large number of fractionating stages and a much more reflux.

3. Basis

The basis for comparison uses a 31.6° API (0.8676 SG) medium density, sweet, moderate sulfur crude (Tia Juana Light) [1]. Tia Juana Light has a density and yield distribution very close to the current (2016) average U.S. crude. The upstream atmospheric tower configuration is the same as in previous work [2] [3] . The upstream atmospheric tower operates at a flash zone pressure of 30 psig and 690°F (308 kPa and 366°C). This gives a 718°F (381°C) cutpoint for the atmospheric tower.

All the vacuum tower configurations use a wet vacuum tower with stripping trays in the bottom of the tower. Vacuum heater, flash zone, and stripping section conditions were kept constant for all cases. Vacuum heater outlet temperature was 775°F (413°C) and vacuum tower cutpoint was 1,046°F (563°C).

All cases kept a constant pressure profile across the wash bed, BPA section and TPA section. Cases that had a fractionation section or a MPA section added used a pressure drop of zero across that section. Of course, real towers do have pressure drop across all sections. The intent of keeping constant pressures was to simplify the analysis of potential heat shifts and fractionation. Evaluation for actual plants must include pressure drop changes when adding pumparounds or fractionation beds.

Performance limits on the vacuum tower have been imposed in actual volumes. Equipment reliability requirements are evaluated in actual volumes. Actual volumetric liquid rate at the bottom of the wash bed (overflash) is kept constant in all case. Yields are shown in standard volumes. Purchasers buy standard volumes. Hence, refinery economics analysis uses standard volumes.

Total distillate yields for all cases are very close. The largest difference is 0.13%. Yield differences arise from changes in wash oil composition and changes in HVGO (BPA) draw temperature. These factors modify wash section fractionation and change the number of standard barrels lost to the bottoms for the same actual volume of overflash at the bottom of the wash bed.

Table 2 shows the results of comparing the different operations. Cases 1-4 focus on improved heat recovery from the base operation. Cases 5-9 examine the results of converting the top product to diesel-boiling range material.

4. Heat Removal – Pumparound and Product

Heat removal in a vacuum tower covers three duties:

- Condensation of the vapor feed.
- Cooling of the vapors down to overhead temperature.
- Condensation of the liquid down to product temperatures.

Net product may leave the vacuum unit as either a hot stream or a cold stream (Figure 10). Hot stream products would leave the unit upstream of the pumparound exchangers. Cold stream products would leave the unit downstream of the pumparound exchangers. Hot-product or cold-product towers do not fundamentally differ in how the vacuum tower works. They do differ in the amount of heat recovery. As product yields shift, the amount of duty in the rundown cooling changes.

Table 2 shows duties for the following:

- PA Q: duty for the pumparound only (hot product tower)
- Prod Q to PA return temperature: extra duty for cooling the product to the pumparound return temperature. This would be the extra duty for a cold product tower.

- Prod Q to final temperature: extra duty from cooling the product from the pumparound return temperature to the final product temperature.
- PA level duty: PA Q plus the Prod Q to PA return temperature.

5. Base Case – Case 1 – Conventional Vacuum Unit – 2 Pumparound, 2 Product

The conventional vacuum unit recovers 15.9 Mbtu/hr (12.5 GJ/hr) TPA level duty and 122.4 Mbut/hr (86.4 GJ/hr) BPA level duty. This base load has more heat recovery into BPA than most typical units. Nevertheless, this heat recovery remains within the feasible range. Units with this split of heat balance between BPA (88%) and TPA (12%) exist.

Figure 11 shows the heat availability for the case. The plot shows total recovered duty versus temperature for a cold product operation. The duty from the pumparound return temperature to the final product temperature is not include ('Prod Q to final temperature' is not included). The desired direction for the heat available curves is shift them up and extend them to the right.

The crude heat required curve has not been added. Attempting to plot vacuum tower heat recovery against crude preheat without examining the atmospheric tower heat recovery would be extremely difficult. In order to focus on vacuum tower effects, the analysis here is being restricted to the generic consequences of an isolated shift in vacuum tower duty and temperature levels.

Economics of vacuum tower operations vary with each specific crude unit. Useful recovered heat generally goes into two areas:

- Preheating crude feed or,
- Steam generation (or boiler feed water preheat).

At current North American pricing (January 2016), a new unit should target a crude preheat temperature of 515°F to 565°F (268°C to 296°C) [3]. Existing units operate in the range of 400°F to 605°F (204°C to 318°C).

6. Fuels Operation and Heat Recovery (Cases 2-4)

Cases 2-4 show the addition of a third pumparound and a third product. Adding an MPA creates an extra level of heat availability. The limiting cases are: all the energy recovered from the MPA is pulled from the TPA (Case 2), or all the heat recovered from the MPA is pulled from the BPA (Case 3). A final example (Case 4) shows an intermediate case where some of the heat comes from the BPA and some from the TPA. Figure 12 plots the heat availability for Case 1 to Case 4.

Adding a pumparound increases cost in the vacuum tower, pumps, external piping, controls, and heat exchangers. Even if economic, the costs of an added pumparound must be compared to the costs of improved heat recovery in other places in the crude unit.

6.1. Case 2 Shifting Duty from TPA into the MPA

Shifting duty from the TPA to the MPA (Case 2) shows the largest gain in the lower temperature region below 300°F (149°C). Above that point, Case 2 has only minor benefits.

In general, most crude units have ample low temperature heat available. Case 2 only helps low temperature heat recovery. Unless the crude unit is running at low energy efficiency, extra low level heat recovery (Case 2) occurs 'below the pinch.' The pinch is where the temperature difference between the heat recovery streams and the feed streams is the smallest. Generally, heat recovery economics are attractive when energy recovery duty is shifted from below the pinch temperature to above the pinch temperature. Even when energy recovery at low temperatures is attractive, an added

MPA must compete with many other ways to recover heat in the crude unit. These other methods will often be more cost effective.

The improved Case 2 heat recovery is also at too low a temperature for effective steam generation or boiler feed water preheat. In addition, the total heat recovered in Case 2 drops. Case 2 operation where TPA duty shifts to an MPA is unlikely to be economic.

6.2. Case 3 Shifting Duty from the BPA into the MPA

Shifting duty from the BPA to the MPA (Case 3) shows three major differences in Figure 12:

- Higher temperatures for the same duty above 450°F (232°C). This improves the heat recovery effectiveness of the hotter heat exchangers.
- Higher total heat recovery duty (the line goes further to the right). This improves total energy recovered.
- Lower temperatures for the same duty below 400°F (204°C). This reduces the heat recovery effectiveness of the colder heat exchangers.

The first two are benefits, the third is a disadvantage. The attractiveness of Case 3 strongly depends upon the balance of possible heat recovery losses in the cold exchangers versus heat recovery gains in the hot exchangers.

For a new unit, Case 3 is not clear cut. In some units, adding a pumparound would be justified, in others it would not be justified. Detailed analysis is required to check the economic benefits of the third pumparound. In general, the larger the capacity of the unit, the more likely the third pumparound would be justified.

Factors that make adding Case 3 to an existing unit more attractive would include:

- Excess surface area in low-temperature exchangers (this reduces the possible energy loss of lower temperatures in the cold exchangers).
- Tight limitations on the fired heater after the heat exchangers (this increases the value of incremental heat recovery at high temperatures).
- Existing pumparound pump capacity limits (this reduces the cost of extra capital for the new MPA pumps).

6.3. Case 4 MPA Duty from a Split Between BPA and TPA

Case 4 takes some of the MPA duty from the TPA and some of the MPA duty from the BPA. This operation improves the temperatures in both the low temperature region and the high temperature region. Around the middle temperature range of 350°F to 400°F (177°C to 204°C) the duty available drops slightly.

The performance of heat exchangers improve for both the cold exchangers and the hot exchangers. The cost is performance of heat exchangers in the middle range degrade. For a well heat-integrated unit, the pinch temperature is likely to be in this range. Additionally, the total heat recovered drops (the end point of the line on Figure 12 shifts left).

For most new units, Case 4 is less attractive than Case 3.

For existing units, Case 4 is most likely to be attractive if the unit is less energy efficient or subject to high fouling rates. Less energy efficient units have higher temperature differences in the heat exchangers, so the small temperature drop in the mid duty section has a smaller cost. High fouling rates make extra temperature driving force in the high temperature range more valuable.

7. Fuels Operation and Diesel Recovery (Cases 5-9)

Cases 5-7 show the addition of a fractionation section to allow for controlling the endpoint of the top product. This allows the top product to easily meet diesel specifications. Two theoretical stages of fractionation were added. For most units, two theoretical stages will capture nearly all of the diesel in the vacuum tower feed [4]. Optimization may show more stages are justified. Nevertheless, two stages is close enough to the optimum answer for accurate comparisons of different vacuum tower configurations. Cases 8 and 9 show attempts to make diesel as a top product with vacuum towers without fractionation sections.

7.1. Cases 5-6 Diesel Product with 2 Pumparounds and 2 Products

Case 5 and Case 6 use a fractionating bed between the TPA and BPA (Figure 3). The bottom liquid from the fractionating bed drops into the BPA section. The TPA product is diesel boiling range material. Case 5 uses yields that match the performance of Case 7. Case 6 has maximum diesel yields at the specification limit for diesel. Case 6 recovers more diesel than Case 5.

Figure 13 compares the heat recovery from Case 5 and Case 6 with the base operation (Case 1). Forcing the TPA to meet diesel distillation specifications drops the TPA draw temperature and reduces the total amount of energy recovery. The lower the distillation temperature of the diesel product, the colder the TPA. Meeting the diesel specification on the TPA also forces lighter material into the BPA. The BPA temperature drops as well. Case 5 shows lower energy recovery and lower temperatures than Case 6.

Economics of diesel recovery depend upon plant specifics. However, for all cases, diesel recovery makes heat integration more difficult, resulting in either lower heat recovery or more capital required for the same heat recovery.

7.2. Case 7 Diesel Product with 2 Pumparounds and 3 Products

Case 7 takes the liquid from the bottom of the fractionation bed and draws it from the tower as a separate product (Figure 4). Creating an extra product stream prevents lighter liquid from dropping into the BPA section. This configuration allows for recovering some diesel without causing BPA heat-integration problems.

In this configuration, there is no fractionation between the combined diesel-LVGO and the HVGO. The maximum diesel yield for a given specification is lower. For comparison, Figure 14 shows the base case (Case 1), Case 5 and Case 7. Case 5 and Case 7 have the same diesel yield.

The diesel recovery option without reflux (Case 7) improves the temperature level of the available duty and slightly reduces the total quantity of energy available. The temperatures improve (rise) due to the production of LVGO at a higher temperature. The LVGO product is available for heat integration even thought the unit has not LVGO pumparound. In existing units with difficult heat integration, using the no reflux configuration can be an attractive option for diesel recovery.

8. Fuels Operation and Diesel Recovery (Cases 8-9) with Conventional Units

Cases 8-9 show conventional fuels vacuum units without fractionation attempting to produce diesel as the top product. Case 8 is the conventional two pumparound tower. Case 9 is the vacuum tower with three pumparounds.

For the base configuration with two pumparounds (Figure 5), converting the TPA into a diesel boiling range specification stream requires shifting duty into the BPA. The BPA condenses more material, making the TPA distillation lighter. This shift increase the BPA rate and decreases the TPA rate. For the base example shown, the TPA rate reaches zero before the TPA distillation matches the required diesel specification. The operation is infeasible.

Some conventional vacuum units can operate with the TPA at a diesel specification. These units combine:

- Low upstream diesel recovery. This increases the rate of diesel entering the vacuum tower.
- Low vacuum unit feed vaporization. This reduces the fraction of heavy material in the BPA.

Case 9 shows meeting a diesel specification on the TPA with three pumparounds (Figure 6). The MPA section creates an effective one-stage of separation between the TPA and BPA. Even one stage of fractionation significantly improves diesel recovery performance.

Figure 15 compares the heat recovery performance of:

- Case 1 (Conventional)
- Case 5 (Fractionation added, TPA-BPA reflux)
- Case 7 (Fractionation added, no TPA-BPA reflux)
- Case 9 (Three pumparound, no internal reflux)

The three pumparound configuration (Case 9) has the best heat recovery of all the diesel recovery configurations. However it has the lowest diesel yield (Table 2).

9. Hydrocracker Feed as Product versus Diesel as Product

Instead of diesel, some fuels vacuum units produce feed for hydrocracking. Hydrocracking catalyst life can vary strongly with the distillation end-point of its feed. In general, hydrocracker feed units have the same conceptual trade-offs as diesel recovery units. The difference between the units is the distillation specification placed on the products. Within capacity and heat balance constraints, many have some ability to shift specifications between diesel operation and hydrocracker feed operation.

10. Conclusions

Conventional vacuum fuels units produce feed streams for downstream conversion units. The downstream units include both FCC and hydrocracking units. Increased diesel demand makes recovering diesel from the vacuum tower more attractive. Multiple vacuum tower configurations can allow the top vacuum tower product to meet a diesel boiling range specification. The two major factors to consider are:

- Increased fractionation will increase diesel yields.
- Increased fractionation will reduce temperatures and make heat recovery more expensive (more capital) or less efficient (less heat recovered).

Trade-offs for adding diesel recovery to both existing plants and to new units must balance the value of diesel recovery against capital costs, yields, and energy recovery.

11. References

[1] R. N. Watkins, *Petroleum Refinery Distillation*. Houston, TX: Gulf Publishing Company, 1973.

[2] A. W. Sloley, "Crude Distillation Sequences," presented at the AIChE Fall National Meeting, San Francisco, CA, 2013.

[3] A. W. Sloley, "Crude Unit Heat Integration - Best Practices," presented at the AIChE Spring Meeting, Austin, TX, 2015.
[4] A. W. Sloley, "Increase diesel recovery," *Hydrocarbon Processing*, vol. 87, no. 6, pp. 85–102, Jun. 2008.

12. Tables and Figures

Table 2E Vacuum Unit Comparisons

Case		1	2	3	4	5	6	7	9
Pumparounds		2	3	3	3	2	2	2	3
Products		2	3	3	3	2	2	3	3
Top prod as diesel						Yes	Yes	Yes	Yes
Diesel at spec						Too Light	Yes	Yes	Yes
MPA Duty from			TPA	BPA	Both PA				
Internal flow to hvgo						Yes	Yes		
Vacuum heater outlet	F	775	775	775	775	775	775	775	775
Vacuum tower cutpoint	F	1,046	1,046	1,050	1,050	1,044	1,045	1,046	1,046
Top PA Duty	Mbtu/hr	11.3	2.2	14.1	2.0	17.6	13.5	9.2	2.3
Mid PA Duty	Mbtu/hr		7.6	6.1	16.7				
Bot PA Duty	Mbtu/hr	77.1	77.1	66.0	66.0	69.6	73.4	77.1	77.1
Top PA Draw Temp	F	290	214	302	209	277	275	259	215
Top PA Return Temp	F	100	100	100	100	100	100	100	100
Mid PA Draw Temp	F		369	482	402			400	371
Mid PA Return Temp	F		200	200	200			200	200
Bot PA Draw Temp	F	558	558	597	597	536	542	558	558
Bot PA Return Temp	F	300	300	300	300	300	300	300	300
Top PA Rate	BPD	9,043	3,099	10,543	2,896	15,316	11,880	9,008	3,189
Mid PA Rate	BPD		6,157	2,787	11,178				
Bot PA Rate	BPD	35,241	35,241	25,676	25,676	35,114	36,075	35,241	35,241
Top Prod	BPD	3,680	583	4,930	449	960	1,380	960	618
Mid Prod	BPD		3,070	2,459	6,904			2,701	3,035
Bot Prod	BPD	20,716	20,716	17,206	17,206	23,283	22,917	20,716	20,716
Top Prod D86 5 vol %	F	560	504	572	501	512	523	515	506
Top Prod D86 95 vol %	F	745	636	745	641	616	638	638	638
Mid Prod D86 5 vol %	F		604	693	623			620	605
Mid Prod D86 95 vol %	F		752	839	800			758	753
Bot Prod D86 5 vol %	F	733	733	771	771	699	709	733	733
Bot Prod D86 95 vol %	F	1,006	1,006	1,018	1,018	1,000	1,001	1,006	1,006
Top Prod Q to TPA ret temp	Mbtu/hr	4.6	0.4	6.6	0.3	1.1	1.6	1.0	0.4
Mid Prod Q to MPA ret temp	Mbtu/hr		3.8	5.4	10.3			4.0	3.8
Bot Prod Q to BPA ret temp	Mbtu/hr	45.3	45.3	44.2	44.2	46.1	46.6	45.3	45.3
Top Prod Q to final temp	Mbtu/hr	0.0	0.0	0.0	0.0	0.0	0.0	0.0	0.0
Mid Prod Q to final temp	Mbtu/hr		1.0	0.8	2.2			0.9	1.0
Bot Prod Q to final temp	Mbtu/hr	17.8	17.8	14.8	14.8	19.9	19.6	17.8	17.8
Top prod final temp	F	100	100	100	100	100	100	100	100
Mid prod final temp	F		150	150	150			150	150
Bot prod final temp	F	180	180	180	180	180	180	180	180
Total distillate	BPD	24,396	24,368	24,594	24,560	24,243	24,297	24,376	24,369
TPA Level Duty	Mbtu/hr	15.9	2.6	20.7	2.3	18.7	15.1	10.2	2.7
MPA Level Duty	Mbtu/hr	0.0	11.3	11.5	27.0	0.0	0.0	4.0	3.8
BPA Level Duty	Mbtu/hr	122.4	122.4	110.2	110.2	115.7	120.0	122.4	122.4
Total Duty	Mbtu/hr	138.4	136.4	142.4	139.6	134.4	135.1	136.6	129.0

Table 2M Vacuum Unit Comparisons

Case		1	2	3	4	5	6	7	9
Pumparounds		2	3	3	3	2	2	2	3
Products		2	3	3	3	2	2	3	3
Top prod as diesel						Yes	Yes	Yes	Yes
Diesel at spec						Too Light	Yes	Yes	Yes
MPA Duty from			TPA	BPA	Both PA				
Internal flow to hvgo						Yes	Yes		
Vacuum heater outlet	C	413	413	413	413	413	413	413	413
Vacuum tower cutpoint	C	563	563	565	565	562	563	563	563
Top PA Duty	GJ/hr	11.9	2.3	14.9	2.1	18.6	14.3	9.7	2.4
Mid PA Duty	GJ/hr		8.0	6.4	17.6				
Bot PA Duty	GJ/hr	81.4	81.4	69.6	69.6	73.4	77.4	81.4	81.4
Top PA Draw Temp	C	143	101	150	98	136	135	126	102
Top PA Return Temp	C	38	38	38	38	38	38	38	38
Mid PA Draw Temp	C		187	250	205			204	188
Mid PA Return Temp	C		93	93	93			93	93
Bot PA Draw Temp	C	292	292	314	314	280	283	292	292
Bot PA Return Temp	C	149	149	149	149	149	149	149	149
Top PA Rate	M3/D	1,013	347	1,181	324	1,715	1,331	1,009	357
Mid PA Rate	M3/D		690	312	1,252				
Bot PA Rate	M3/D	3,947	3,947	2,876	2,876	3,933	4,040	3,947	3,947
Top Prod	M3/D	412	65	552	50	108	155	108	69
Mid Prod	M3/D		344	275	773			302	340
Bot Prod	M3/D	2,320	2,320	1,927	1,927	2,608	2,567	2,320	2,320
Top Prod D86 5 vol %	C	293	262	300	260	267	273	268	263
Top Prod D86 95 vol %	C	396	336	396	338	325	337	337	337
Mid Prod D86 5 vol %	C		318	367	329			327	319
Mid Prod D86 95 vol %	C		400	448	427			403	400
Bot Prod D86 5 vol %	C	389	389	410	410	371	376	389	389
Bot Prod D86 95 vol %	C	541	541	548	548	538	538	541	541
Top Prod Q to TPA ret temp	GJ/hr	0.5	0.0	0.7	0.0	0.1	0.2	0.1	0.0
Mid Prod Q to MPA ret temp	GJ/hr		0.4	0.6	1.2			0.4	0.4
Bot Prod Q to BPA ret temp	GJ/hr	5.1	5.1	5.0	5.0	5.2	5.2	5.1	5.1
Top Prod Q to final temp	GJ/hr	0.0	0.0	0.0	0.0	0.0	0.0	0.0	0.0
Mid Prod Q to final temp	GJ/hr		0.1	0.1	0.3			0.1	0.1
Bot Prod Q to final temp	GJ/hr	2.0	2.0	1.7	1.7	2.2	2.2	2.0	2.0
Top prod final temp	C	38	38	38	38	38	38	38	38
Mid prod final temp	C		66	66	66			66	66
Bot prod final temp	C	82	82	82	82	82	82	82	82
Total distillate	M3/D	2,732	2,729	2,755	2,751	2,715	2,721	2,730	2,729
TPA Level Duty	GJ/hr	12.5	2.4	15.6	2.1	18.7	14.5	9.8	2.5
MPA Level Duty	GJ/hr	0.0	8.4	7.0	18.8	0.0	0.0	0.4	0.4
BPA Level Duty	GJ/hr	86.4	86.4	74.6	74.6	78.6	82.6	86.4	86.4
Total Duty	GJ/hr	98.9	97.2	97.2	95.5	97.3	97.1	96.7	89.3

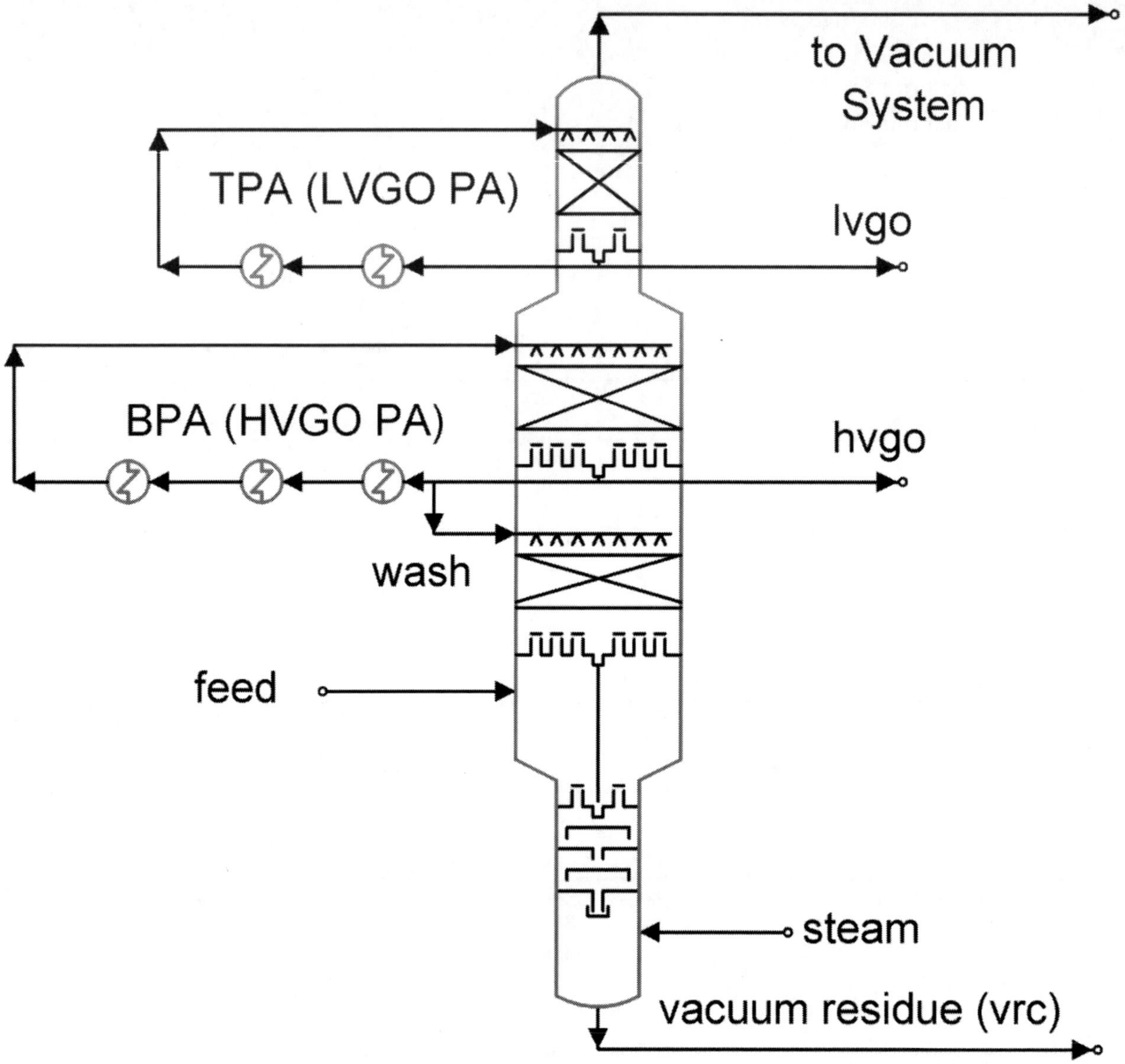

Figure 1 – Case 1: 2 Pumparound, 2 Product, Conventional

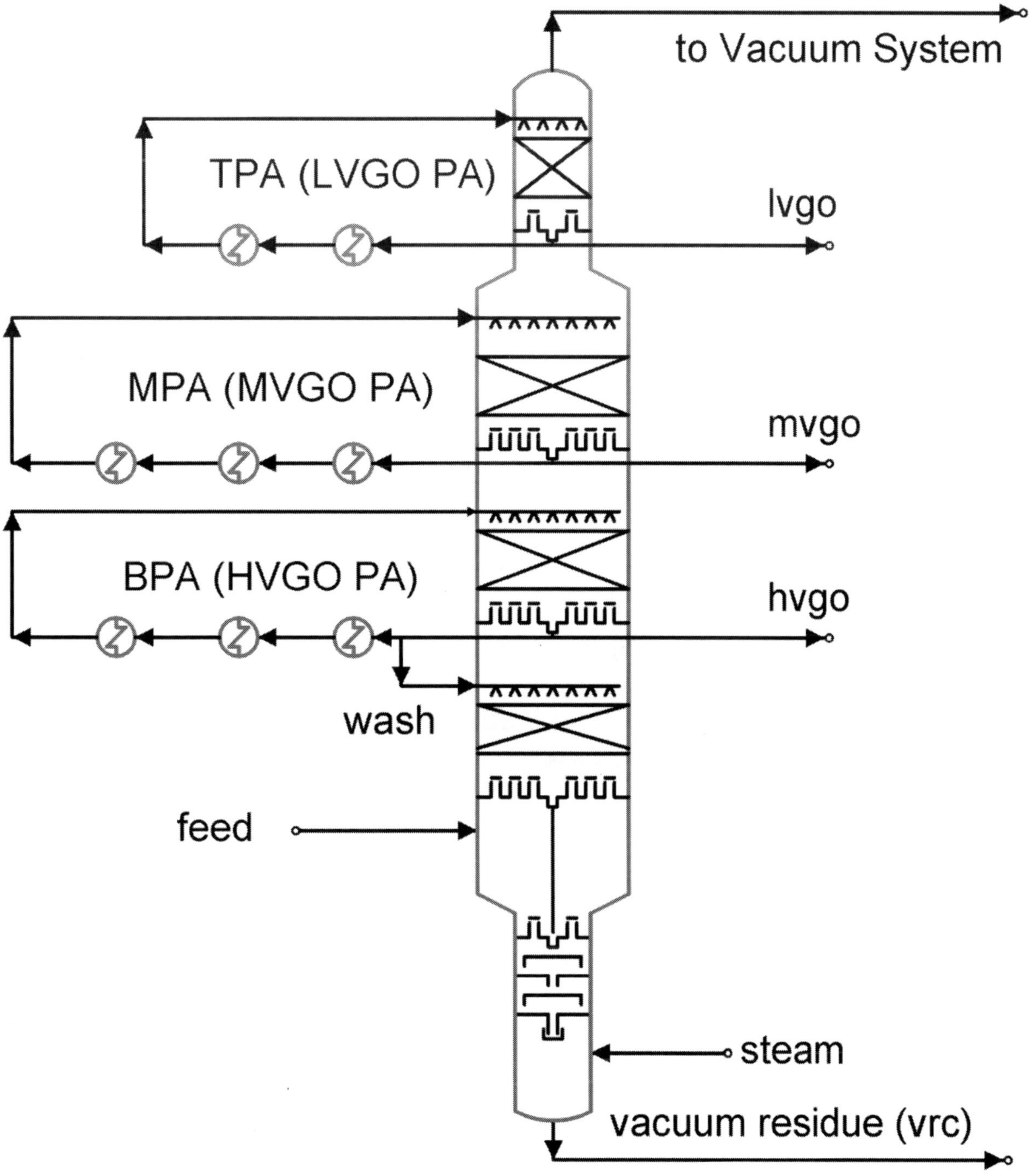

Figure 2 – Case 2, 3, 4: 3 Pumparound, 3 Product

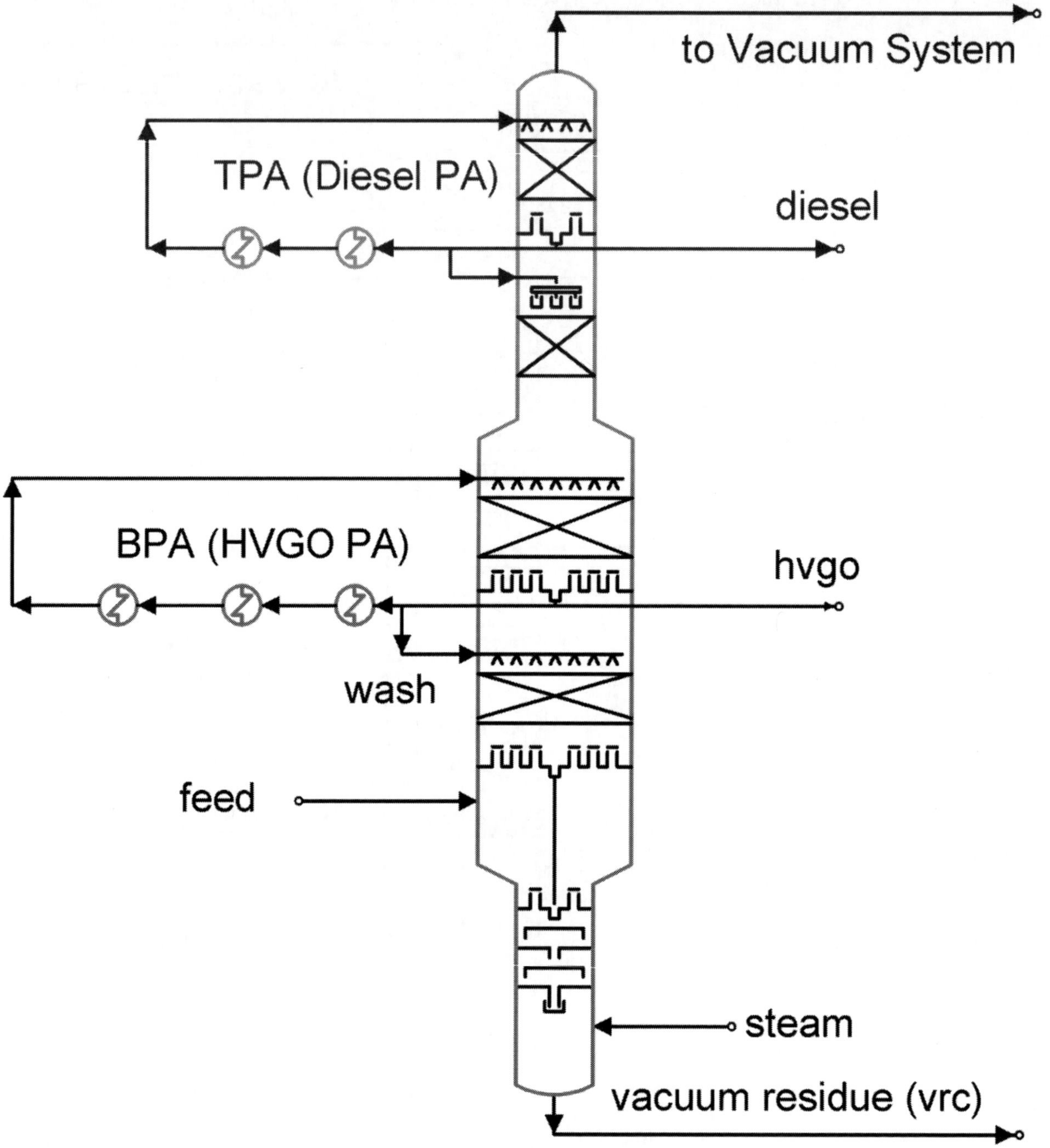

Figure 3 – Case 5, 6: 2 Pumparound, 2 Product, Diesel Operation

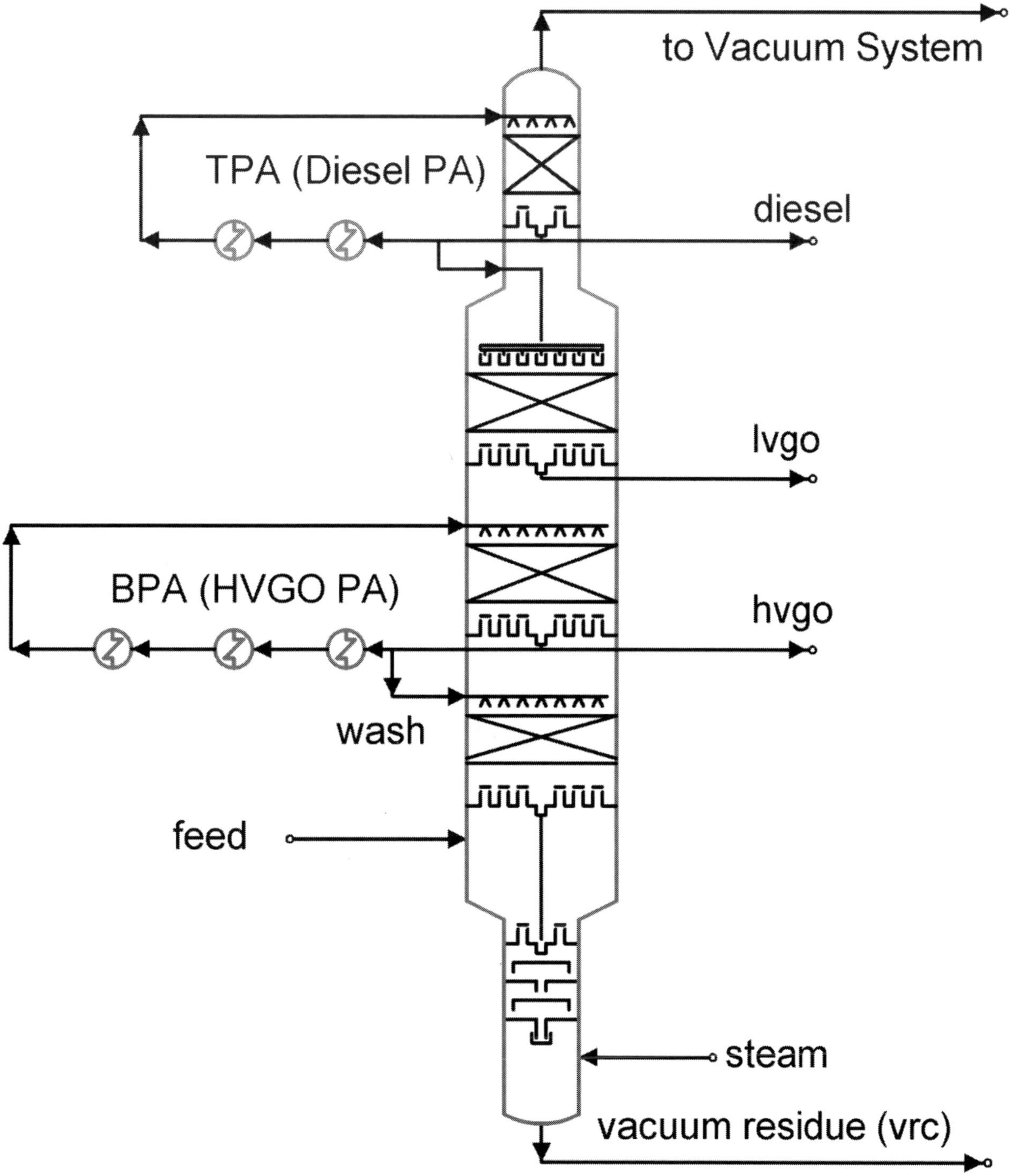

Figure 4 – Case 7: 2 Pumparound, 3 Product – Top Product as Diesel

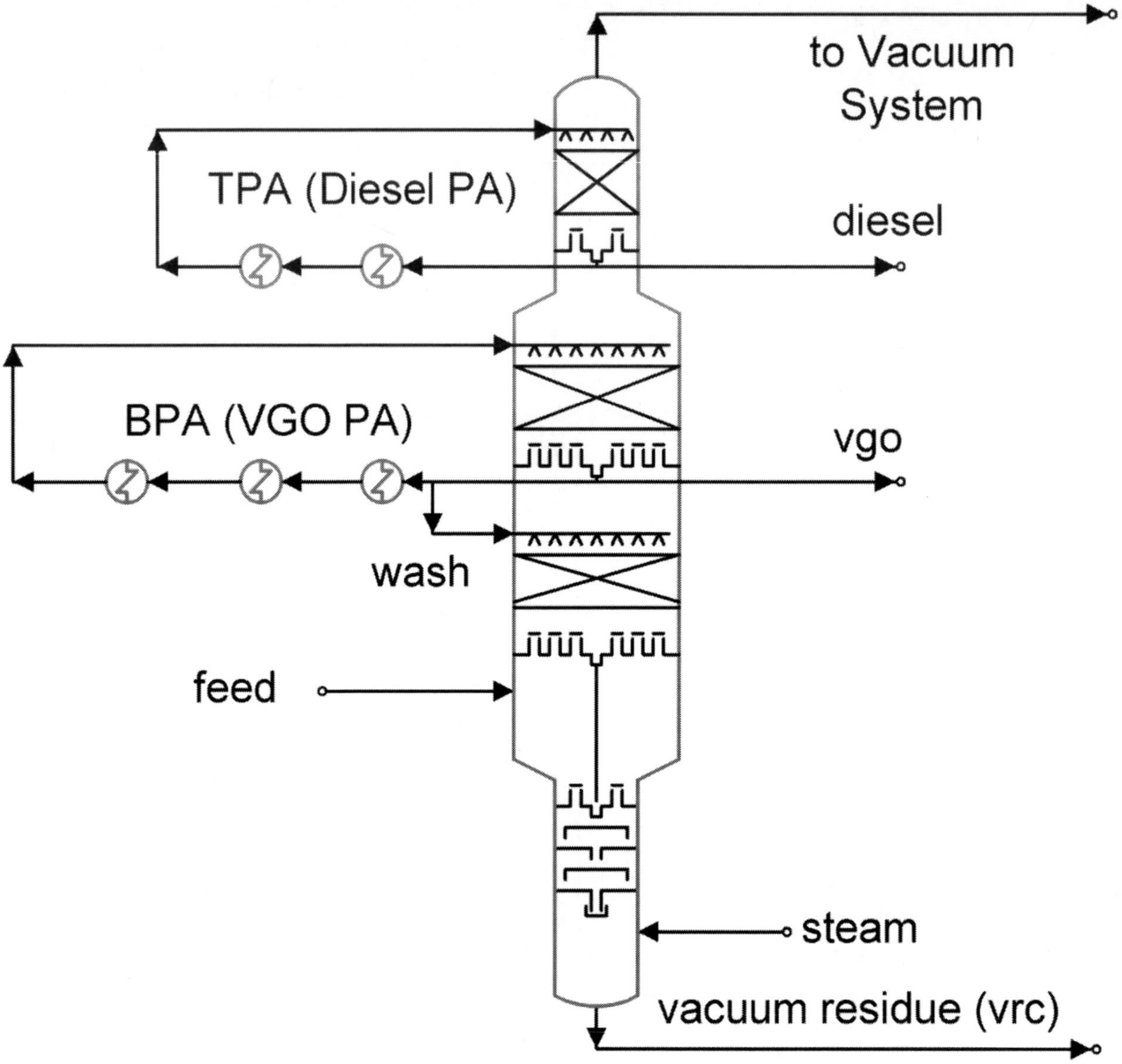

Figure 5 – Case 8: 2 Pumparound, 2 Product – Top Product as Diesel, No Internal Fractionation

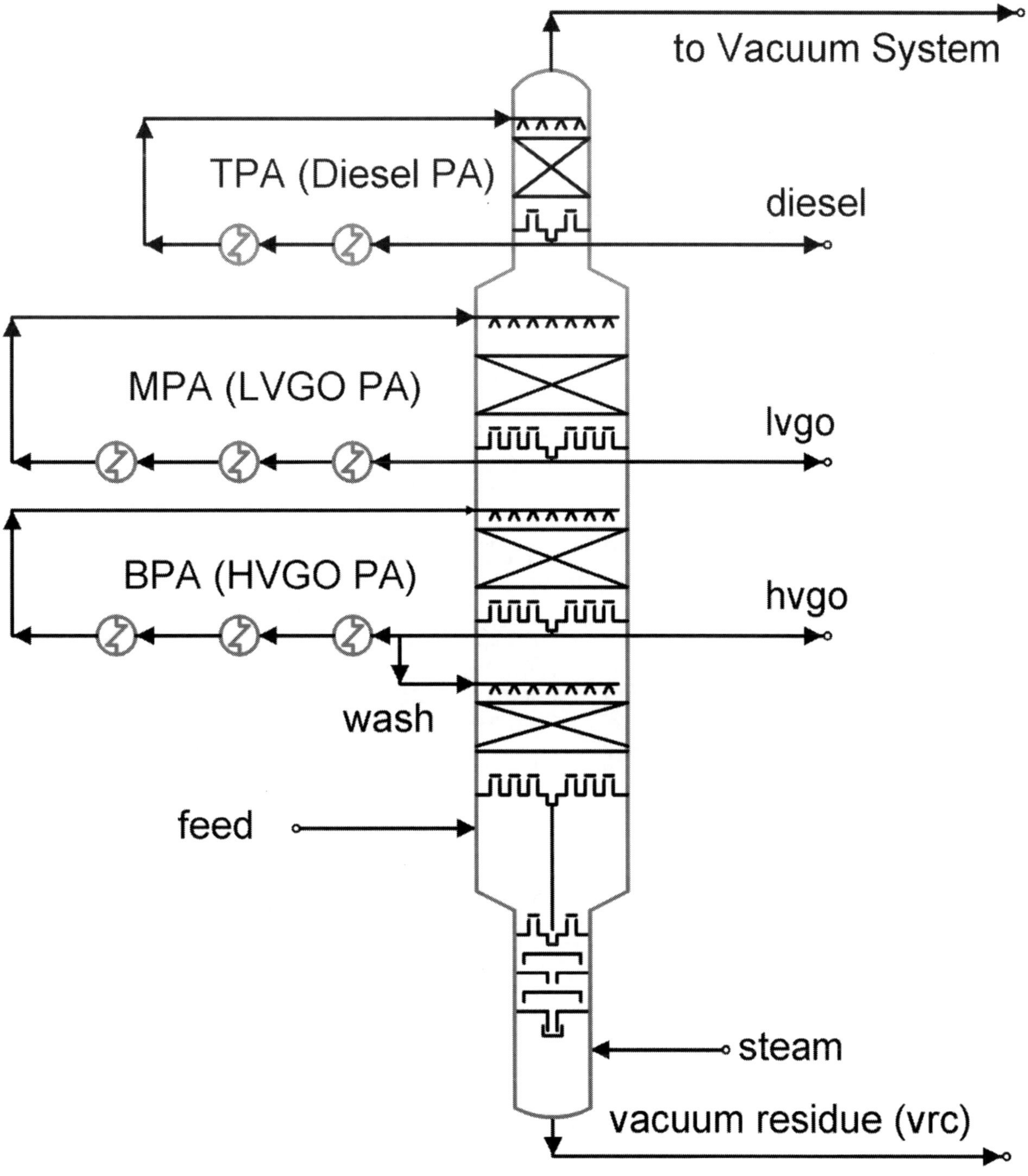

Figure 6 – Case 9: 3 Pumparound, 3 Product – Top Product as Diesel, No Internal Fractionation

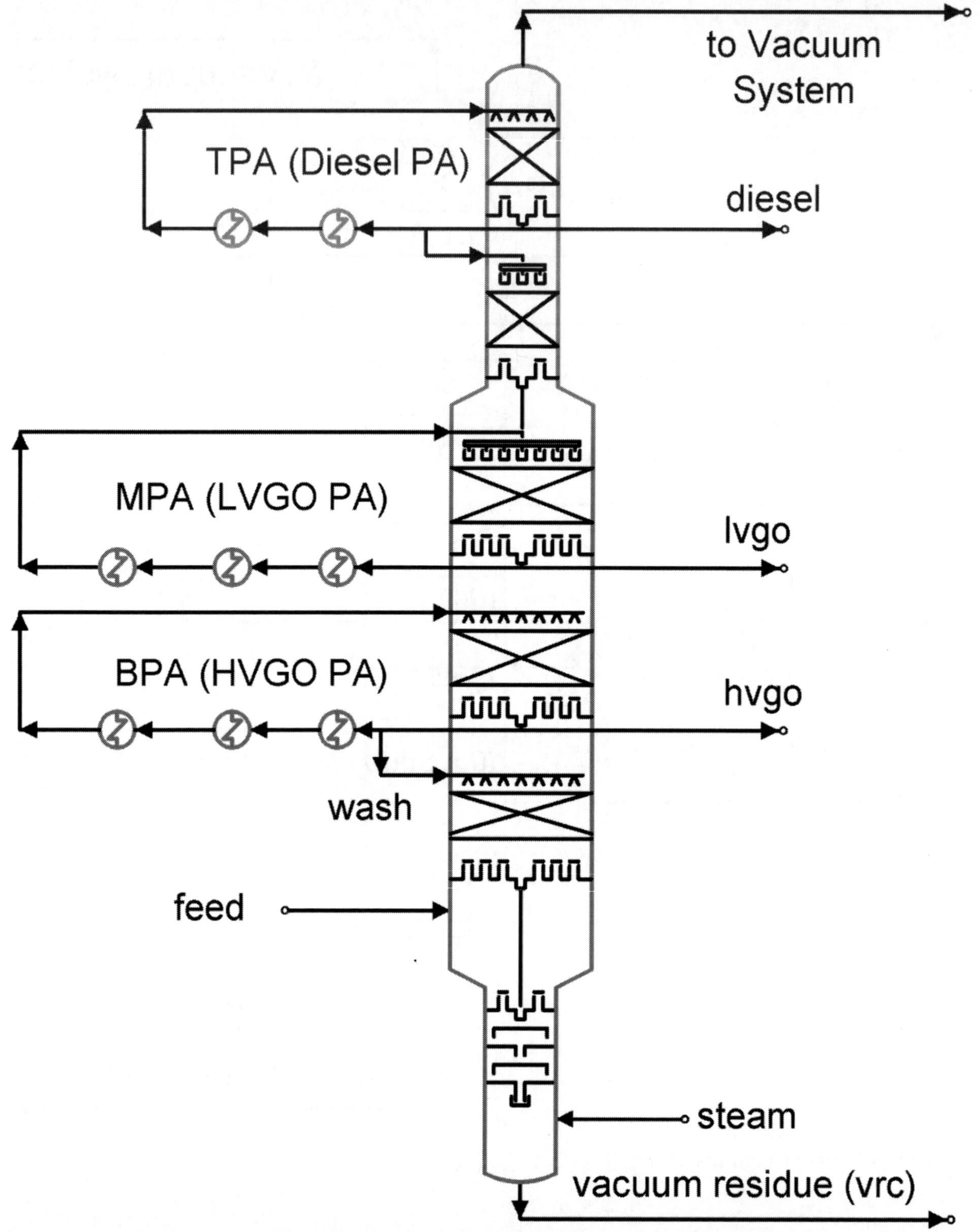

Figure 7 –3 Pumparound, 3 Product – Top Product as Diesel, Internal Fractionation

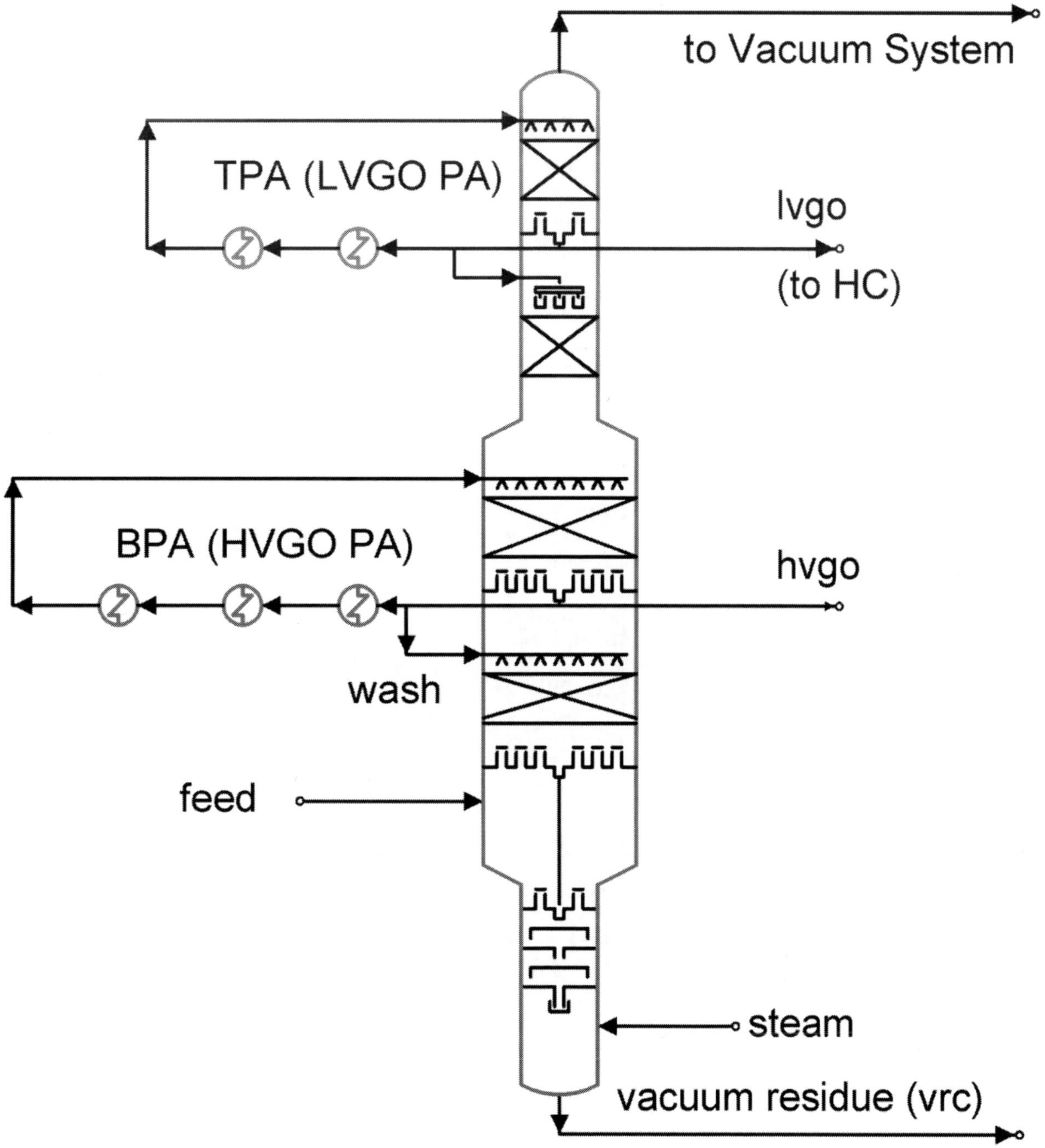

Figure 8 – 2 Pumparound, 2 Product – Hydrocracker Feed Operation, Internal Fractionation

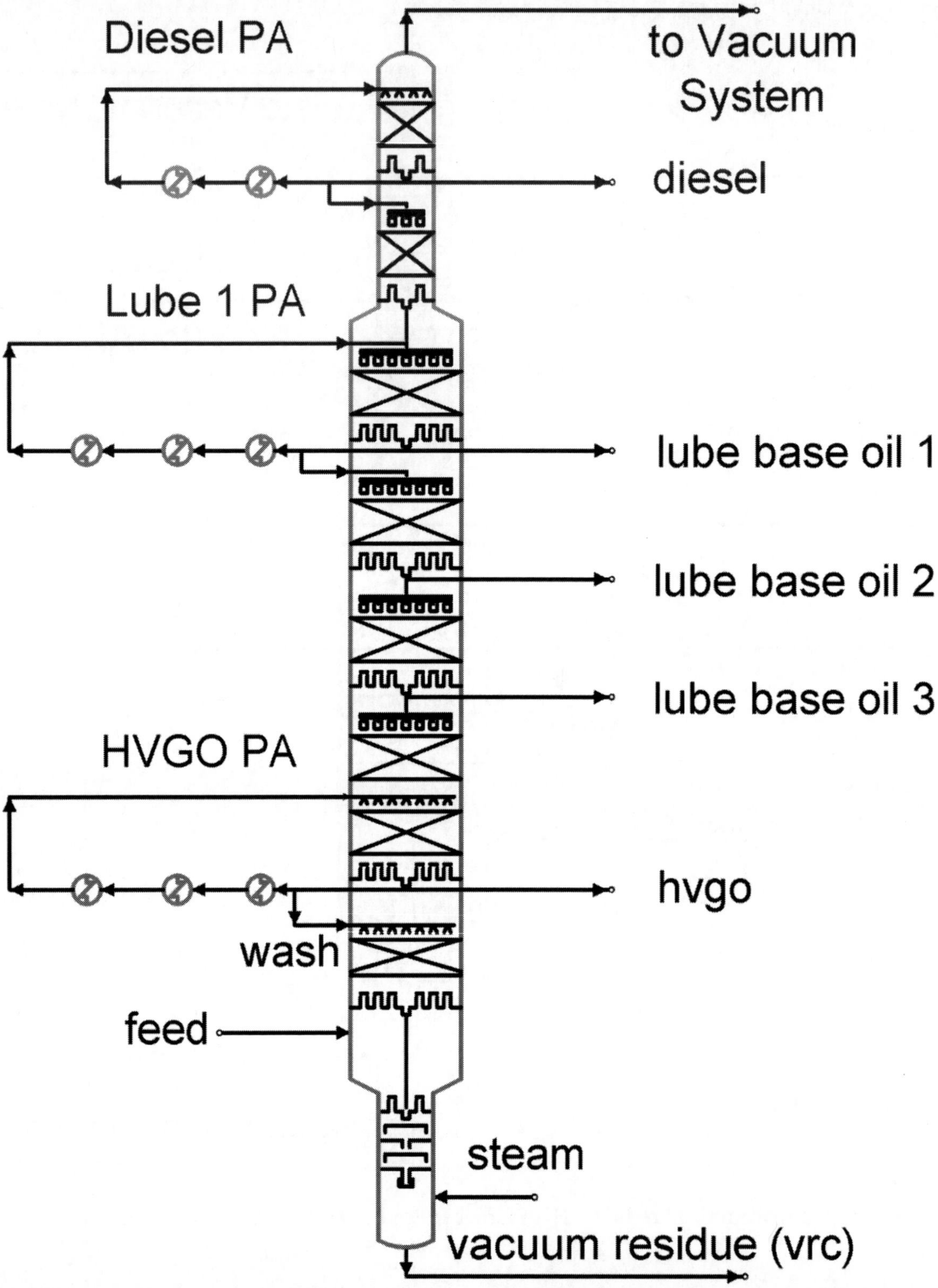

Figure 9 – Lubricants Vacuum Tower

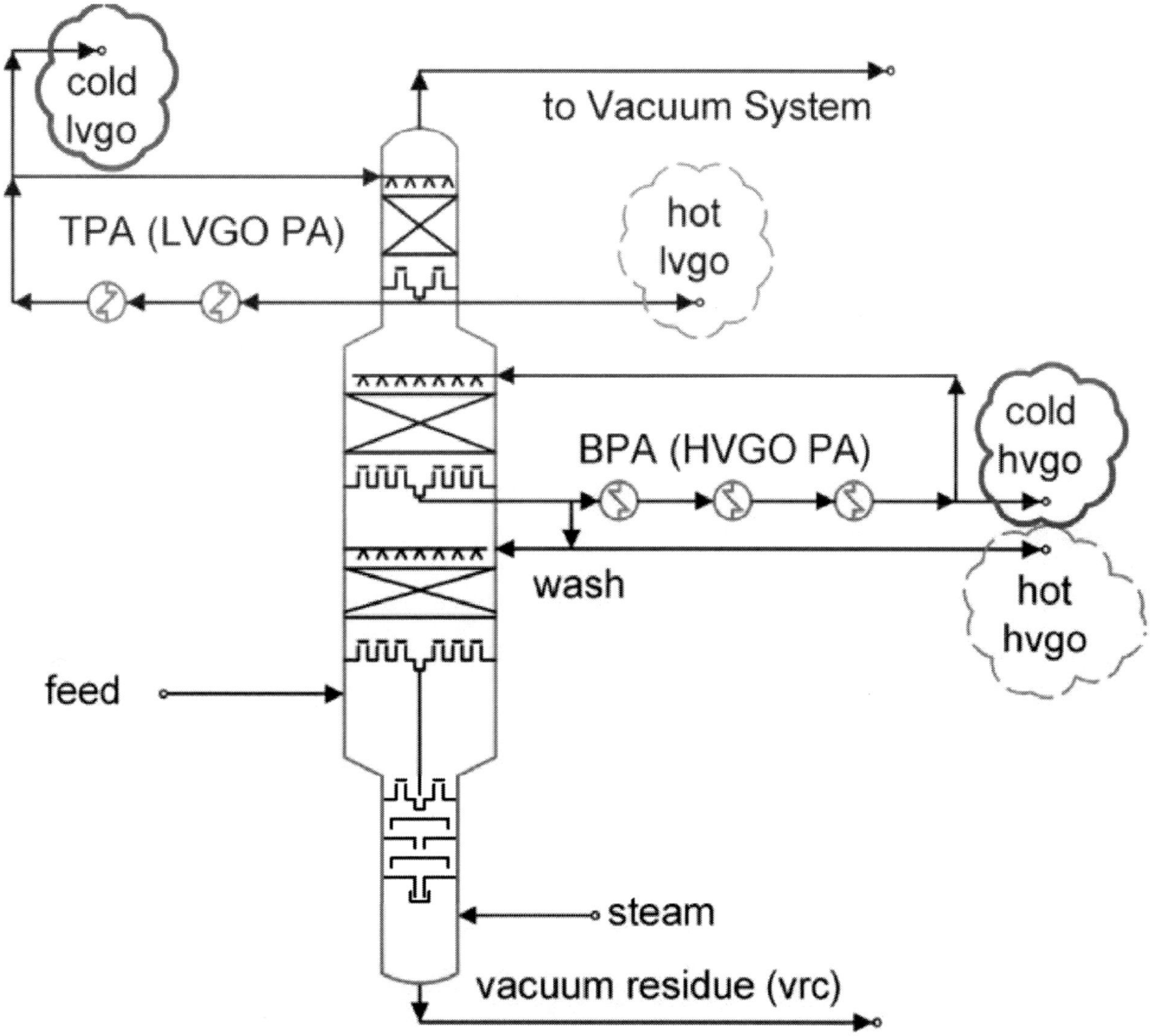

Figure 10 Hot-product Versus Cold-Product

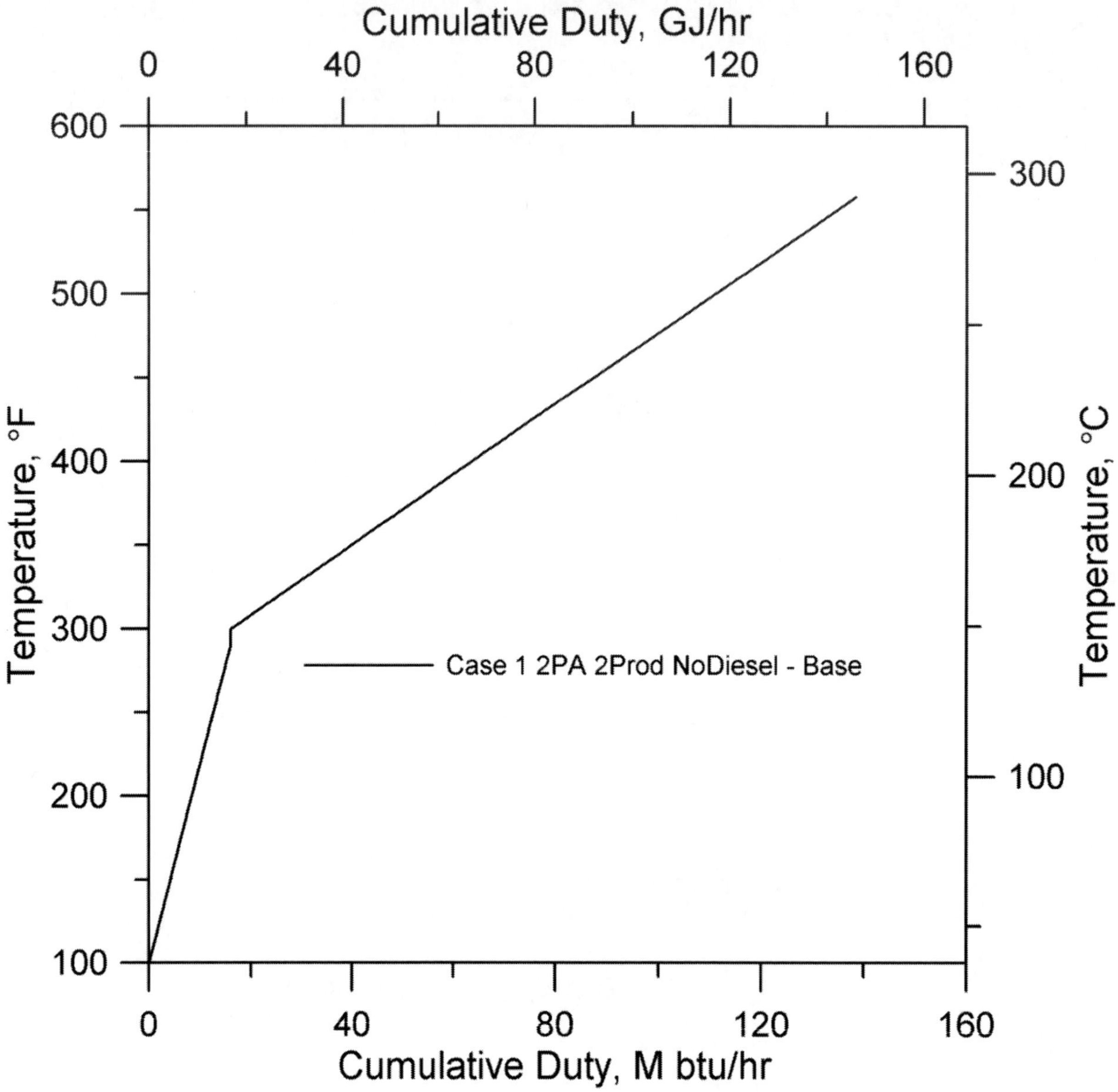

Figure 11 Case 1 (Base) Cumulative Duty Recovered Versus Temperature

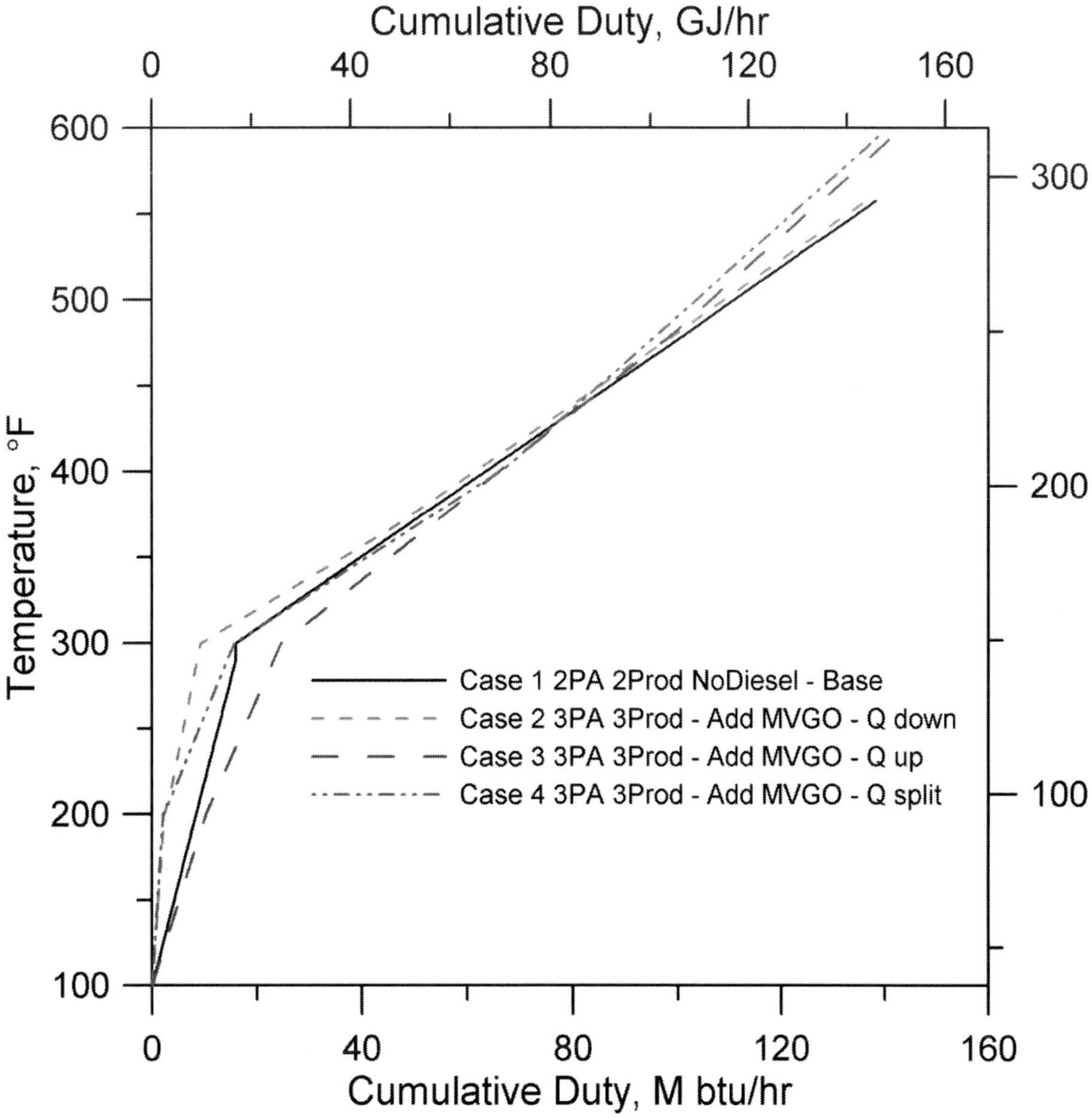

Figure 12 Cumulative Duty Recovered Versus Temperature – Adding a Pumparound

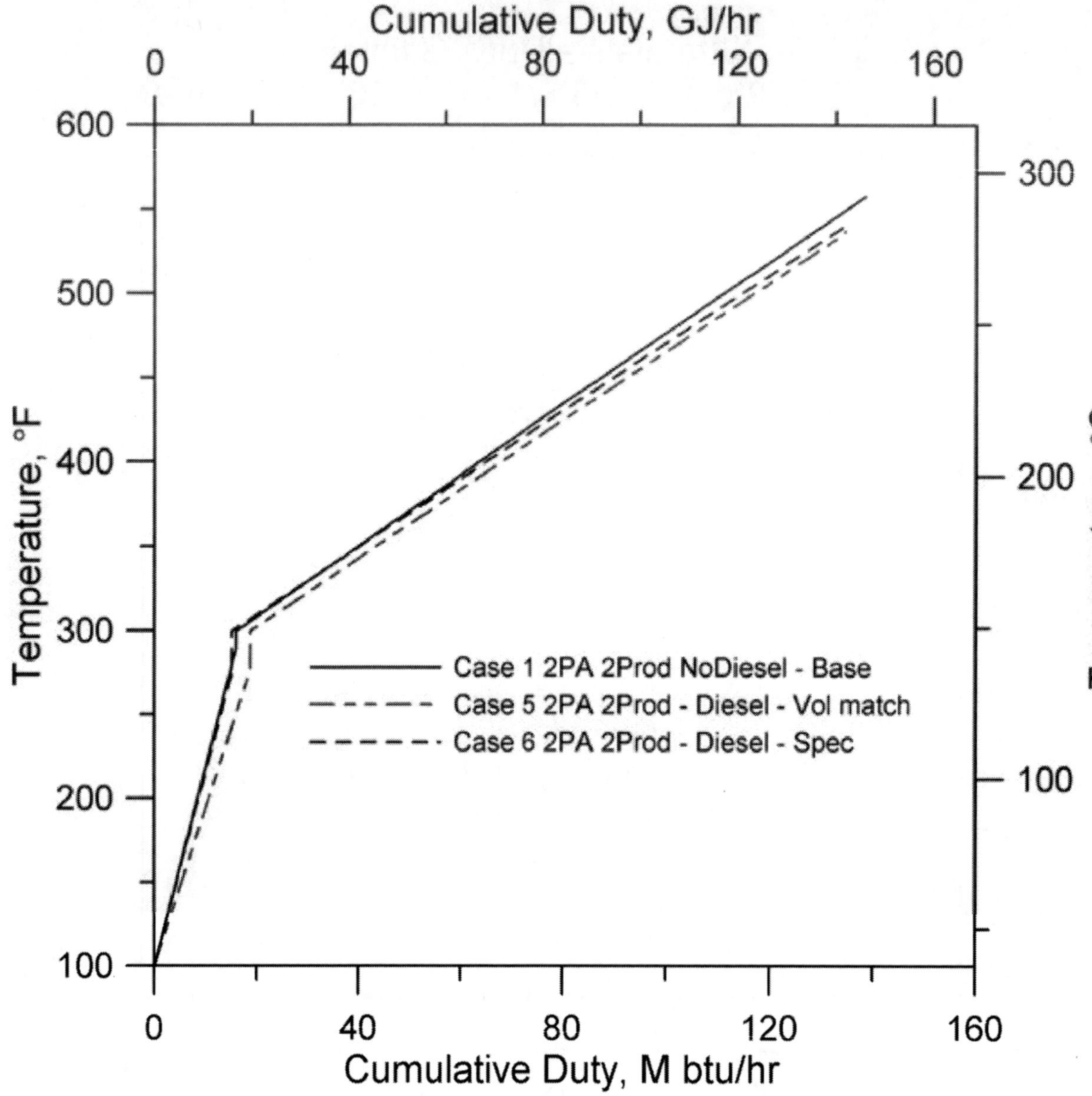

Figure 13 Cumulative Duty Recovered Versus Temperature – Diesel Recovery

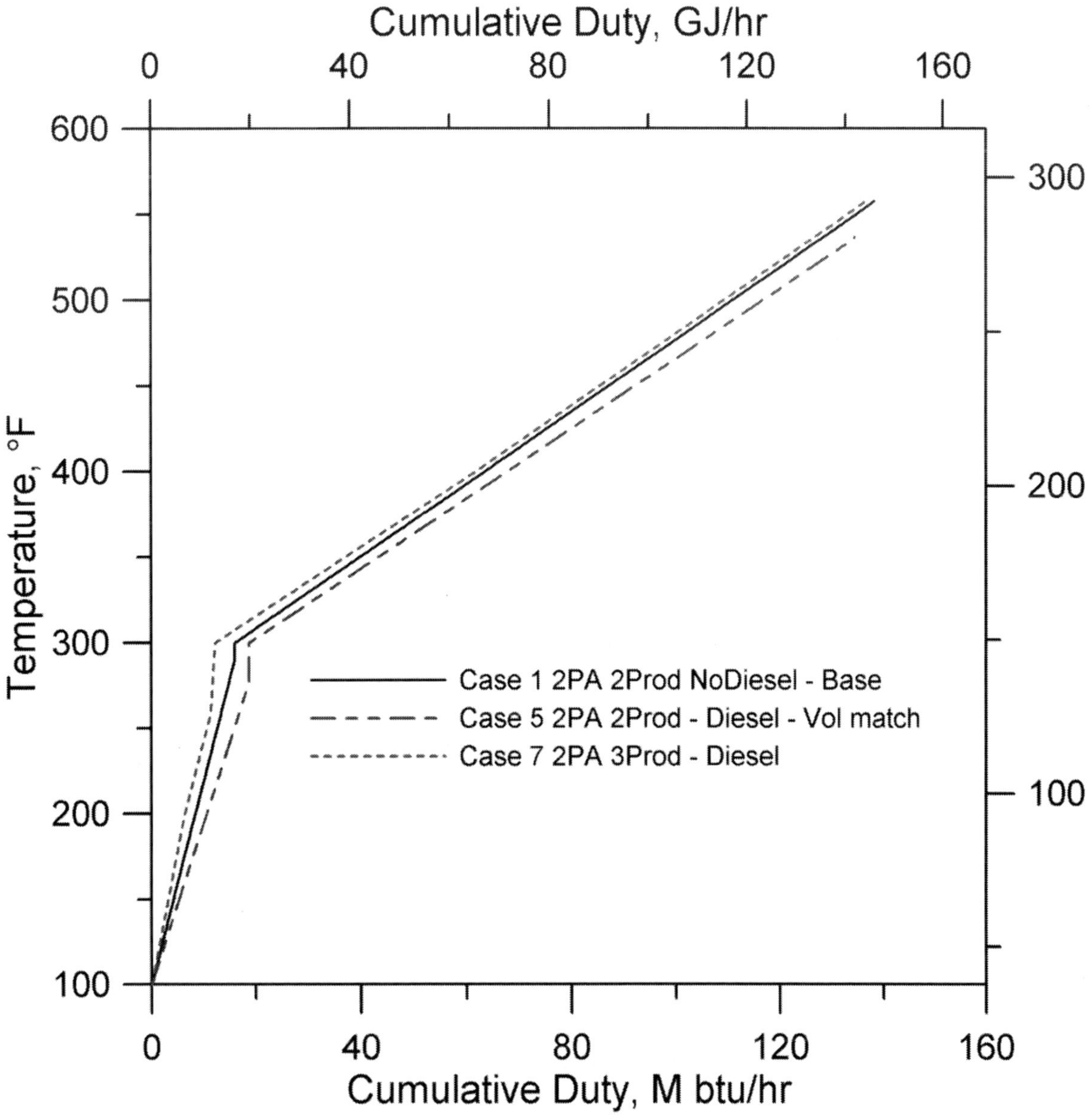

Figure 14 Cumulative Duty Recovered Versus Temperature – Diesel Reflux Options

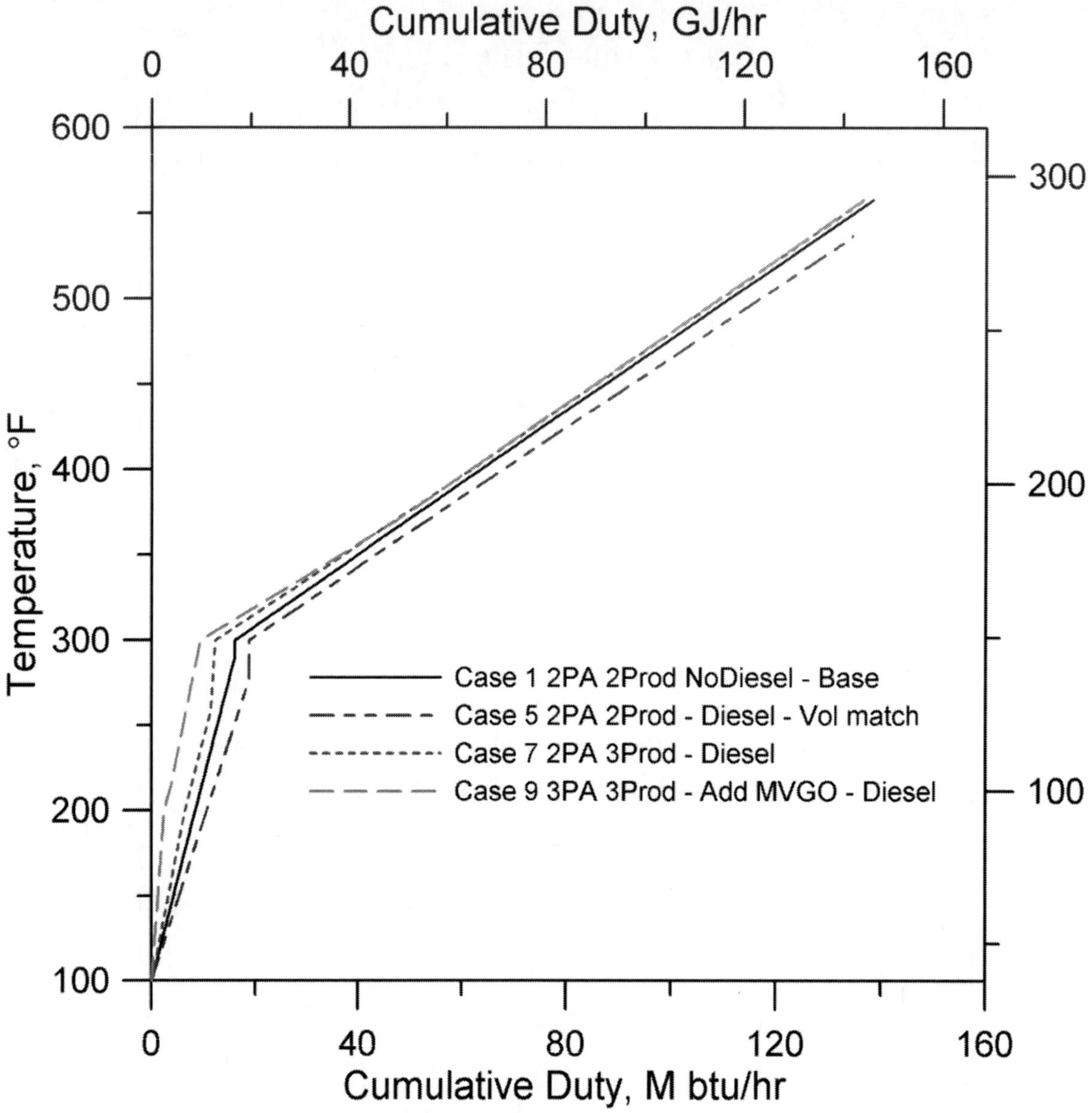

Figure 15 Cumulative Duty Recovered Versus Temperature – Diesel Options Expanded

13. Appendix – Glossary, Definitions, Abbreviations and Conventions

Abbreviations and terms in text and figures that may not have been defined or that may be non-standard usage.

These definitions are tailored to the most common industry use of the term in crude distillation and petroleum refining. Many plants use their own variations on these definitions and other terms in place of those listed. Other technical fields may have different meanings for these terms.

Term or abbreviation	Definition
API degrees, API°	See API Gravity
API Gravity	A measure of liquid density defined by the American Petroleum Institute. The scale was derived from the reading marks on a specific type of measurement instrument. $API^{O} = \frac{141.5}{SG} - 131.5$ $SG = \frac{141.5}{API^{O} + 131.5}$ Where SG is specific gravity at 60°F (15.6°C). Water has an API gravity of 10. An API gravity lower than 10 is denser than water, an API gravity higher than 10 is less dense than water.
ASTM Distillation	Old usage. ASTM Distillation quoted without a descriptive qualifier refers to an ASTM D86 distillation. This is a batch distillation without reflux.
Boiling range	See distillation range.
BPA	Bottom pumparound.
bpd	Barrels per day, standard volume barrels at 60°F (15.6°C)
CP	Cutpoint
Cutpoint	The ASTM D2892 distillation temperature that would achieve the same split between overhead products and bottoms products that the unit achieves. The D2892 distillation basis is atmospheric equivalent temperature. Typical usage is volume percent.
Distillation range	Range of AEBP boiling temperatures for a stream. Typically expressed as an ASTM D-2892 equivalent (true boiling range) or an ASTM D-86 equivalent (ASTM boiling range) on a 5-95 volume percent basis. Other basis may be used.
Flash zone	Feed entry point into a main fractionator.
MB	Million btu/hr (abbreviated to fit in figures), mmbtu/hr
MBtu/hr	Million btu/hr, mmbtu/hr
mmbtu/hr	Million btu/hr, industry standard usage remains the old form using a

	lower-case m for thousand. 'mm' equals thousand X thousand to get million.
MPA	Middle pumparound.
Specific gravity	Density of a stream at 60°F (15.56°C) compared to the density of water at 60°F (15.56°C).
T_{XX}	Temperature at which XX volume (or weight) percent of liquid is distilled.
TBP	True boiling point. Distillation curve based on ASTM D2892 or an equivalent.
TPA	Top pumparound.
Transfer line	Pipe connecting the outlet of a fired heater to the downstream main fractionator.

Figure conventions;

- Lower case text and abbreviations in figures refer to a stream.
- Upper case text and abbreviations in figures refer to an equipment item.

Paper 27b

Atmospheric Crude Distillation

Equipment State of the Art

Andrew W. Sloley

Advisian
Energy Center II
575 N Dairy Ashford St
Houston, TX 77079

Prepared for Presentation at the
American Institute of Chemical Engineers
2016 Spring Meeting
Houston, TX April 10-14, 2016

UNPUBLISHED

1. Summary

Atmospheric crude distillation is the first processing step in the oil refinery. Overall distillation performance sets feed rates and properties to downstream units. Improved separation increases refinery profits and flexibility.

This paper provides an overview of the state-of-the art on atmospheric crude distillation equipment. The major focus is on towers with trays. However, some comments are included on specific situations where packing may be preferred.

Specific topics covered include:

- Service types (fractionating, heat-transfer, de-entrainment, stripping)
- Vapor and liquid rate zones
- Conventional versus high-capacity trays
- Blowing in wash trays
- Mechanical reliability
- Stripping trays and packing

In many ways refinery atmospheric crude distillation towers fit conventional mass-transfer equipment requirements. However, specific parts of the atmospheric crude distillation service merit special attention. The discussion highlights areas requiring special care, areas where conventional equipment fits, and situations well suited for modern high-capacity column internals.

Services commonly include heat transfer (pumparounds), fractionation, de-entrainment (feed entry) and stripping trays. In modern atmospheric towers the dominant choices revolve around trays. For a given capacity, a tower with trays usually has a lower investment cost than a tower with packing, even if the packed tower might have a smaller diameter. An exception is the wash zone of the atmospheric crude column. For that service, grid or structured packing often gives better overall economics.

High-capacity trays have been a major recent development. They often allow for significant capacity increases in existing towers. For new units conventional trays with sloped downcomers are usually the favored choice for minimum investment.

Process upsets often subject the stripping section of these towers to large mechanical forces. Pressure relief design (trap door) trays and mechanically reinforced trays are often favored for this service.

2. Atmospheric Tower Description

Figure 1 shows the conventional layout of an atmospheric crude tower. Two examples of major physical components present in typical towers for this service are illustrated. The first example has a packed wash section. By choice, the packing should be structured packing, grid or a combination of the two. The second example has trays throughout the column.

Historically most atmospheric crude towers use the second configuration: entirely trayed. However, the first configuration, the tower with the packed wash section, has significant technical advantages that often make it more profitable even though investment is higher.

3. Predominant Equipment

3.1. Trays

Trays are the predominant fractionation equipment in atmospheric crude towers. Economics are the driving force behind this choice. For a given capacity, even though a conventional tower with trays has a larger diameter than a tower with packing, the overall installed cost of the tower with trays (including auxiliaries and internals) is lower than the cost of a tower with packing.

Functionally, trays have three operating areas:

- Active area: the region where the vapor and liquid mix. Mass transfer occurs in this area.
- Disengagement area: the region above the tray where liquid droplets disengage from the rising vapor.
- Downcomer: the region where vapor bubbles disengage from the liquid. The downcomer also accumulates a liquid head that provides the driving force for flow down the column. The downcomer must provide enough inventory to overcome the total of:
    - Vapor pressure drop trough the tray
    - Head losses for liquid flowing down the tower
    - Hold-up provided by the outlet weir (if any)
    - Crest height over the outlet weir (if any)

Figure 2 illustrates these sections for a conventional tray.

Three other factors favor trays in atmospheric tower services:

- Atmospheric towers have multiple side draw products and multiple vapor feeds from sidestrippers. Sumps can be easily added to trays for partial liquid draws. The modest vapor rates from the side-strippers in this service usually allow for the use of tower nozzles without internals. The trays have sufficient pressure drop to adequately distribute the vapor. Even if high rates are present in the side-stripper returns, relatively simple internal pipes can distribute the vapor.
- Liquid draws for pumparounds are relatively straightforward. Again simple sump and piping arrangements are adequate.
- The trays in the top of the column often suffer from salt deposition. Water injection may clean out soluble salts. This is more effective for cleaning trays than packing.

3.2. Equipment Capacity Evaluation

Conventional trays have multiple mechanical features that determine the trays' upper capacity, lower capacity, and efficiency. For atmospheric crude columns the major mechanical choices are tower diameter, tray spacing, and number of liquid flow paths for the tray. As a rule-of-thumb, Table 1 shows the basic relations between these variables.

Table 1 Major Tray Capacity Trade-offs

Feature	Liquid Handling	Vapor Handling
Tray spacing, more distance	Increases maximum rate No change on minimum rate	Increases maximum rate No change on minimum rate
Tray spacing, less distance	Decreases capacity No change on minimum rate	Decreases capacity No change on minimum rate
Tray diameter, more diameter	Increases maximum rate Increases minimum rate	Increases maximum rate Increases minimum rate
Tray diameter, less diameter	Decreases maximum rate Decreases minimum rate	Decreases maximum rate Decreases minimum rate
Tray passes, more	Increases maximum rate	Approximately no change in rates
Tray passes, less	Decreases maximum rate	Approximately no change in rates

Tray design and performance is highly complex. Many specific situations can create exceptions to these rules. In certain cases, what look like minor modifications to a tray layout may cause large changes in tray performance. Engineers should consult detailed references to tray operation when evaluating specific circumstances [1] [2].

3.3. Active Area Devices

Current industry practice uses both sieve trays and valve trays in atmospheric tower service. Based on surveys by the author, ~75% of the trays in atmospheric columns, excluding stripping trays, are valve trays. The trend over the last 10 years has been an increasing number of fixed valves. Nevertheless, movable valves remain the most common valve-type. Sieve trays cost less but have lower vapor handling capacity. Correctly applied sieve trays and valve trays have similar capacity and efficiency for the different services.

A few units remain in service with bubble-cap trays and tunnel-cap trays. Due to higher pressure drop, lower capacity, and higher cost these trays are much less common today compared to 30 years ago. The benefit of bubble-cap and tunnel-cap trays was capability to run at very low liquid rates.

3.4. Flexibility of Multiple-Pass Trays

Multiple pass trays divide the liquid flow into different flow paths. Figure 3 shows one-pass and two-pass trays. Figure 4 shows three-pass and four-pass trays. Multiple flow paths have two significant benefits:

- Increased liquid rate per flow path decreases weir load (volumetric flow per unit length of weir). Lower weir loads reduce tray pressure drop and increase tray capacity.
- Decreased flow path length across the active area. If the flow path length becomes too long, performance may suffer from vapor cross-flow channeling [3] [4]. For most towers, flow path lengths of over 8-10 ft (2.4-3.0 m) create situations where cross-flow channeling may occur. In specific cases, lower flow path lengths may create a problem.

The author rarely uses three-pass trays. They have a non-symmetrical layout and may be difficult to design. Other options, including some of the multiple downcomer trays (§3.5) often perform more effectively than three-pass trays.

Increasing the number of flow paths increases maximum liquid handling capacity. Many towers use multiple flow paths to handle pumparound liquid rates.

Increasing the number of flow paths also increases the minimum vapor rate required. At low vapor rates, some of the tray sections may remain inactive [5]. Inactive tray sections reduce heat-transfer and mass-transfer efficiency. Preventing inactive sections requires a minimum number of valves to open [6]. Table 2 summarizes guidelines for this.

Table 2 Vapor Turn-Down for Valve Trays

Flow paths	Minimum Number of Valves Open %
2	40
3	60
4	80

The minimum number of valves open does not directly correlate with vapor rate turndown. A valve tray may have 100% of its valves open at much less than 100% of the tray's vapor handling capacity. Vapor rate flexibility can be improved by using valves with multiple weights and using caged valves. Both steps increase the tray pressure drop. The higher the tray pressure drop, the higher the flexibility.

3.5. High-Capacity Trays

Current high-capacity tray technology uses one or more of the following to increase tray capacity:

- Increased weir length to reduce tray pressure drop. This increases both vapor and liquid handling capacity (Figure 5).
- Directional valves to use the vapor to push liquid across the tray (Figure 6).
- Hanging downcomers to increase the active area on the tray. This increases vapor handling rate (Figure 7).

- Centrifugal devices to increase vapor-liquid separation.
- Other de-entrainment devices to improve vapor-liquid separation.

Techniques are listed in chronological order of development. The first three have extensive application in atmospheric crude towers. When done correctly, they may significantly improve tower capacity. The last two, centrifugal separators and use of de-entrainment devices, are relatively rare due to their expense. In general, high capacity devices have lower operating flexibility than conventional trays [7].

4. Equipment Services

Equipment inside the atmospheric tower can be classified into the unit operation function (heat transfer...) or the service (diesel pumparound...).

The basic equipment functions in the refinery atmospheric column are:

- Heat-transfer (always present)
- De-entrainment (always present)
- Mass-transfer (always present)
- Stripping (nearly always present)
- Auxiliary equipment (always present)
    - Structural components (beams, rings)
- Auxiliary equipment (often present)
    - Collector trays
    - Feed entry devices

Figure 8 illustrates the location of the functions in an atmospheric tower:

- Mass-transfer – fractionation between products
- Heat-transfer – direct contact heat exchange for the pumparounds
- De-entrainment – removal of liquid droplets from rising vapor
- Stripping – recovery of light material from tower bottoms

In a refinery atmospheric tower, most of the internals are in mass-transfer operation. The other operations, while important, use up less of the tower.

5. Mass-Transfer

Mass-transfer equipment has the intent of fractionating between adjacent product cuts. The refinery atmospheric column creates the feeds to the downstream units. Some flexibility is possible on most of the products. Nevertheless, fractionation requirements are critical for the atmospheric column. The trays must work properly and have a reliable efficiency.

5.1. Tray Efficiency – General

Tray efficiency is normally stated as a percent of the effectiveness of a theoretical plate. A tray with the same effectiveness as a single theoretical plate has 100% efficieny. Tray efficiency is determined by the effectiveness of the vapor-liquid mixing on the tray and how much liquid or vapor bypasses the mixing. A partial list of mechanical and hydraulic factors that affect tray efficiency includes:

- System physical properties
- Absolute vapor and liquid rates

- Vapor-liquid ratio on the tray
- Backmixing on the tray
- Mechanical layout including
  - Device type (sieve hole, fixed-valve, movable valve, bubble-cap, other)
  - Device (sieve hole or valve) size
  - Flow path length
  - Downcomer inlet and outlet velocities
  - Downcomer weir height
  - Weir geometry (setback, picketed, other)
- Condition (fouled, corroded, leaking, blocked)

Different methods of defining tray efficiency have been used. The most common definition is the Murphree tray efficiency [8] [9]. Murphree tray efficiency is used here.

Many correlations are available for calculating tray efficiency. The author has always taken the position that the more ways are in use to calculate something, the more likely it is that none of them work very well. The reason someone develops another correlation (or is willing to pay for it to be done) is that the previous methods do not give useful results. The author's library of papers on tray efficiency has roughly 75 different methods and variations attempting to create a unified approach to tray efficiency calculation. Most of them only apply to specific physical conditions, tray diameters, and limited equipment operating characteristics. None of them work well as a general method.

Tray efficiency is best determined from operating data. For atmospheric distillation column fractionating trays, the most important factor determining tray efficiency is the volumetric ratio of vapor and liquid on the tray. This assumes well-designed, properly installed, and maintained trays.

Discussions of tray efficiency often use detailed correlations combining physical properties such as viscosity, diffusivity, and composition gradients. However, for the mixtures of hydrocarbons present in crude oils at a vapor-liquid equilibrium (boiling-condensing), the differences due to these factors are minor. Only liquid-phase viscosity variation has any significant effect. For the heavier products in the column (going down the column), the viscosity increases, reducing tray efficiency.

5.2. Tray Efficiency – Liquid and Vapor Hydraulics

The major factor determining tray efficiency in atmospheric crude distillation is the liquid-vapor volume ratio on the tray. Figure 9 shows a conventional plot of vapor-liquid profiles in an optimized atmospheric column [10]. The conventional plot gives a rough idea of the vapor-liquid loads. Descending the column, the liquid loads in the fractionation sections generally drop. This results from two factors:

- Energy integration favors recovering heat at as high a temperature as possible. This shifts condensation into the lower parts of the column.
- Product draws take liquid from the column.

The example shown has large liquid rates between trays 21-32 and trays 3-8. This is due to large economic incentives for maximizing the middle distillate (kerosene plus diesel) yields versus other products. But for the typical tower, liquid rates will more closely align with those shown on trays 40-35 and trays 17-12.

While Figure 9 gives a general idea of the vapor and liquid loads, it does not give full insight into the tray hydraulics. Figure 10 gives a better comparison. Figure 10 modifies the presentation in four ways:

- The rates are converted to wet rates. This includes steam present in the vapor. This gives the actual vapor load required for tray hydraulics.
- The rates are converted to actual volumes.
- The vapor has been changed from net vapor (vapor leaving the stage) to entering vapor (vapor entering the stage). Fractionation calculations use net vapor and net liquid rates. This is the vapor leaving the stage and liquid leaving the stage. For proper evaluation of tray pressure balance the required rates are entering vapor and leaving liquid.
- The results are shown as a ratio of liquid-to-vapor on the tray.

All fractionation sections have less than 1% of the system volume as liquid. In fact, liquid fractions drop to as low as 0.2%. Very low liquid rates shift the tray operating regime from the froth regime into a spray regime [11] [12] [13]. Spray regimes tend to have lower efficiencies. The section on wash services (§7.2) includes a more detailed discussion of some problems with low liquid rates and operation in the spray regime.

5.3. Tray Efficiency - Recommendations

Based on plant operating data the following efficiencies are recommended for the services listed:

Table 3 Recommended Fractionation Service Efficiencies [2]

Service	Based on Tray Type	Efficiency, %
Light naphtha-heavy naphtha	Valve	75
Heavy naphtha-kerosene	Valve	70
Kerosene-Diesel	Valve	70
Diesel-Gas Oil	Valve	65
Wash	Valve	50
Stripping	Sieve	30
Side Stripping	Valve	25

Observed efficiency is typically within 10% of these values. The efficiencies quoted are for typical operating conditions and reflect conventional liquid-vapor ratios and physical properties for these services. Wash, stripping and sidestripping services are included for completeness.

Operating data, when available, should be used to calculate actual efficiencies for existing columns.

6. Heat-Transfer Services

The heat-transfer service in the refinery atmospheric tower takes sub-cooled pumparound return liquid and condenses the rising vapors. The heat-transfer returns the pumparound liquid to saturated conditions. As they lose heat, the vapors condense and create the net product liquid. Industrial units have been built with from zero (spray chamber) to 10 trays in pumparound sections.

Tray calculations for heat-transfer are complex [14] [15] [16]. The dominant factors are the relative vapor and liquid rates and mixing energy. These set the surface area and vapor bubble size. Experience has shown that 2-3 trays will meet heat-transfer requirements in nearly all pumparounds. Most units use three trays in each pumparound. Revamps that add a pumparound to an existing tower often use only two trays to minimize tower changes.

7. De-Entrainment Services

The atmospheric tower feed enters in the tower flash zone. The high-velocity feed creates small liquid droplets. The feed entry device coupled with the lower column vapor velocity allow the larger droplets to fall towards the tower bottom. However, the smaller droplets still entrain up the tower.

The section of the tower between the lowest product side draw and the feed is often called the wash zone. The name wash zone recognizes that the major purpose of this section of the tower is to de-entrain liquid droplets from the vapor leaving the flash zone.

7.1. Fractionation versus De-entrainment in the Wash Section

Either fractionation or de-entrainment performance may set the liquid rate required in this section. Figure 11 shows two atmospheric crude towers. One tower has AGO as the lowest side-draw product. The other tower has diesel as the lowest side draw product.

AGO normally has no distillation specification on the heavy end. Downstream processing economics favor making the product as heavy as possible (maximizing product yields). Entrainment of atmospheric residue contaminants limits the product quality. The trays, or packing, between the flash zone and the AGO are purely a de-entrainment device.

Diesel has a distillation specification on the heavy end [17]. Specific plants may vary from this because of blending and other constraints, but the atmospheric crude tower must produce a product stream close to the specification values. This distillation specification is in addition to the de-entrainment requirements. In a tower with diesel as the lowest side-draw, either the fractionation requirements or the de-entrainment requirements set the liquid load in the wash section. In nearly all cases, the fractionation load sets a higher liquid requirement.

7.2. Minimum Liquid Load for Trays versus Packing

One lower limit for tray capacity is normally quantified as a minimum liquid load on the outlet weir. At low liquid rates with low vapor density and high liquid density, tray operations change from continuous liquid-phase (froth) to continuous vapor-phase (spray) [18]. The lower liquid limit quoted before this transition causes efficiency to fall is 1 gpm/inch of weir length [6] (1.49 lpm/cm).

When the vapor phase is continuous, the liquid no longer flows smoothly over the outlet weir. Figure 12 illustrates this by showing three zones on a tray in spray operation. In all areas the rising vapor creates drops of liquid spray over the tray deck. In the inlet area, spray back against the downcomer is collected and falls back to the tray. In the center of the tray, the spray droplets fall back down to the tray. Close to the outlet, the vapor projects some of the spray droplets over the downcomer. The droplets over the downcomer fall to the tray below. The tray functions in a diffusion-like operation with the driving force for flow coming from a mass-density gradient across the tray. This operation is called tray blowing [19]. When trays enter blowing operation both fractionation and de-entrainment effectiveness drop.

Figure 1 shows atmospheric crude columns with both a packed wash section a trayed wash section. The tower with the trayed wash section shows one-pass trays. Using one-pass trays is an attempt to reduce the weir length and prevent blowing. However, if the diameter is too large, the flow gradient across the tray can create cross-flow channeling and other problems.

The tray with the lowest liquid load is the tray immediately above the flash zone. Table 4 illustrates the effect of some different options to resolve tray limits. The top half of Table 4 shows the effect on yields for a constant feed temperature. The bottom half of Table 4 shows the change in required feed temperature for constant yields. The results are at a constant tower pressure and stripping steam-to-bottoms ratio.

Table 4 Trays or Packing in Wash Zone Service

Packing					110% of min liquid
Tray passes		2	1	1	1
Liquid at weir	gpm/in	1.00	1.00	0.50	0.30
Weir length	Inches	188 *	110	110	110
Flow path length	Inches	76	152	152	152
Overflash	Vol % of crude	4.87	2.86	1.43	0.86
Constant heat input					
Distillate yield	Vol %	56.91	58.48	59.63	60.10
Feed Temp	°F	690.0	690.0	690.0	690.
Heat duty	M btu/hr	222.6	222.6	222.6	222.6
Stripping Stm	Lbs/hr	17,990	17,320	16,840	16,640
Bottoms Temp	°F	668.2	669.9	670.9	671.2
Constant Yield					
Distillate yield	Vol %	60.10	60.10	60.10	60.10
Feed Temp	°F	715.0	702.7	693.8	690.0
Heat duty	M btu/hr	248.4	235.7	226.5	222.6
Stripping Stm	Lbs/hr	16,640	16,640	16,640	16,640
Bottoms Temp	°F	692.8	682.5	676.7	671.2

* bottom tray has outboard downcomers.

Table 4 shows four cases;

- A two-pass tray at 1 gpm/in (1.49 lpm/cm). This tray has a high liquid flow requirement. Overflash is the amount of liquid that must flash in the tower over the rate of tower products to produce the liquid required for the wash section. Numbers for this are often quoted, but the author finds the concept of little use. Liquid rates depend highly on the tray configuration. However, the value here of 4.87% is very close to oft-quoted values of 5%. All the other trays have better performance.
- A one-pass tray at 1 gpm/in (1.49 lpm/cm).
- A one-pass tray with 0.50 gpm/in (0.75 lpm/cm) weir load. This is around the lower limit for even mediocre de-entrainment on trays.
- A one-pass tray with 0.30 gpm/in (0.60 lpm/cm) weir load. This was chosen for matching performance to a tower with a packed wash section. For a conventional tray this is likely to have ineffective de-entrainment.

Minimum liquid requirements on the bottom tray have significant yield or operating effects. For all liquid rates, the two pass tray example has a 71% higher liquid demand (188" weir length compared to 110" weir length). Tray design can optimize these rates; nevertheless, the possible improvements all have limits. Some approaches include:

- Smaller tower diameter by using a larger tray spacing, 30" instead of 24", for example.

- Parallel baffles in the wash section (see §8 on stripping trays for examples) to reduce effective weir lengths.
- Splash baffles [5]
- Picket-fence weirs [20]

Of these, picket-fence weirs are the most common. Figure 13 shows a tray with a picket-fence weir. The pickets act as deflectors that prevent too much liquid from spray above the downcomer. Effective outlet weir load increases. The liquid mass inventory on the tray goes up and both mass-transfer and de-entrainment improve.

The normal limit for structured packing or grid in wash service is a minimum liquid rate of 0.15 gpm/ft^2 (6.11 lpm/m^2) liquid load at the bottom of the bed. For a 16' (4.88 m) diameter tower this gives a liquid load of 30.1 gpm (114 lpm). The final case in Table 4 (right column) has a liquid load equal to 110% of the minimum packing load. At this liquid rate the de-entrainment effectiveness exceeds the performance of trays operating with 1 gpm/in (1.49 lpm/cm) of weir length.

Further optimization might include using wash sections of reduced diameter. This decreases the liquid load requirement and improves unit operation.

Wash services in atmospheric columns have easier service requirements than wash services in vacuum columns. A full discussion of wash section performance with packing can be found in references on vacuum towers [21].

7.3. Wash Oil Distributors

Wash oil distributors merit some additional comments. Vendor recommendations vary. The two major options are either spray nozzle distributors or trough distributors. Both present challenges. The ideal de-entrainment device would do no heat transfer and no mass transfer. Neither device comes is clearly superior.

The author typically recommends spray nozzle distributors. Spray distributors have the major benefit of low initial cost and minimal maintenance cost and time and have reasonable performance. Sprays are very effective heat-transfer devices (bad), but they give low mass-transfer efficiency in the bed (good). These characteristics compete against each other. Finally, low residence times in the distributor pipes inside the tower tend to reduce coking.

Trough distributors have the benefit of eliminating the liquid sprayed onto the vessel wall, and reduce splash back from the liquid entering the packing surface. They do no heat transfer. However, their longer residence time increases the likelihood of coking, especially at high temperature operation.

Spray distributors and trough distributors have similar risk profiles for plugging from scale, dirt, and other solids. The balance of pressure drop used across the orifice versus the number of distribution points results in distributors with roughly equal plugging risk due to scale. Even with liquid distribution baffles or drip point multipliers, hole sizes are small in trough distributors.

Systems that pump liquid back to the tower for the wash distributor should use parallel basket strainers to remove scale. Piping downstream of the basket strainers should be 5-Cr alloy or better to prevent scale created in the pipe plugging the distributor.

7.4. Liquid Collection Beneath Wash Bed

A liquid collector should be installed beneath the wash bed. The liquid collector is typically a chimney tray. It performs the following:

- Acts as a vapor distributor to the packed bed above
- Collects liquid to bypass the feed zone below, avoiding quenching the flash zone and reducing unit yield
- Reduces probability of liquid re-entrainment at high vapor rates

8. Stripping Services

Nearly all atmospheric columns have a stripping section below the feed. Liquid in the feed is collected and stripped with steam. Stripping allows for reduced flash zone operating temperature at the same yield, reduces light material entering the downstream vacuum tower, and provides flash control of the bottoms product. Trays are the most common device used.

8.1. Tray Process Design and Layout

Stripping trays in this service are normally sieve trays with large-diameter holes (>= 0.75 inches, 19 mm). Occasionally conventional 0.5 inch (12.7 mm) diameter holes are used. Some units have used fixed-valves for the stripping trays, but that has few benefits and increases the risk of plugging.

Vapor rates change dramatically across each tray. It is extremely difficult to create a useful tray layout that will work across the entire stripping section. The bottom tray has such a low vapor rate that the low hole area it requires would cause the top tray to have excessive pressure drop and flood. The top tray has such a high vapor rate that the bottom tray with that hole area would weep and have extremely low efficiency. Multiple tray layouts are required.

Various parameters need to be evaluated to set the stripping tray layout. These include minimum dry tray pressure drop, weeping, hole spacing, and total tray pressure drop. Even with customized tray designs, the trays will have low efficiency. At best, tray efficiency will be 25-30%.

Table 5 illustrates the difficulty. It includes two cases for the example unit:

- Low steam: 2. 5 lbs of steam per standard barrel of bottoms (0.75 weight percent steam to bottoms rate).
- High steam: 10 lbs of steam per standard barrel of bottoms (2.97 weight percent steam to bottoms rate).

Though these cases are labeled low-steam and high-steam, most commercial units run rates less than the low-steam rate here. Typical rates are often as low as 1.0 lbs of steam per barrel of bottoms (~0.30 weight percent steam to bottoms rate). The low-steam rate defined here is the upper end of most commercial practice. The rate of 10 lbs of steam per barrel is approaching an optimum design for a unit with 12-16 real trays (3-4 theoretical trays). The economic optimum for specific plants will vary based on the economic structure for the plant.

The basis for Table 5 is 8 trays with 25% tray efficiency. Tray design analysis showed the following zones for effective tray design;

- Trays 8-5, top 4 trays, one design.
- Trays 2-4, middle trays, one design.
- Tray 1, bottom tray, one design.

Table 5 Tray Limits for Example Stripping Section (Two-Pass Trays)

Tray	Tower diameter In (m)	Limited by	Operating Problems
Low Steam Rate	90 (2.29)		
8-5		Liquid weir load, high	
2-4		Liquid weir load, high	
1		Liquid weir load, high Hole area, low (3.3%) Hole pitch, high	Weeping
High Steam Rate	126 (3.20)		
8-5		Liquid weir load, at limit	
2-4		Liquid weir load, at limit	
1		Hole area, low (3.3%) Hole pitch, high	Weeping

Note that the diameters shown are 90" (2.29m) and 126" (3.20m). These would be reduced diameter sections below the feed zone. The diameter just above the feed zone for this tower is 192" (4.88m).

Industrial practice takes four approaches to dealing with reduced diameter sections:

- Use very large downcomers to reduce the tray active area appropriately
- Reduce the diameter of the bottom section of the overall tower
- Install a reduced diameter can section inside a full-sized tower shell
- Use channel baffles to reduce the effective active area

Channel baffles are often used in conjuction with reduced diameter sections in order to optimize active area and weir load. Figure 14 shows a channel baffle tray. The side baffles (channels) isolate a portion on the outside of the tray. All the vapor and liquid flow through the center section only.

8.2. Tray Mechanical Design

Operating conditions subject the stripping trays to severe service conditions. Tray failure occurs routinely. The two major sources of tray failure are pressure surge due to flashing of water and upsets due to high liquid level. Unit startup creates most of the problems. Some units have never started up successfully with the stripping trays in place. A systems approach is needed to keep the stripping trays intact. Steps include configuration of external piping to reduce the risk of water getting to the tower in the stripping steam, reliable level measurement, sound operating procedures and mechanical upgrades to the trays.

Conventional tray design uses two design criteria. The first is a point load of 300 lbs (136 kg) to meet installation requirements. The second is a distributed load of 60 lbs per ft^2 (293 kg/m^2) to meet operating requirements. These criteria are insufficient for severe mechanical services. A distributed load design of 288 lbs per ft^2 (1,406 kg/m^2) is required for the stripping section. A combination of strengthened construction and special features including through-bolting, shear clips, truss-lugs and extra-strength components can meet this requirement [22].

8.3. Alternative Tray Mechanical Configurations – The Trap-Door Tray

One alternative to handling pressure surges is the trap-door tray. Figure 15 and Figure 16 show the trap-door tray configuration [23]. The trap-door tray was invented and first applied in an atmospheric crude column [24] suffering from tray damage. In service, a chain with a weight attached runs between the trap-door and the tray deck. The chain prevents the panel from excessive opening and the weight pulls the panel closed after it opens.

8.4. Packed Stripping Sections

Packed stripping sections have been used in atmospheric crude towers. A smooth-surface structured packing is the best choice for this, although it is an ambitious service. The benefit is the ability of the packing to handle higher steam rates, have an increased number of stages, and to easily deal with the varying vapor load in the stripping section. The major risk is plugging of the packing or distributors due to coking or solids. Distributor plugging has been a larger source of problems than packing coking. Packed stripping sections work best with clean crudes.

Reports of using random packing in atmospheric tower stripping sections date back to 1979 at Guam Refining [25]. Documented cases of structured packing in atmospheric tower stripping sections date back to 1987 [26]. In the author's experience some units with random packing in the stripping section have operated over 20 years with minimal problems. Other units with packing in the residue stripping services have had annual coking events. Before taking this path, all benefits and costs should be weighed.

9. Auxiliary Equipment

In atmospheric crude towers auxiliary equipment may include collector trays, water draw trays, feed inlet devices, and other equipment. Appropriate references should be consulted for these items [1] [21] [27].

10. Conclusions

Distillation equipment design for atmospheric crude distillation is all about trays. With one exception trays have the appropriate capacity, flexibility, and cost performance to make them the favored distillation internal in this service. Both valve trays and sieve trays work effectively. In the wash service, just above the feed zone, structured packing or grid is often the superior technical choice.

Tray design for this service fits well into the conventional tray category. Revamping units for higher throughput favors high-capacity trays. The stripping section at the bottom of the tower requires careful attention to mechanical fabrication details to keep the trays intact and in place.

11. References

[1] H. Z. Kister, *Distillation design*. McGraw-Hill, Inc., 1992.

[2] A. W. Sloley, "Distillation design: trays," in *Encyclopedia of Chemical Processing*, vol. 2, 5 vols., S. Lee, Ed. New York, NY: Marcel Dekker, 2005, pp. 749–764.

[3] H. Z. Kister, "Can valve trays experience vapour cross flow channeling?," *The Chem Eng*, Jun. 1993.

[4] H. Z. Kister, K. F. Larson, and Madsen, P. E., "Vapor cross-flow channeling on sieve trays: fact or myth?," *Chem Eng Prog*, vol. 88, no. 11, pp. 86–93, Nov. 1992.

[5] A. W. Sloley and P. B. Fleming, "Successfully downsize trayed columns," *Chem Eng Prog*, vol. 90, no. 3, pp. 39–47, Mar. 1994.
[6] "Bulletin 681R1 44 frequently-asked questions and answers about trays and packing." Glitsch, Inc., 1983.
[7] A. W. Sloley, "Should you switch to high-capacity trays?," *Chem Eng Prog*, vol. 95, no. 1, pp. 23–35, Jan. 1999.
[8] E. V. Murphree, "Rectifying column calculations," *Ind Eng Chem*, vol. 17, no. 7, pp. 747–750, Jul. 1925.
[9] E. V. Murphree, "Graphical rectifying column calculations," *Ind Eng Chem*, vol. 17, no. 9, pp. 960–964, Sep. 1925.
[10] A. W. Sloley, "Atmospheric distillation process - fundamental concepts," presented at the AIChE Spring Meeting, New Orleans, LA, 2014.
[11] M. J. Lockett, "The froth to spray transition on sieve trays," *Trans IChemE*, vol. 59, pp. 26–34, 1981.
[12] M. Prado, "The bubble-to-spray transition on sieve trays: mechanisms of the phase inversion," University of Texas, Austin, TX, 1986.
[13] M. Prado, K. L. Johnson, and J. R. Fair, "Bubble-to-spray transition on sieve trays," *Chem Eng Prog*, vol. 83, no. 3, pp. 32–40, Mar. 1987.
[14] J. R. Fair, "Direct contact gas-liquid heat exchanger for energy recovery," *Trans ASME*, vol. 112, no. 8, pp. 216–222, Aug. 1990.
[15] J. R. Fair, "Process heat transfer by direct fluid-phase contact," presented at the Process Heat Transfer Symposium, 12th National Heat Transfer Conference, Tulsa, OK, 1971, vol. 118, pp. 1–11.
[16] T. P. Ognisty and Y. S. Kao, "Performance of fractionation devices in direct contact heat transfer," presented at the PACHEC IV Pacific Area Chemical Engineering Conference, Acapulco, Mexico, 1988.
[17] "D975-15b Standard specification for diesel fuel oils," ASTM International, West Conshohocken, PA, 2015.
[18] H. Z. Kister, "Design and layout for sieve and valve trays (Column internals 3)," *Chem Eng*, vol. 87, no. 18, pp. 119–123, Sep. 1980.
[19] A. W. Sloley, "Improve tray operations," *Hyd Proc*, vol. 80, no. 6, pp. 85–86, Jun. 2001.
[20] D. R. Summers and A. W. Sloley, "How to handle low-liquid loadings," *Hyd Proc*, vol. 86, no. 1, pp. 67–78, Jan. 2007.
[21] A. W. Sloley, "Refinery vacuum towers - state of the art," presented at the AIChE National Meeting, San Francisco, CA, 2013.
[22] G. H. Shiveler, "Use heavy-duty trays for severe services," *Chem Eng Prog*, vol. 91, no. 8, pp. 72–81, Aug. 1995.
[23] A. R. DiNicolantonio and H. R. Theysen, "Hinged pressure relief tray," 4,133,852, 09-Jan-1979.
[24] A. R. DiNicolantonio, "Private communication," 1997.
[25] L. D. Nicholson and H. A. Erskine, "Crude-unit revamp cuts resid yield," *OGJ*, vol. 78, no. 12, pp. 102–117, Mar. 1980.
[26] P. Roy and A. C. Mercer, "The use of structured packing in a crude oil atmospheric distillation column," presented at the Distillation and Absorption, Brighton, UK, 1987, vol. 104, pp. A103–A114.
[27] A. W. Sloley, "Don't get drawn into distillation difficulties," *Chem Eng Prog*, vol. 94, no. 6, pp. 63–78, Jun. 1998.

[28] G. J. Keller, "Apparatus for liquid and vapor or gas mass transfer," 3,729,179, April 24, 29173.
[29] W. Bruckert, "Vapor-liquid contact system and method," 3,410,540, 12-Nov-1968.
[30] J. Forgrieve, "Commerical jet-tray fractionators," in *Proceedings of the International Symposium on Distillation*, Brighton, UK, 1960, pp. 185–189.
[31] J. O. Nye, "Distillation tray/downcomer," 5,242,628, 07-Sep-1993.
[32] K. T. Chuang and A. E. Everatt, "Gas-liquid contacting apparatus," 4,504,426, 12-Mar-1985.

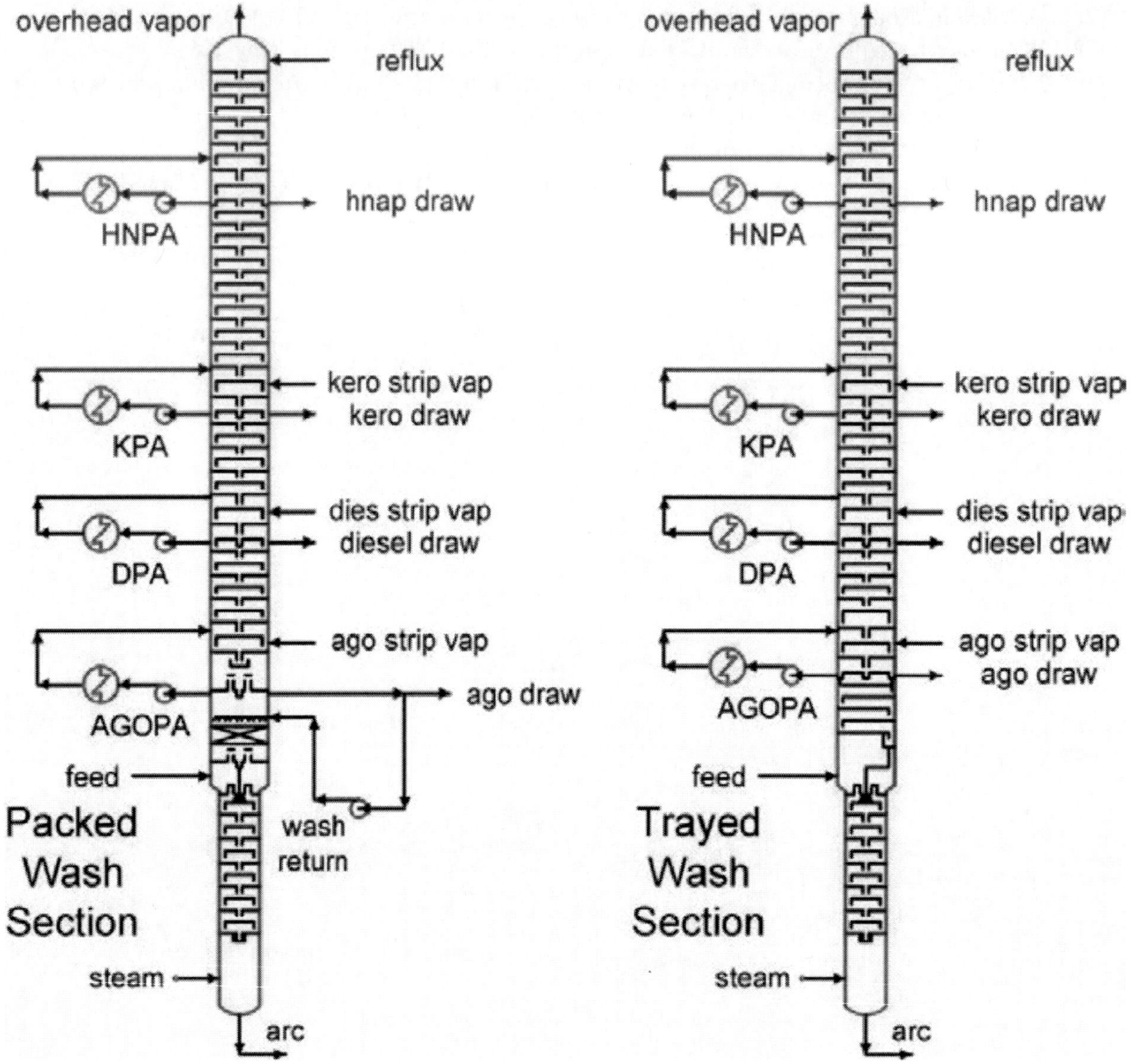

Figure 1 Atmospheric Crude Distillation Tower Configuration Examples

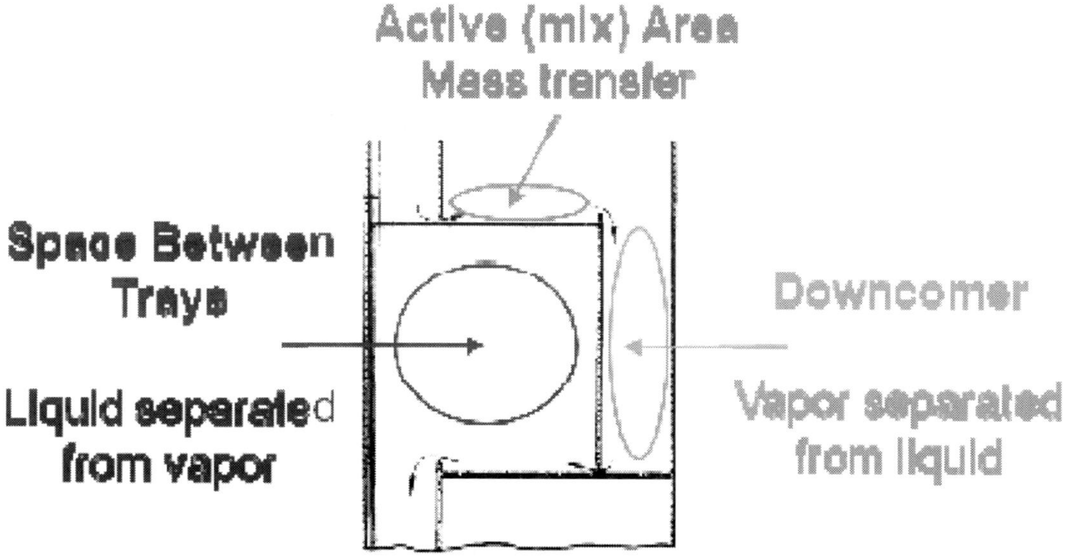

Figure 2 Functional Areas of a Conventional Tray

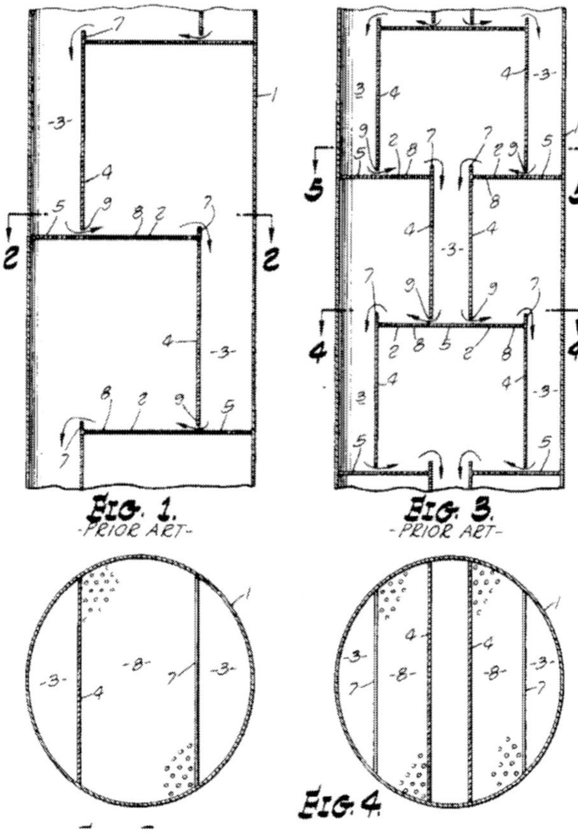

Figure 3 One-Pass and Two-Pass Trays, Plan and Elevation Views, from [28]

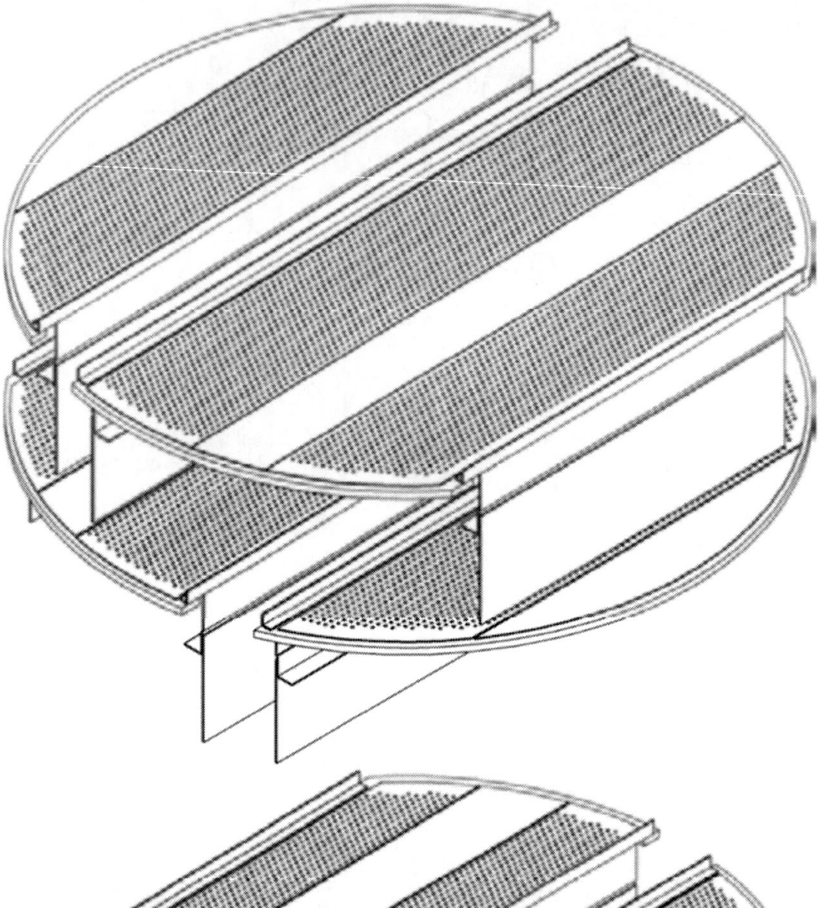

Figure 4 Three-Pass (top) and Four-Pass Trays (bottom), Isometric views

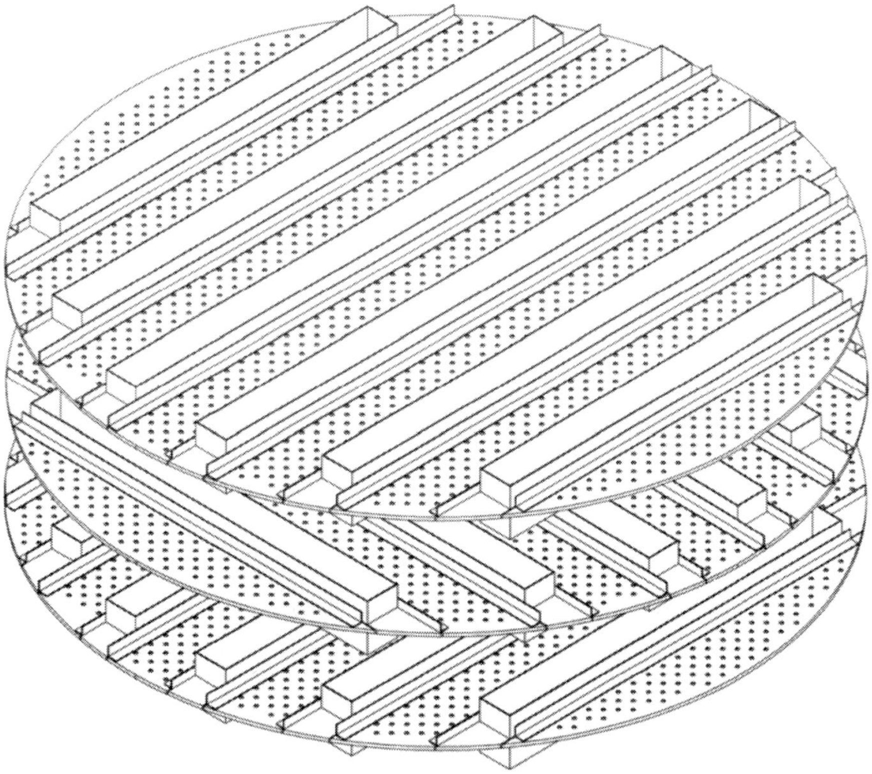

Figure 5 High-Capacity Tray With Extended Downcomer Length [29]

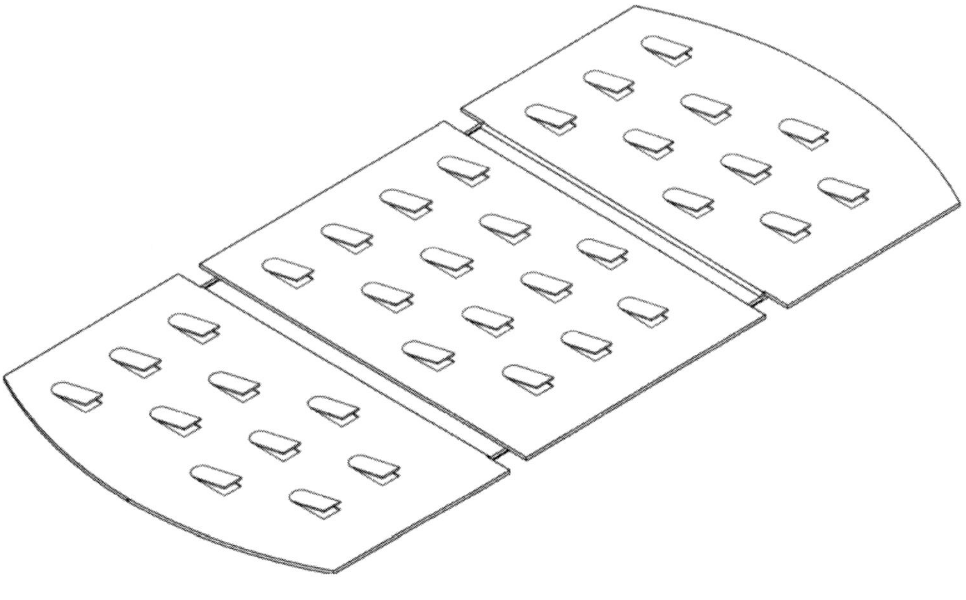

Figure 6 Jet Tray Deck [30]

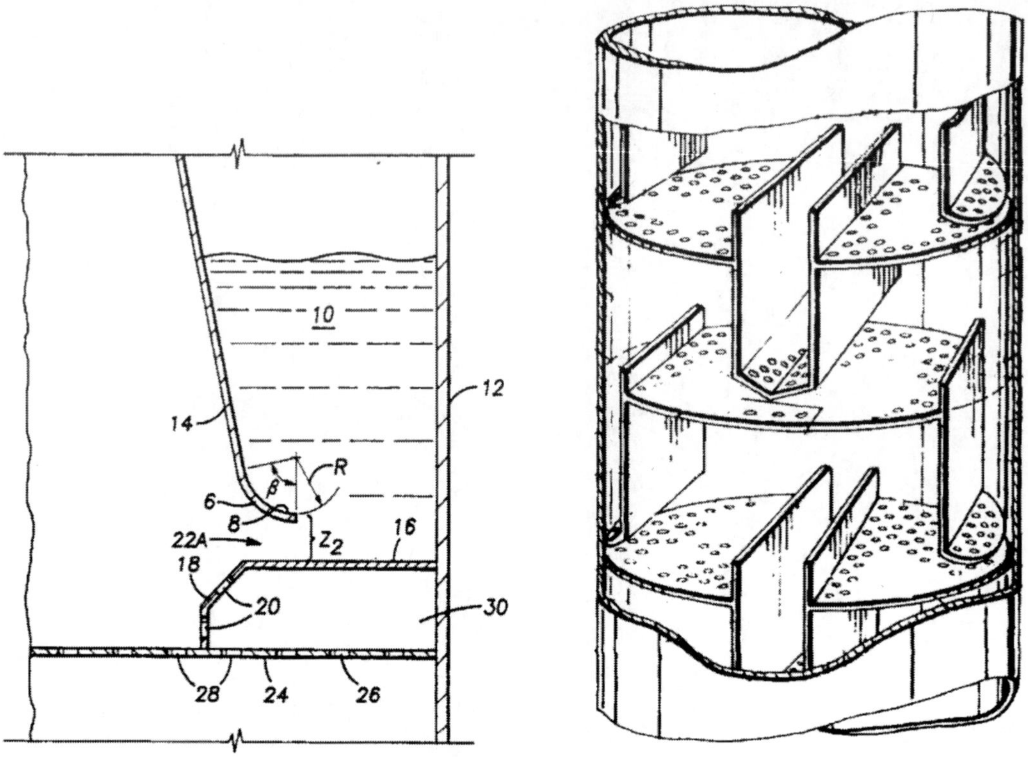

Figure 7 Hanging Downcomers, [31] left and [32] right. Left section shows a raised tray component (#16) with side vapor injection from holes (#20).

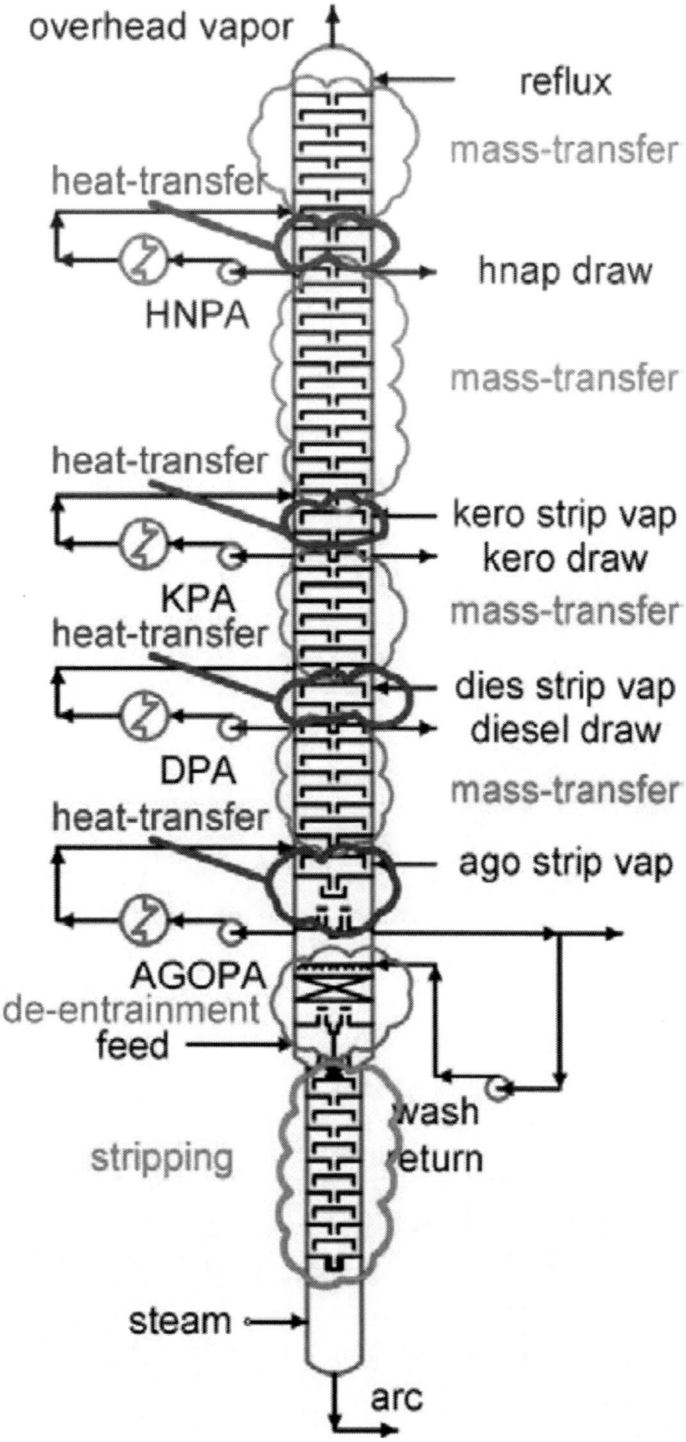

Figure 8 Atmospheric Tower Internals Services

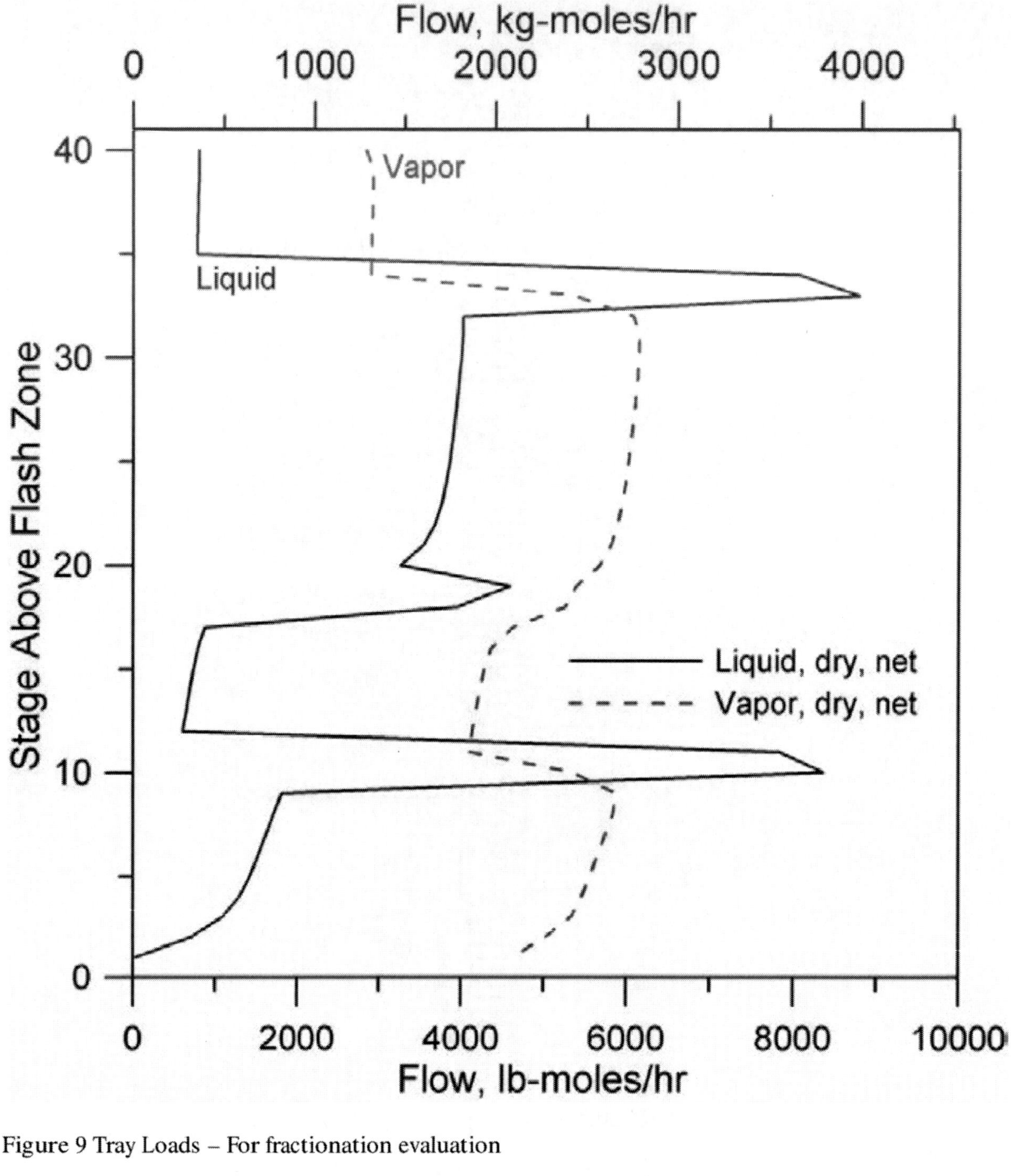

Figure 9 Tray Loads – For fractionation evaluation

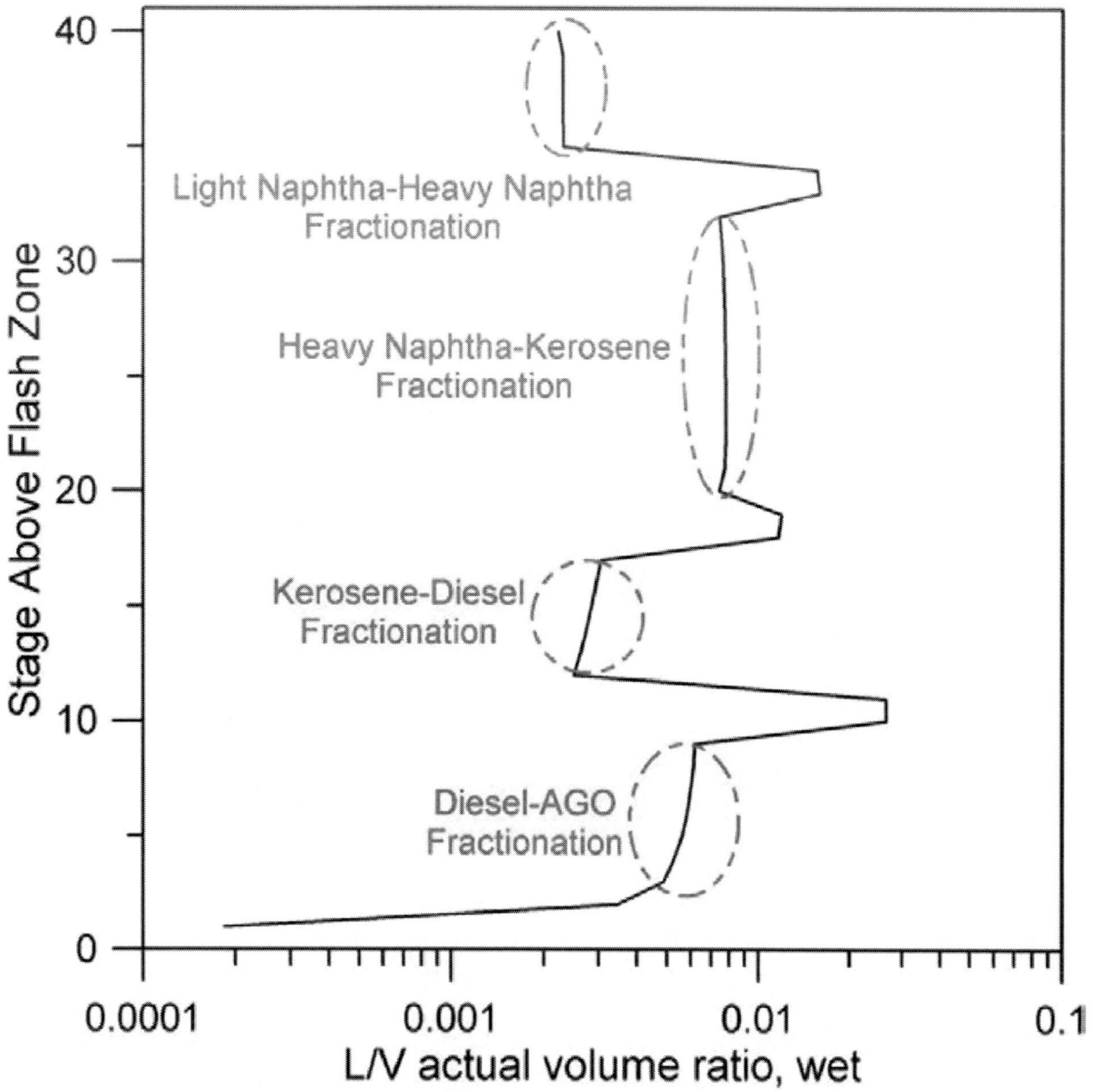

Figure 10 Tray Loadings, Volume Ratios

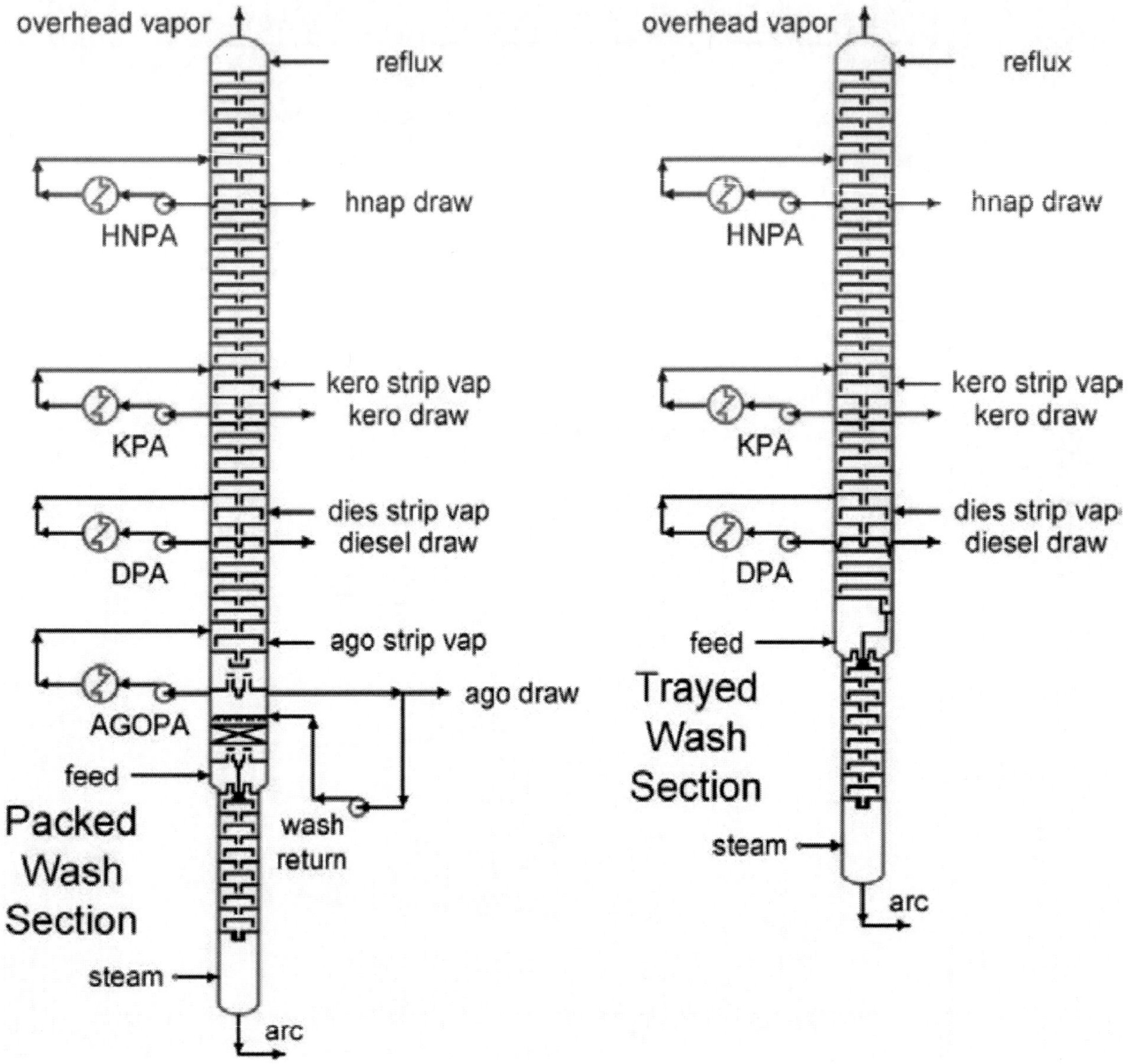

Figure 11 AGO versus Diesel as Lowest Side-Draw

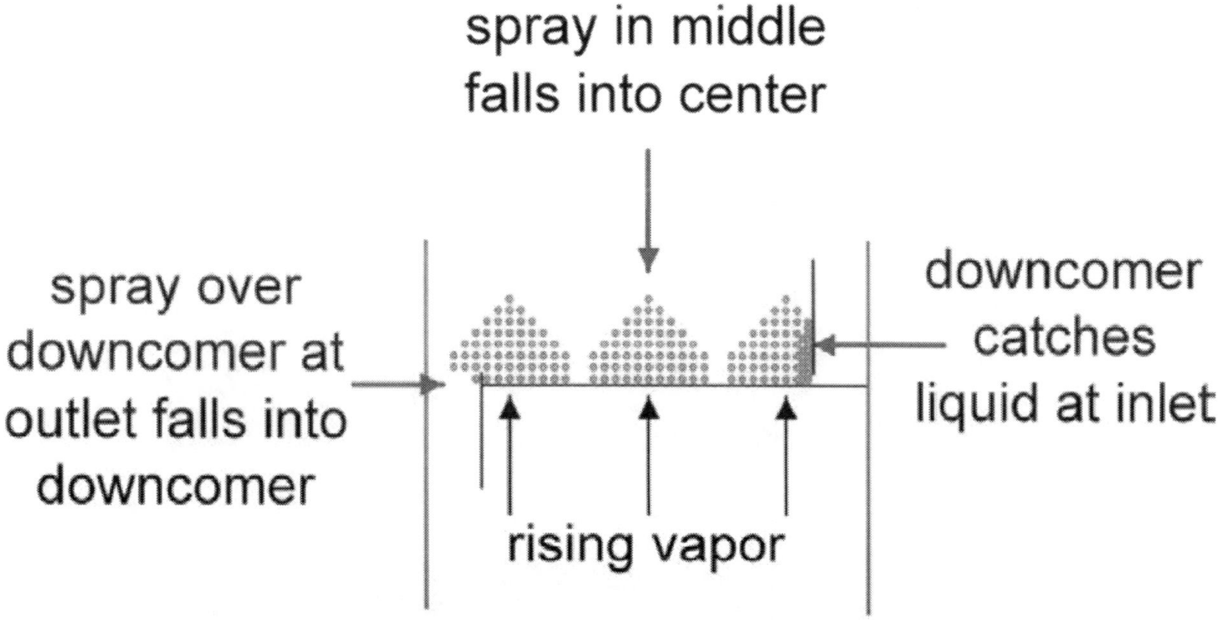

Figure 12 Blowing Operation in Spray Regime

Figure 13 Tray with Picket-Fence Weir (Courtesy Sulzer Chemtech Ltd.)

Figure 14 Tray with Channel Baffles (Courtesy Sulzer Chemtech Ltd.)

Figure 15 Tray with Channel Baffles and Trap Doors (Courtesy Sulzer Chemtech Ltd.)

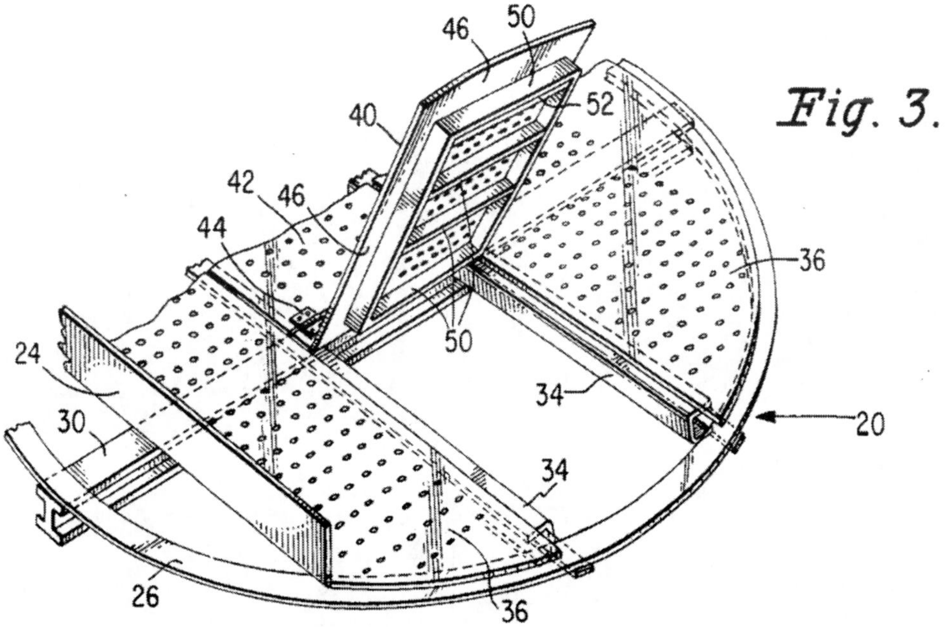

Figure 16 Trap Door Tray for Severe Service (from [23]).

Balanced Distillation Equipment Design for Crude Distillation Units

Sounho Lee – GTC Technology US LLC
2016 AIChE Spring Meeting / Distillation Symposium / "Distillation in Refining Session"
Houston TX April 11, 2016

GTC Technology
→ Engineered to Innovate

Crude Distillation Unit Distillation Equipment

GTC Technology
→ Engineered to Innovate

- Fouling and corrosion: major issues
- Distillation equipment fouling resistance & efficiency: inverse relationship
- Balancing / optimization between fouling resistance and efficiency requirement
- Poor process and/or equipment design know-how often downgrade column performance and reduce unit run length
- Two (2) actual retrofit case studies
  - Crude distillation units

AIChE 2016 Spring Meeting / Distillation Symposium

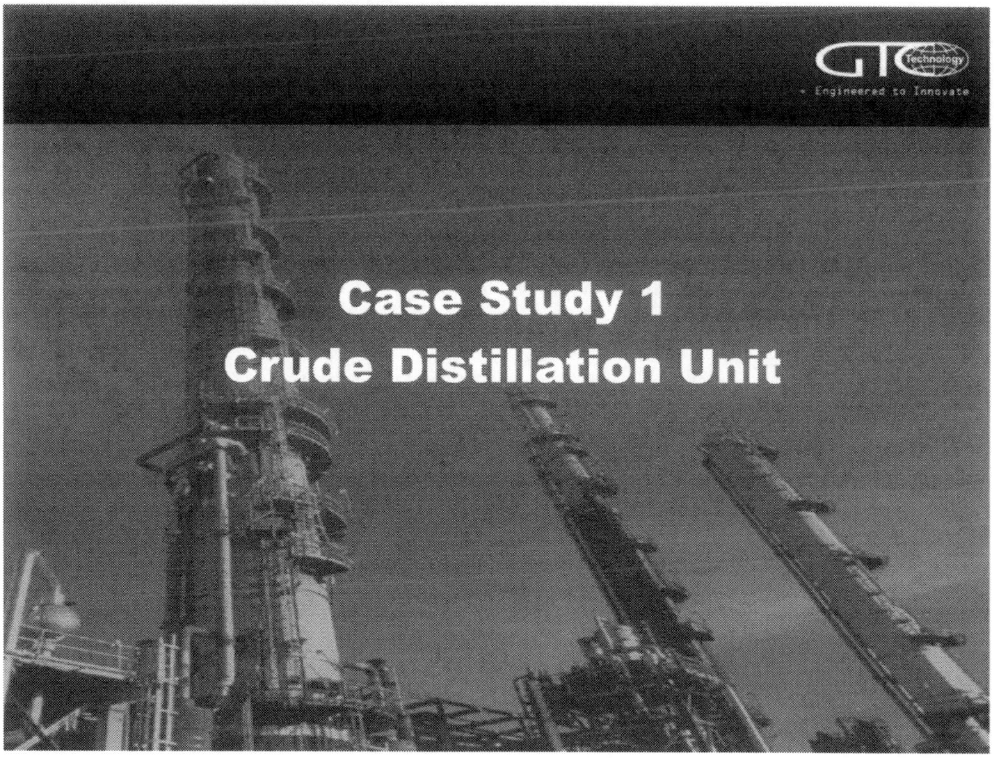

Case Study 1- Crude Distillation Unit
Unit Configuration

GTC Technology – Engineered to Innovate

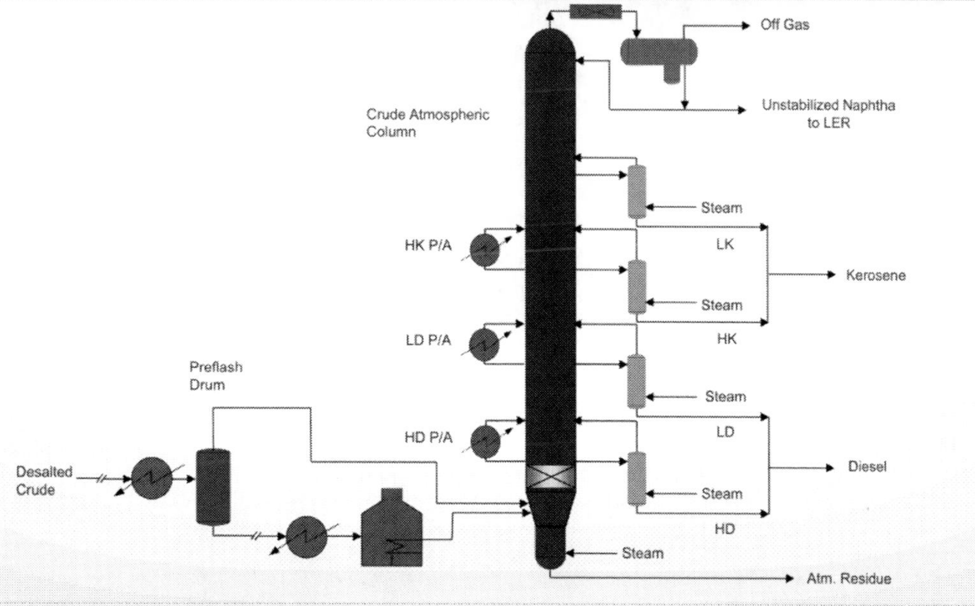

AIChE 2016 Spring Meeting / Distillation Symposium

Case Study 1- Crude Distillation Unit Limitation

- Valve / perforation hole wear & corrosion
  - Naphtha / light kerosene fractionation section
- Fouled trays
  - Light kerosene/heavy kerosene fractionation section
  - Heavy kerosene/light diesel fractionation section
- Side product draw limitation
  - Limited product yields : kerosene & diesel
  - Varied crude slate composition

Case Study 1- Crude Distillation Unit Original Side Draw Configuration

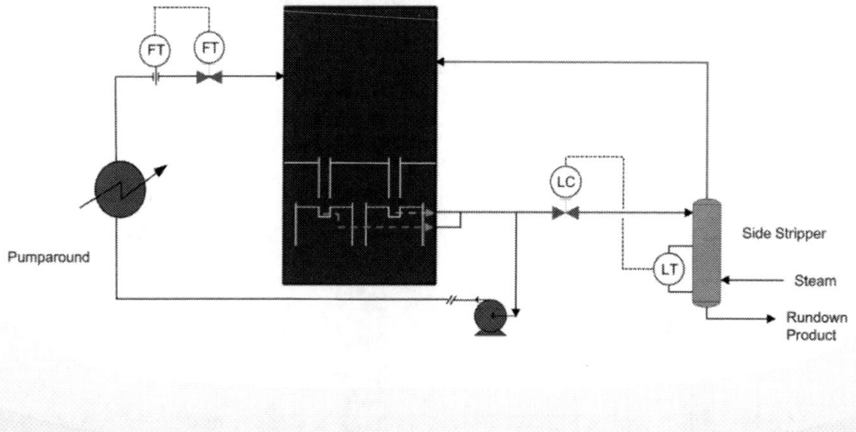

Case Study 1- Crude Distillation Unit Modification

- Fixed valve tray implementation
  - Improving fouling/wear resistance
- Intermediate product/pumparound draw:
  - Original: withdrawal from fractionating tray directly
  - Modification: conversion to chimney tray
- Losing one fractionating tray for each
  - Light kerosene/heavy kerosene fractionation section
  - Heavy kerosene/light diesel fractionation section
  - Light diesel / heavy diesel fractionation section
- Sensitivity Analysis
  - Heavy kerosene/light diesel fractionation section

Case Study 1- Crude Distillation Unit Sensitivity Analysis

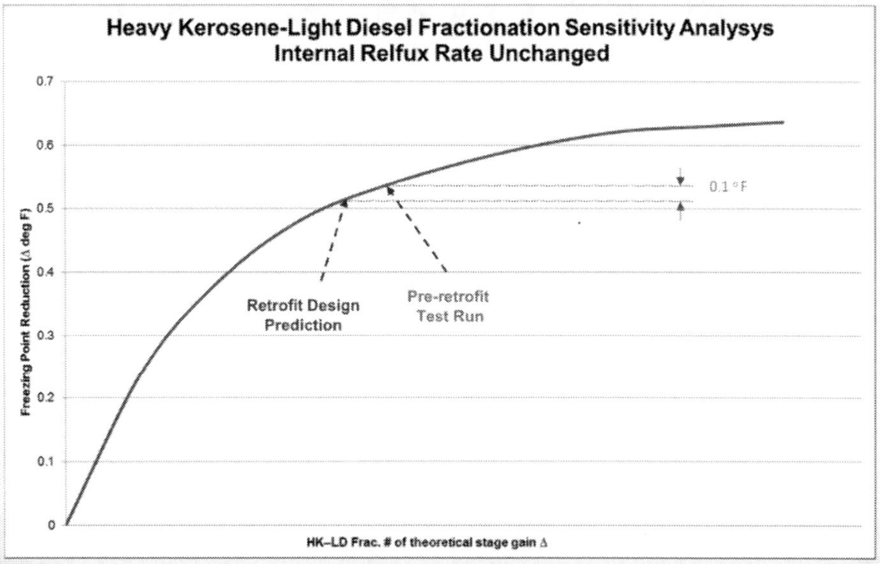

Case Study 1- Crude Distillation Unit Sensitivity Analysis

GTC Technology – Engineered to Innovate

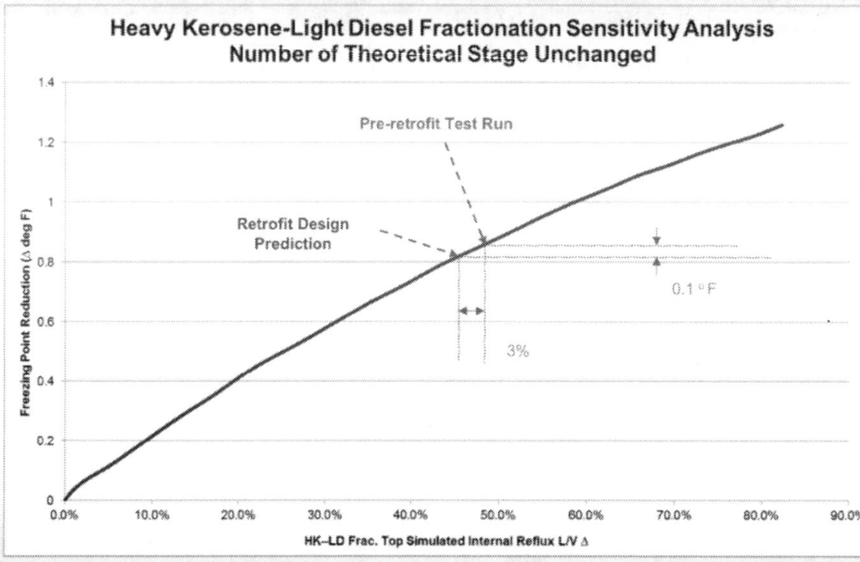

AIChE 2016 Spring Meeting / Distillation Symposium

Case Study 1- Crude Distillation Unit Performance Summary

GTC Technology – Engineered to Innovate

Parameter	Pre-retrofit Test Run	Post-retrofit Test Run
Crude Charge (BPD)	BASE	+ Δ 12 %
Unstabilized Naphtha Yield (LV%)	BASE	+ Δ 6.0 %
Kerosene Yield (LV%)	BASE	+ Δ 19.5 %
Diesel Yield (LV%)	BASE	+ Δ 12.1 %
Naphtha – Light Kerosene Gap, °F	BASE	+ Δ 1.0 °F
Heavy Kerosene – Light Diesel Gap, °F	BASE	- Δ 1.0 °F
Kerosene Flash Point, °F	BASE	Δ 0.0 °F
Kerosene Freezing Point, °F	BASE	+ Δ 0.2 °F
Heat removal – Overhead condenser (% of total BTU/hr)	72 %	62 %
Heat removal – Heavy Kerosene P/A (% of total BTU/hr)	4 %	6 %
Heat removal – Light Diesel P/A (% of total BTU/hr)	7 %	10 %
Heat removal – Heavy Diesel P/A (% of total BTU/hr)	16 %	21 %
HK–LD Frac. Top Simulated Internal Reflux L/V (wt)	BASE	- Δ 15 %

AIChE 2016 Spring Meeting / Distillation Symposium

Case Study 1- Crude Distillation Unit Post Retrofit Sensitivity Analysis

GTC Technology – Engineered to Innovate

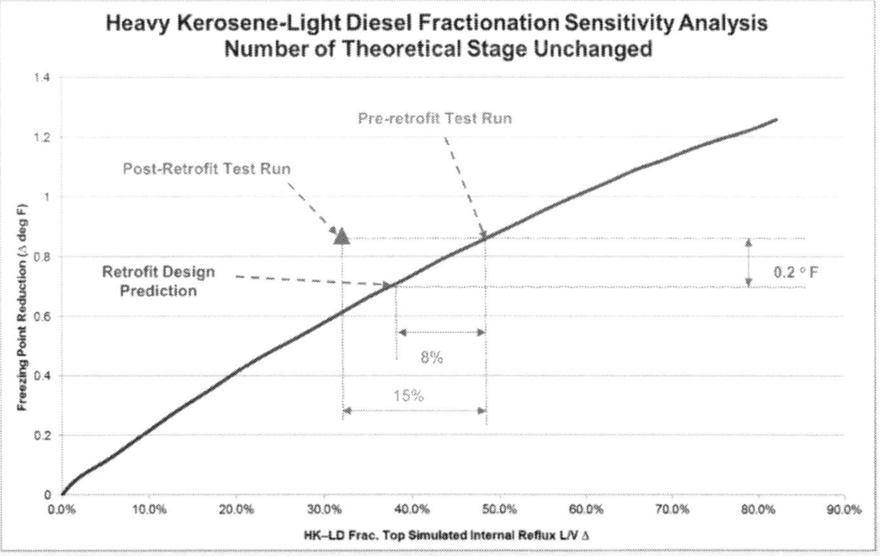

AIChE 2016 Spring Meeting / Distillation Symposium

Case Study 1- Crude Distillation Unit Fractionation Efficiency Comparison

GTC Technology – Engineered to Innovate

Heavy Kerosene / Light Diesel Fractionation	Pre-retrofit Test Run	Retrofit Prediction	Post-retrofit Test Run
Kerosene Freezing Point, °F	BASE	+ Δ 0.1 °F	+ Δ 0.2 °F
HK–LD Frac. Top Simulated Internal Reflux L/V (wt)	BASE	- Δ 3 %	- Δ 15 %
Simulated Section Efficiency [note 1]	~55%	~50%	~63%

Note
1. Overall efficiency (# of theoretical stage ÷ # of fractionating tray)

- New fractionating trays
  - Fixed valve
  - Equipped the performance enhancing features

AIChE 2016 Spring Meeting / Distillation Symposium

Case Study 2- Crude Distillation Unit
Unit Configuration

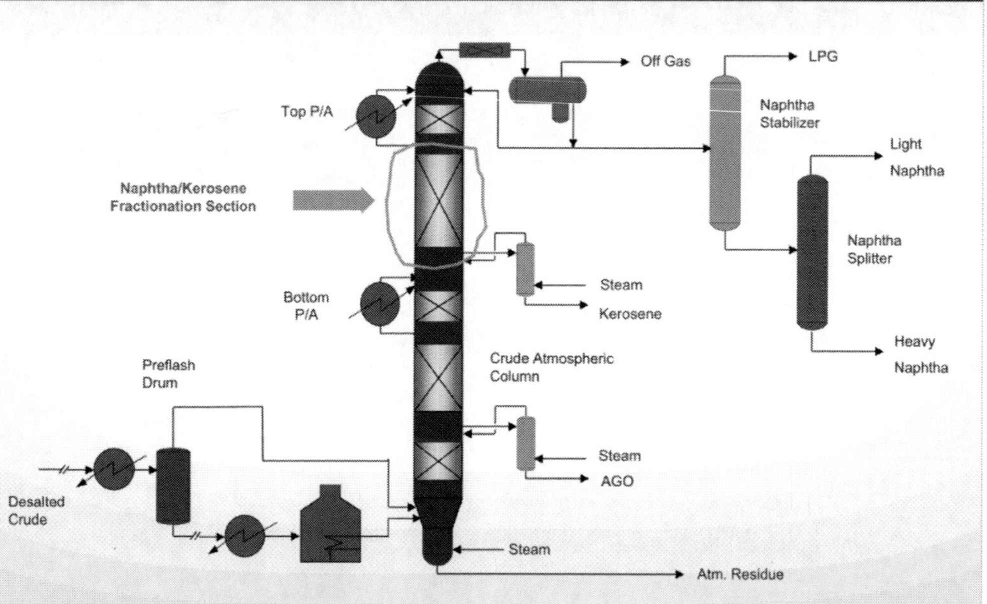

AIChE 2016 Spring Meeting / Distillation Symposium

Case Study 2- Crude Distillation Unit Limitation

GTC Technology • Engineered to Innovate

- Downgraded degree of fractionation between naphtha & kerosene
  - After 4 month operation
- Unit trouble
  - Kerosene yield limitation
  - Downstream reformer performance
- Liquid distributor fouling
  - Naphtha/kerosene fractionation

AIChE 2016 Spring Meeting / Distillation Symposium

Case Study 2- Crude Distillation Unit Root Cause Identification

GTC Technology • Engineered to Innovate

- Ammonium salt formation
  - Transferred to naphtha/kerosene fractionation section
- Limited column inside wall cladding
  - Rust from the column shell (carbon steel section)
- Original liquid distributor design for naphtha/kerosene fractionation
  - High drip point density
  - Small drip hole size

AIChE 2016 Spring Meeting / Distillation Symposium

Case Study 2- Crude Distillation Unit Gravity Distributor Optimization

GTC Technology – Engineered to Innovate

$$H = \left(\frac{L_v}{k \times N \times HA} \right)^2$$

H : Liquid Height ("Head") above Round Hole
L_v : Liquid Volumetric Flow
N : Number of Holes (Drip Point Density)
HA : Individual Hole Area (Drip Hole Size)
K : Orifice Coefficient

- Drip point density vs. Drip hole size
  - Higher drip point density reduces individual hole size @ total hole area
- Distributor operating range
  - Adequate liquid head should be maintained @ turndown rate

AIChE 2016 Spring Meeting / Distillation Symposium

Case Study 2- Crude Distillation Unit Distributor Modification Summary

GTC Technology – Engineered to Innovate

Naphtha/Kerosene Fractionation Packing	Unit	Original	Modification
Nominal Crimp Size	inch	1"	Re-used
Nominal Specific Surface Area	ft^2/ft^3 (m^2/m^3)	38 (125)	Re-used
Inclination Angle	°	45°	Re-used

Naphtha/Kerosene Fractionation Distributor	Unit	Original	Modification
Distributor Drip point density	dp/ft^2	16	4.6
Distributor Operating Range	-	2.5 : 1	1.7 : 1
Drip Hole Elevation note 1	inch	1.5"	3"
Distributor Drip Hole size	inch	3 / 16"	7 / 16"

Note
1. From bottom of trough

AIChE 2016 Spring Meeting / Distillation Symposium

Case Study 2- Crude Distillation Unit Performance Summary

GTC Technology – Engineered to Innovate

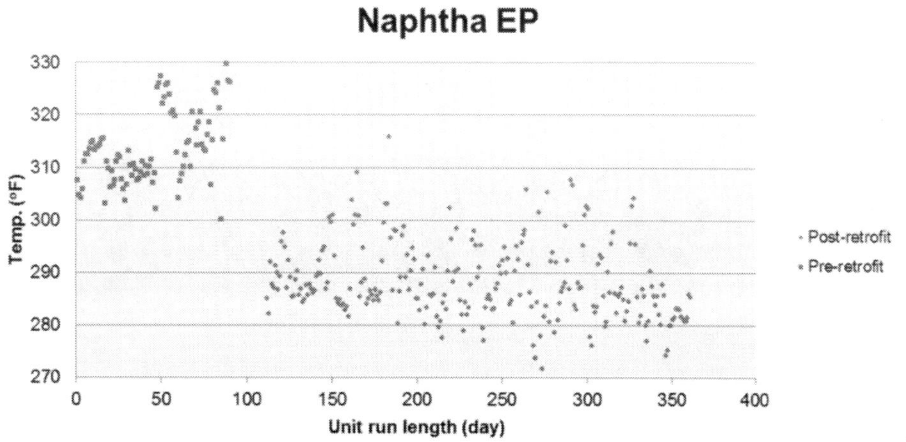

AIChE 2016 Spring Meeting / Distillation Symposium

Case Study 2- Crude Distillation Unit Performance Summary

GTC Technology – Engineered to Innovate

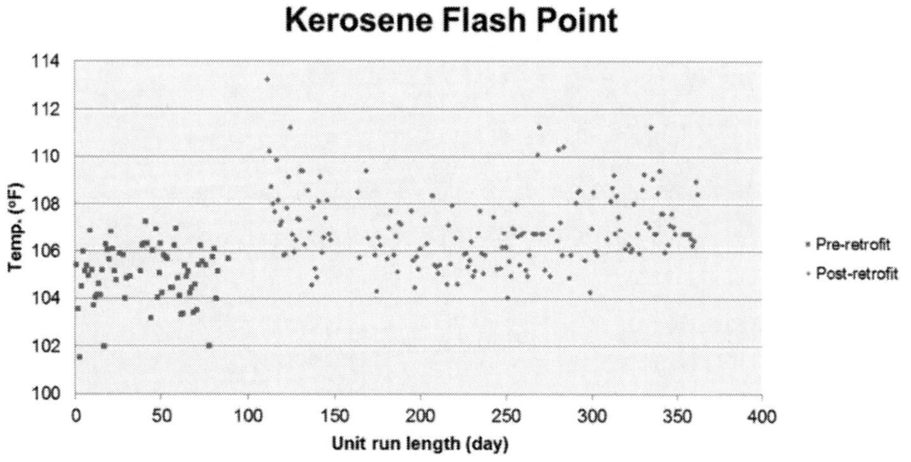

AIChE 2016 Spring Meeting / Distillation Symposium

Thank you

Engineered to Innovate

Thank you

443916 Rigorous Simulation of a Crude Distillation Unit with Heat Exchanger Network

Abstract:

Crude distillation is one of the most important operations in oil refineries since it provides feed to downstream units for further processing. Because of the nature of the present complex and highly integrated refineries, it is critical to understand the operation of the crude distillation unit (CDU) in order to accurately predict the performance and thereby plan operations of the downstream units in response to market demand of the different products.

The present study focuses on simulation of a CDU using a commercially available process simulator Petro-SIM. A detailed model has been developed and validated using refinery operating data. A sensitivity analysis of the model has also been conducted for incorporating different property packages, and Peng-Robinson Equation of State has been found to provide excellent predictions for the CDU performance. Parametric studies have been performed for key operating scenarios in order to understand the relationship among process parameters, feed qualities and product profiles and properties. Different scenarios were studied to demonstrate effects of draw rates of different products on the respective cut properties. Such a robust predictive model opens up opportunities for CDU operational optimization, contributing to an overall refinery profit margin enhancement. The CDU consumes a significant amount of energy. In order to conduct energy optimization studies, e.g. using pinch analysis technique, a detailed accurate model such as the one being presented would be highly useful.

MANCHESTER 1824

The University of Manchester

process integration limited

PIL: The experts in process improvement technologies

Retrofit and off-line operational optimisation of heat-integrated crude oil distillation systems

11/04/2016

16AIChE Spring Meeting, Houston, TX

Contact:
Lluvia Ochoa-Estopier
Tel: +44-7411-946-473
lluvia.ochoa-estopier@processint.com
www.processint.com

www.processint.com | +44 161 974 0090 | Station House, Stamford New Road, Altrincham, Cheshire, CW5 7BH, UK

Contents

1. Knowledge Transfer Partnership program
2. Overview of Process Integration Limited (PIL)
3. Background
4. Methodology
5. Case Study
6. Conclusions and future work

PIL: The experts in process improvement technologies

Slide Number: 2

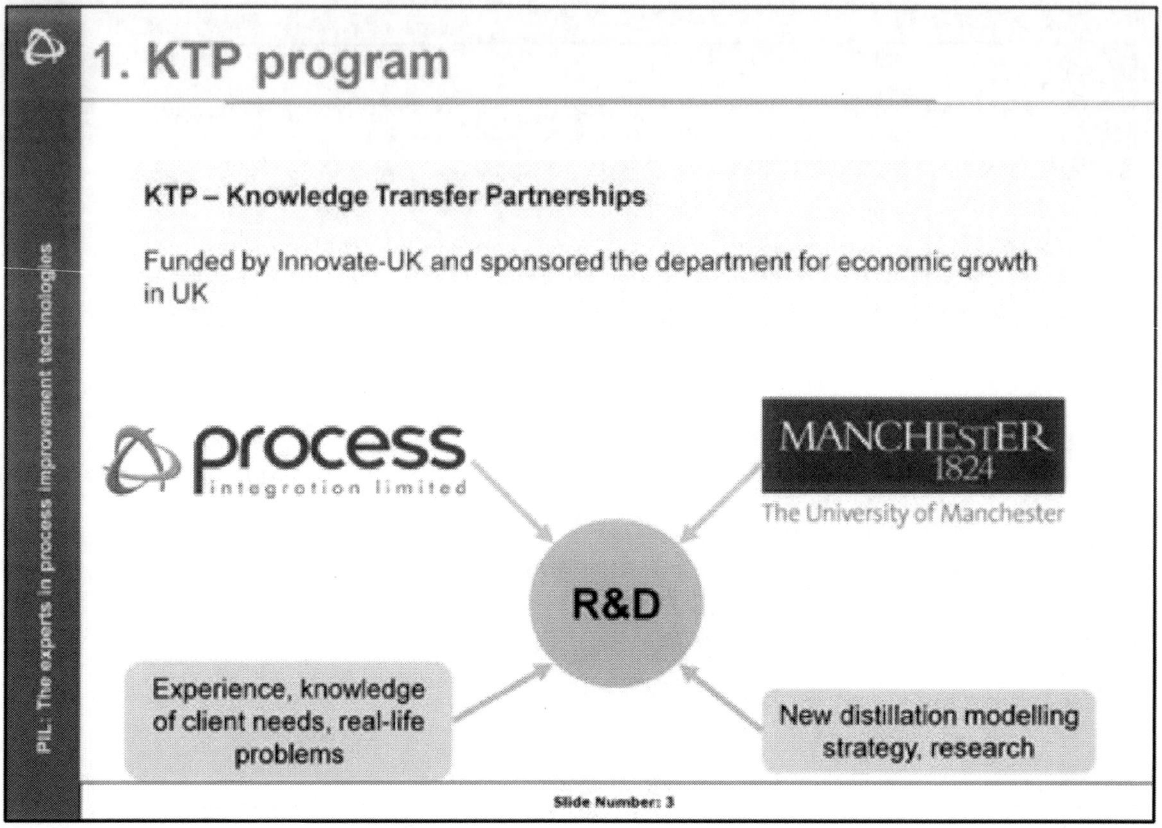

This research work is part of a 30-month Knowledge Transfer Partnership (KTP) program sponsored by the department of economic growth in UK.

"Knowledge Transfer Partnerships (KTP) is Europe's leading programme helping businesses to improve their competitiveness and productivity through the better use of knowledge, technology and skills that reside within the UK knowledge base." (https://connect.innovateuk.org/web/ktp, 17/03/2016)

The objective of this KTP is to develop and embed new mathematical approaches for the optimisation of refinery crude units into commercial tools and deliver a step change in the company's consultancy and software portfolio.

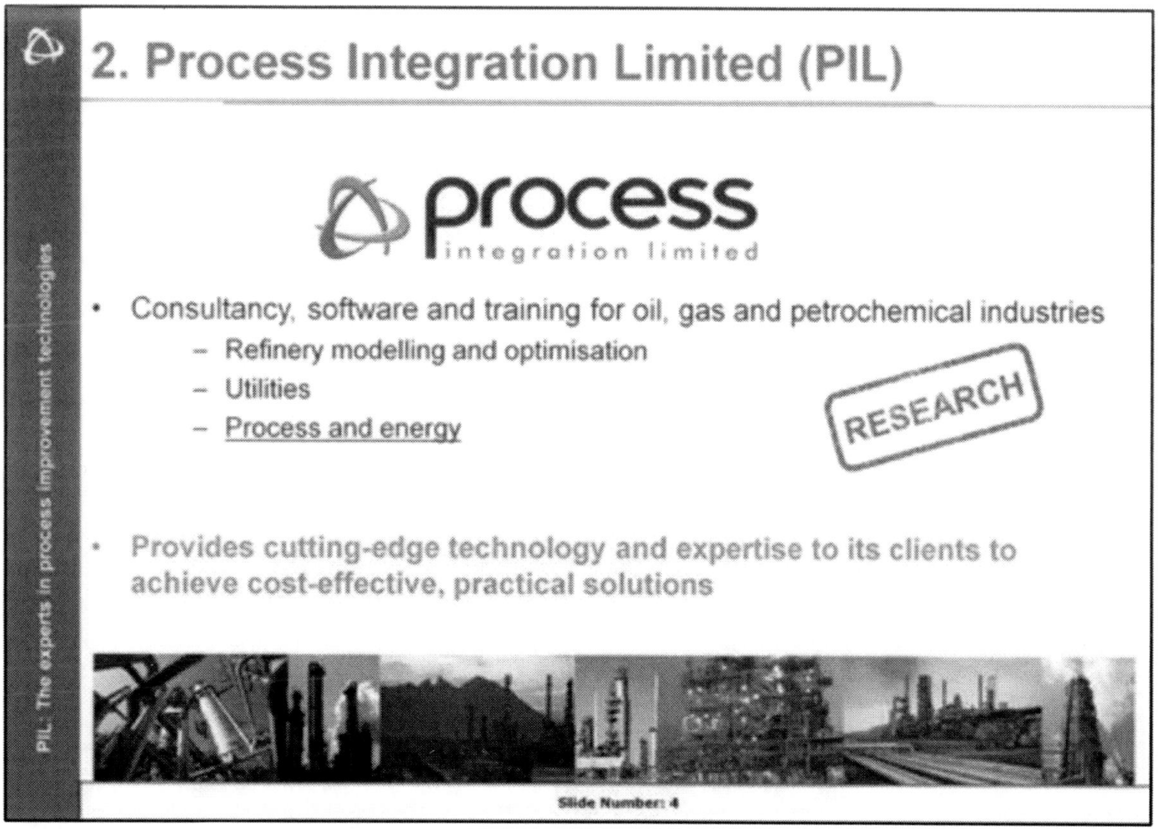

"Process Integration Limited (PIL) provides state-of-the-art process improvement technologies for the oil, gas, petrochemical and chemical industries by identifying innovative improvements in operation and design. The company's technology and know-how can effectively be leveraged to:

Increase energy efficiency
Decrease emissions
Improve yields
Increase capacity"

(http://www.processint.com/about-us, 17/03/2016)

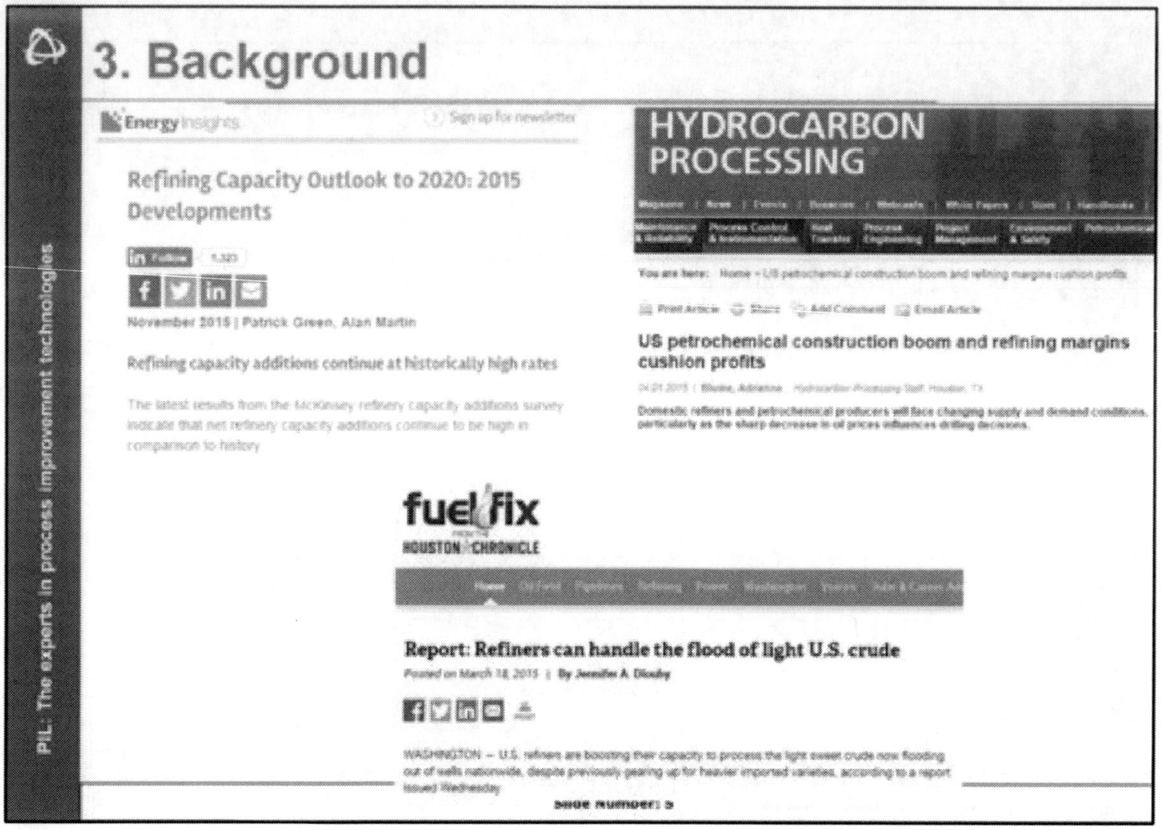

Refiners today are facing challenges related to costs, global and local demand, changes in crude oil feedstock, changes crude oil prices etc. The key to success is to maintain a cost-effective operation in these changing scenarios.

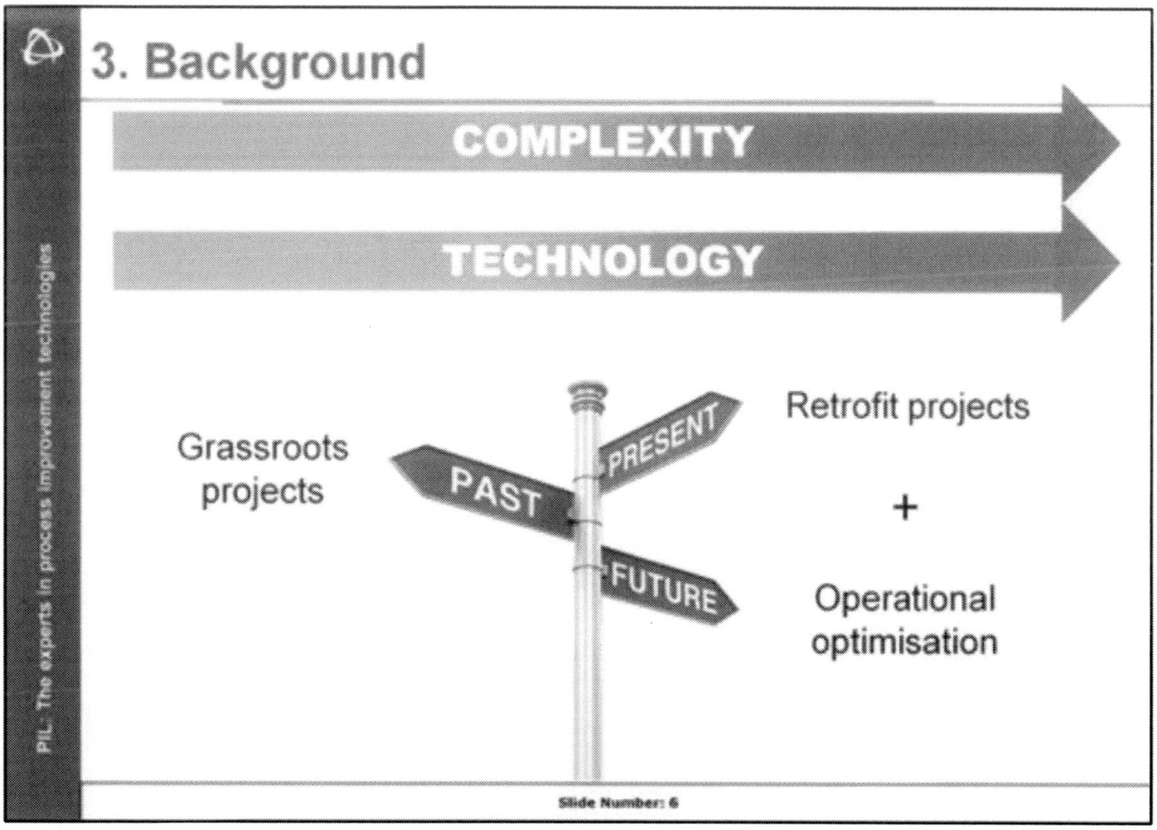

Refineries that develop research and implement projects to process crude oil in a cost-effective way, while meeting strict environmental regulations and product quality requirements, ensure a competitive position in the global market. In this context, the crude oil distillation process is key to improve the economic and environmental performance of refineries.

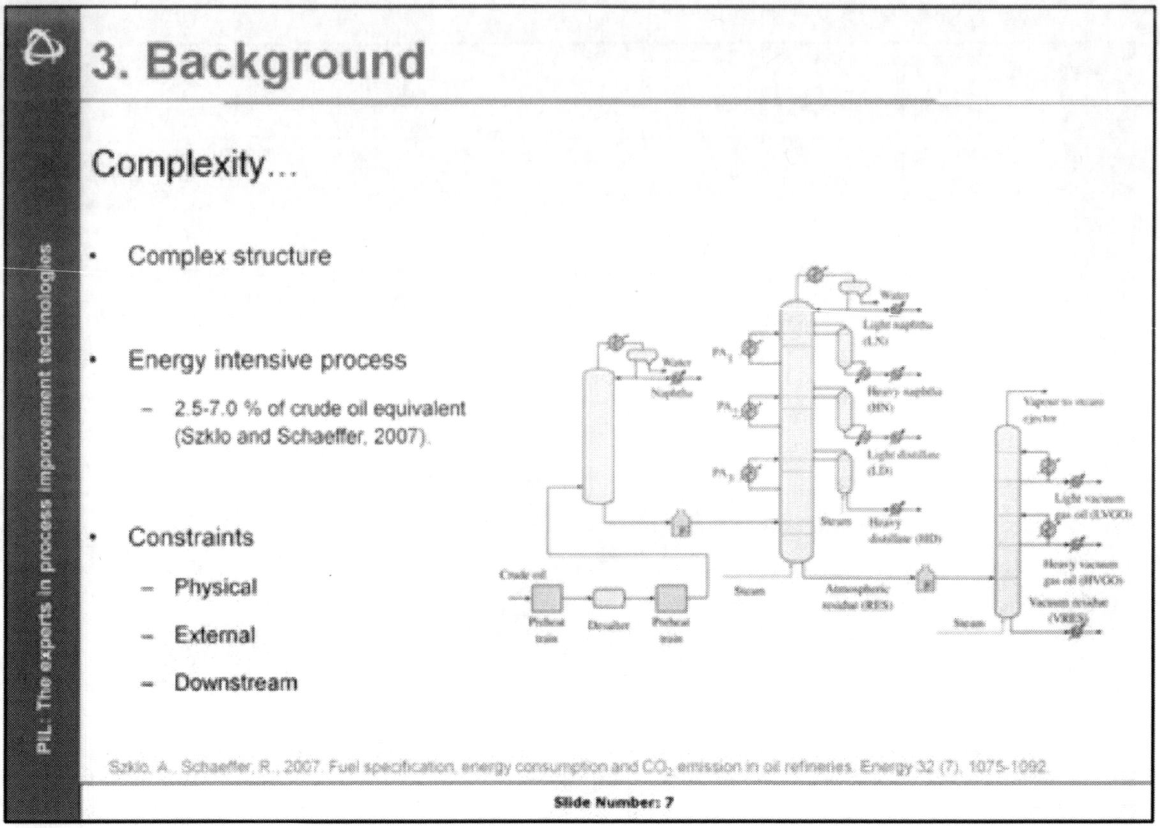

Energy intensive:

- Environmentally challenging
- Economically challenging

Complex structure

- Expensive to modify
- Strong interactions between the column and HEN
- Many degrees of freedom for design
- Practical constraints

Feedstock characteristics

- Variable conditions
- Large number of components

Several products

- Product specifications

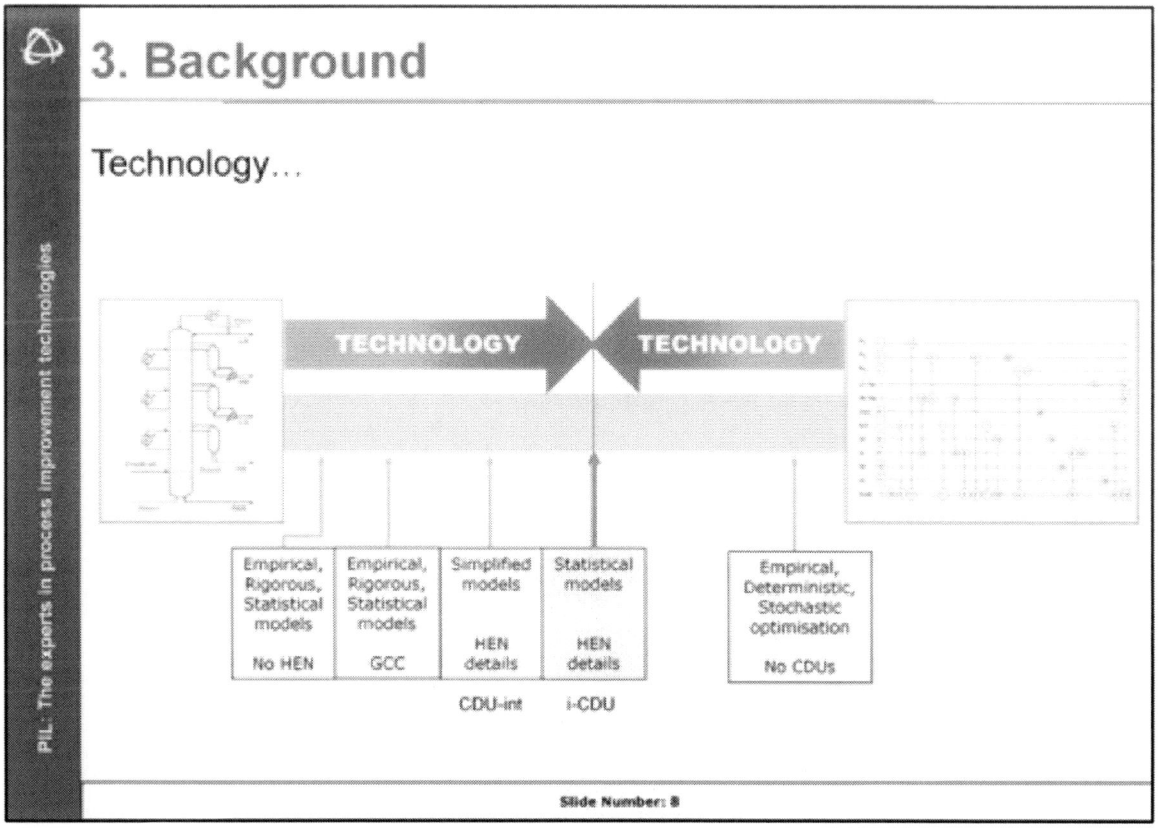

Over time, new computational tools have facilitated the implementation of optimisation-based design approaches, e.g. grass-roots design, retrofit and operational optimisation approaches. However, to develop such tools, several challenges need to be addressed. For heat-integrated crude oil distillation systems, the main challenge is finding models that are accurate enough to represent interactions between the crude oil distillation unit (CDU) and HEN, but at the same time are robust and computationally efficient enough to be implemented in optimisation frameworks.

Rigorous, simplified, hybrid and statistical models have been employed to represent crude oil distillation units. Each type of model has characteristics that define its suitability for a given design application. In particular, available degrees of freedom and model robustness are very important to select the best distillation model for each design purpose. For grass-roots design and retrofit, simplified models are most suitable, while for operational optimisation, statistical models are most suitable.

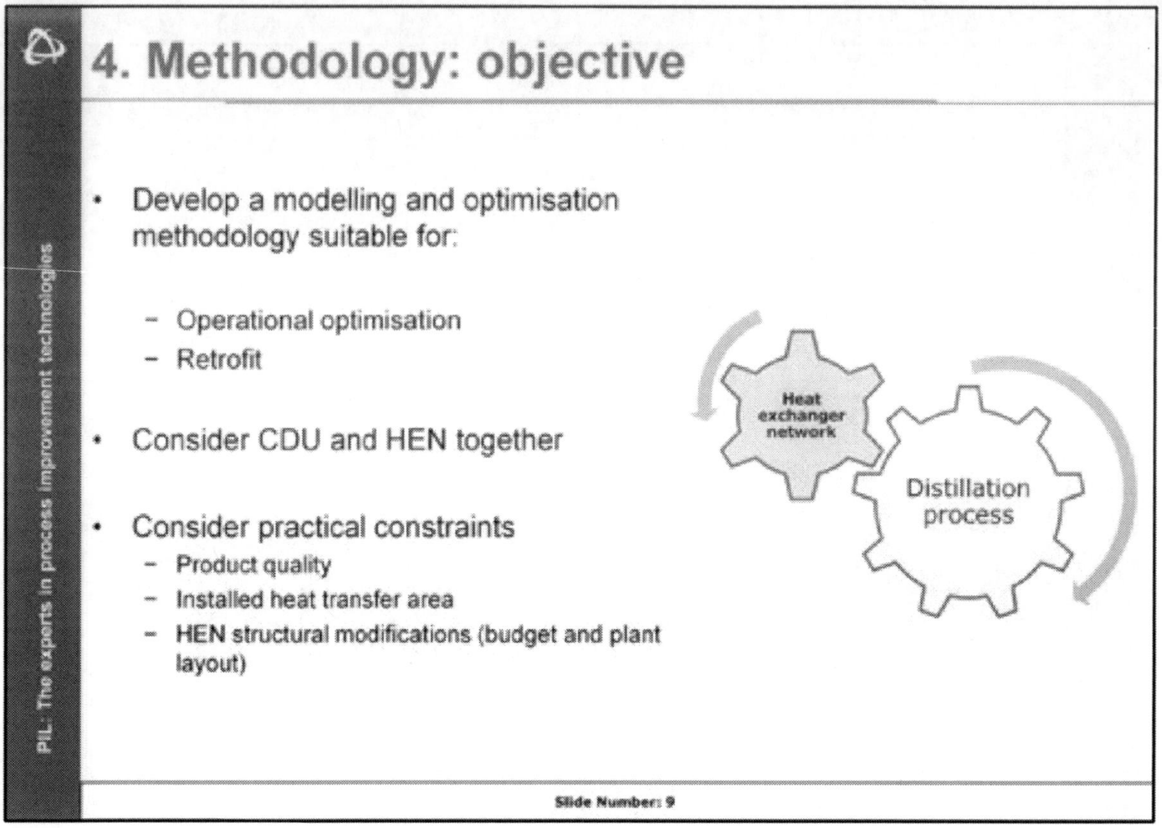

As discussed previously, many challenges arise in the design of heat-integrated crude oil distillation systems. In this project, a new optimisation approach is proposed to address some of these challenges. The new approach considers the distillation process and heat exchanger network together, which allows interactions within the overall system to be exploited. Operational optimisation is proposed for the distillation process, while retrofit of the heat exchanger network is allowed to accommodate the new operating scenarios. The new approach aims to produce improved designs in terms of product yields and operating costs, while taking into account process constraints (e.g. product quality, bounds on operational variables, etc.).

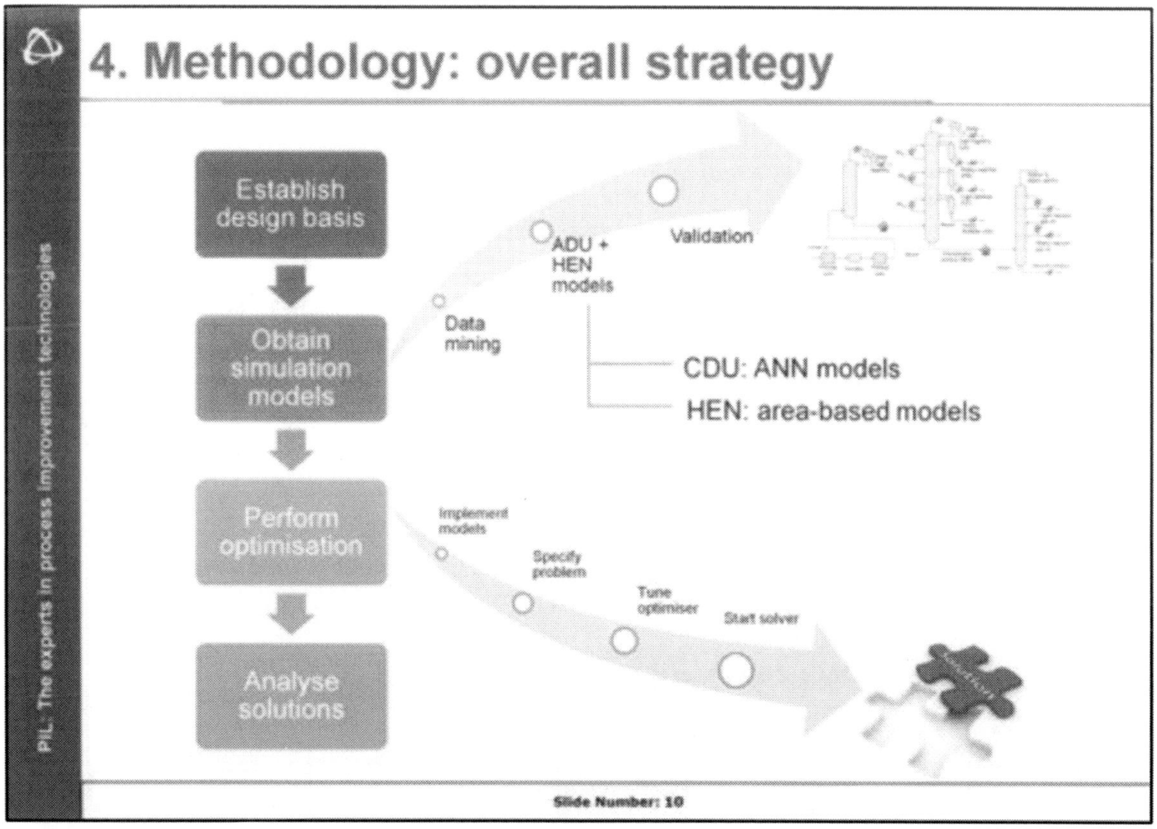

A new optimisation framework for heat-integrated crude oil distillation units using Artificial Neural Networks (ANNs) is proposed. In this approach, a statistical model for the atmospheric tower is constructed from data generated in rigorous models simulations. To represent the HEN, mass and energy balances are solved. The heat-integrated system is incorporated in an optimisation scheme that can optimise product profit and operating costs, while proposing retrofit modifications for the HEN. The optimisation problem is formulated to consider the dominant operational degrees of freedom (such as pump-around specifications or product yields) and constraints (e.g. product quality, heat exchanger area).

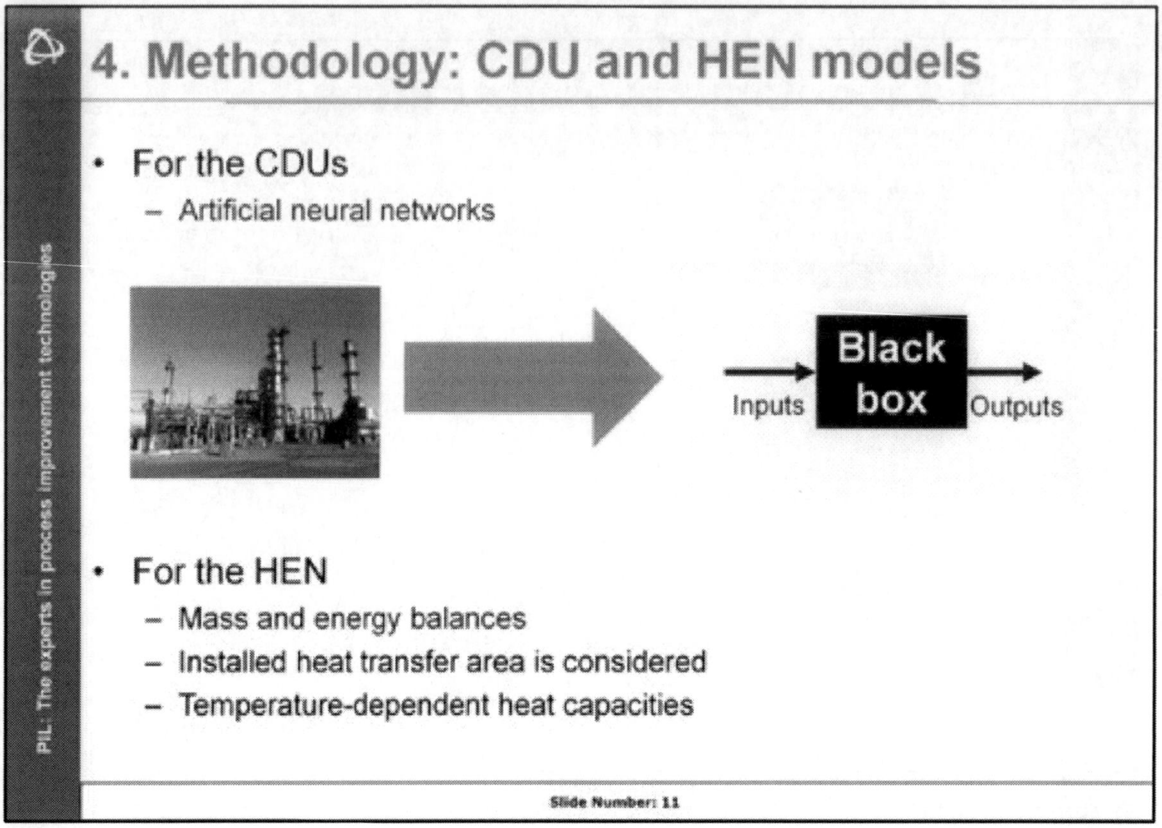

The main advantage of statistical models, such as ANNs, consists in their capability of capturing complex relations between great number of variables. They allow the designer to easily model systems that are often nonlinear, with strong interactions between their elements, such as CDUs. Another advantage is that they can build models from real plant data and are self-adapting to new information. Nevertheless, because ANNs are constructed using stochastic relations, their performance is entirely dependent on the quality of the information supplied to build the model, that is, the ANN is as good as the data used to represent the system.

iHeat is used to simulate the HEN. iHeat calculates the utility requirements and outlet temperatures of every heat exchange unit, splitter and mixer. iHeat also ensures that the calculated temperatures do not violate the stream energy balance, minimum temperature approach and installed heat transfer area constraints.

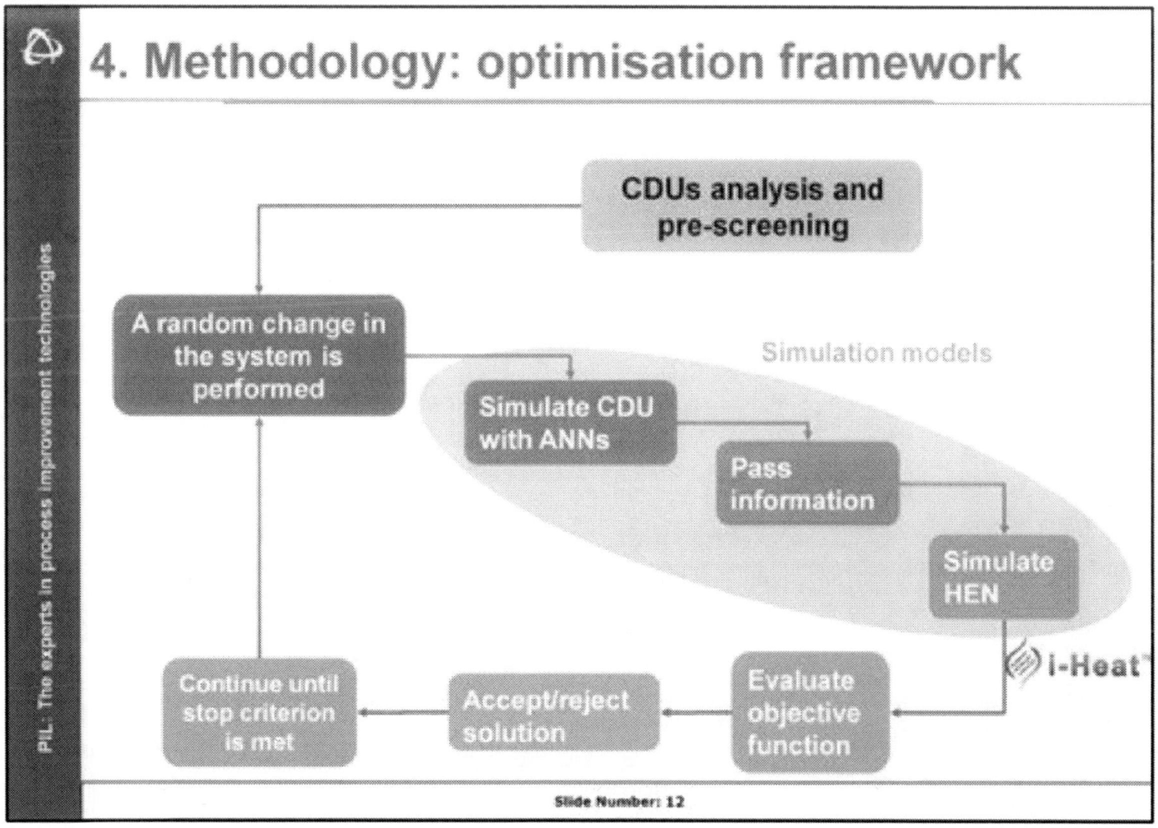

1. The optimiser starts with a feasible set of CDU operating conditions and HEN structure. The objective function for the initial distillation system is calculated.

2. For each iteration, the algorithm randomly performs a change in the distillation system. Changes can be implemented either on the operating conditions of the CDU or the HEN structure.

3. Once the CDU or HEN move is applied, the distillation system is simulated to compute the objective function. The CDU is simulated using ANNs while the HEN is simulated using i-Heat.

4. For a feasible set of operating conditions and HEN structure, the optimisation algorithm accepts or rejects each design of each iteration based on the value of the objective function. If the design is not feasible (i.e. violates constraints), the design is automatically rejected by the optimiser.

5. Steps 2 and 4 are repeated until the optimisation ends.

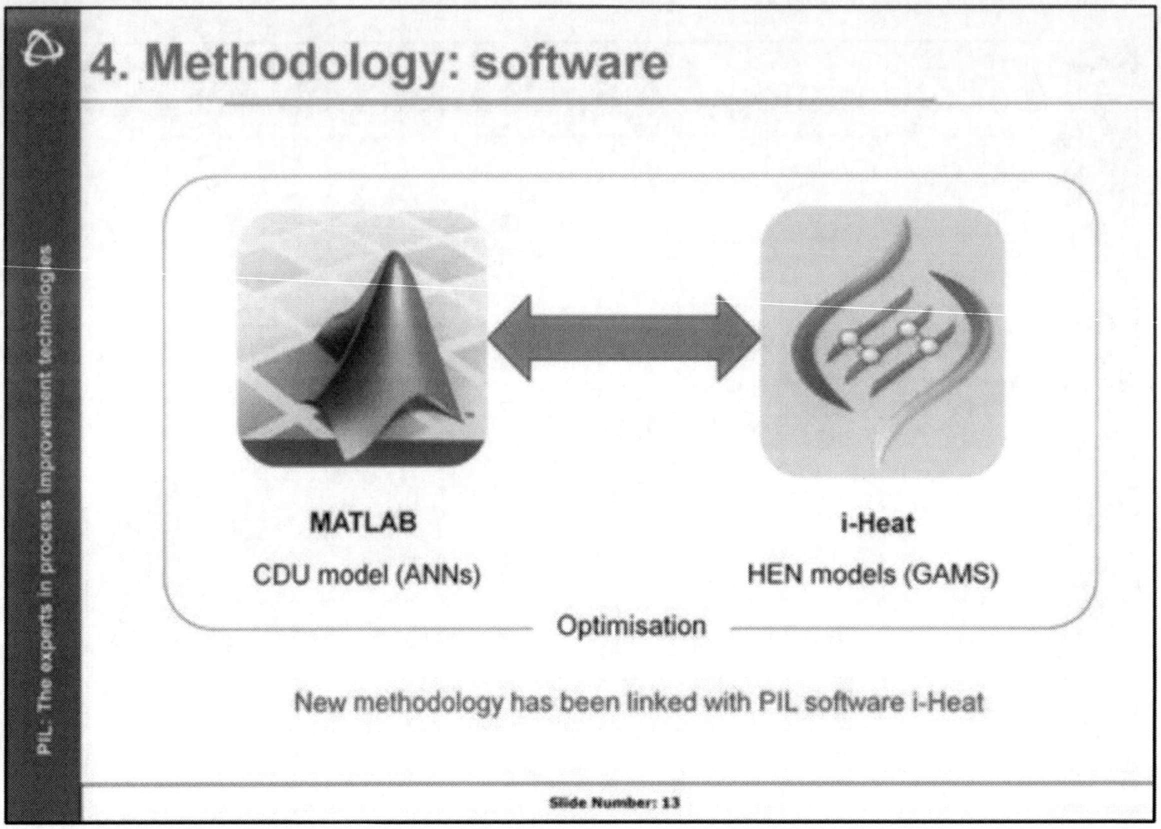

This slide show the software used to simulate the distillation system.

The CDU is simulated using ANNs developed in MATLAB while the HEN is simulated using i-Heat. MATLAB and i-Heat are linked during optimisation.

PIL: The experts in process improvement technologies

4. Methodology: challenges and solutions

Challenge	Solution
Consider CDU and HEN together	Methodology based on the work of Chen (2008) and Ochoa-Estopier (2014)
Detailed CDU models are highly non-linear and hard to converge	Build a simple CDU model using statistical models
HEN models are highly combinatorial and non-linear	Use simulated annealing optimisation to design the HEN and the CDU
CDU product quality needs to be kept	Screen scenarios that meet specifications using sensitivity analysis and neural networks
HEN installed areas need to be taken into account	HEN model considers heat transfer areas and are kept during optimisation
Pressure and temperature-dependent thermal properties	Simple first- and second-order models to represent this behaviour

Chen, L., 2008. Heat-Integrated Crude Oil Distillation Design. PhD Thesis. The University of Manchester, Manchester, UK.

Ochoa-Estopier, L. M., 2014. Optimisation of Existing Heat-Integrated Crude Oil Distillation Systems. PhD Thesis. The University of Manchester, Manchester, UK.

Slide Number: 14

5. Case Study

- **Modified from a Chinese refinery**
  - Reduction in energy consumption
  - Product yield improvement
- **Distillation system**
  - 2 prefractionators
  - 2 atmospheric units
  - 1 vacuum unit
  - 40 optimisation variables
  - 18 product quality constraints
- **HEN**
  - 50 process-to-process heat exchangers
  - 20 utility heat exchangers
  - Many practical constraints (e.g. installed area, EMAT, target temperatures)

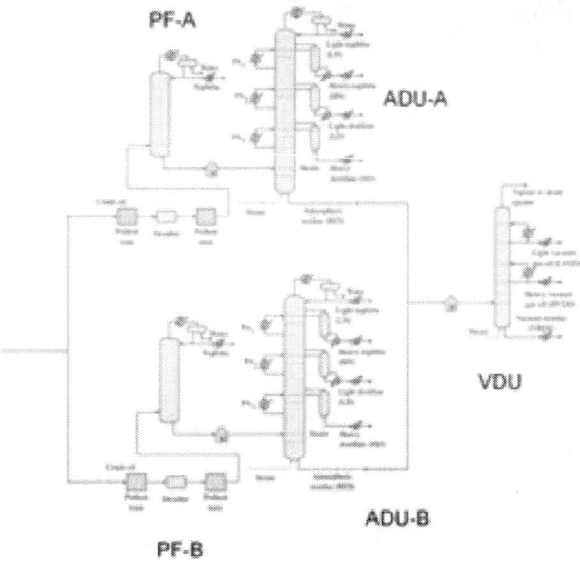

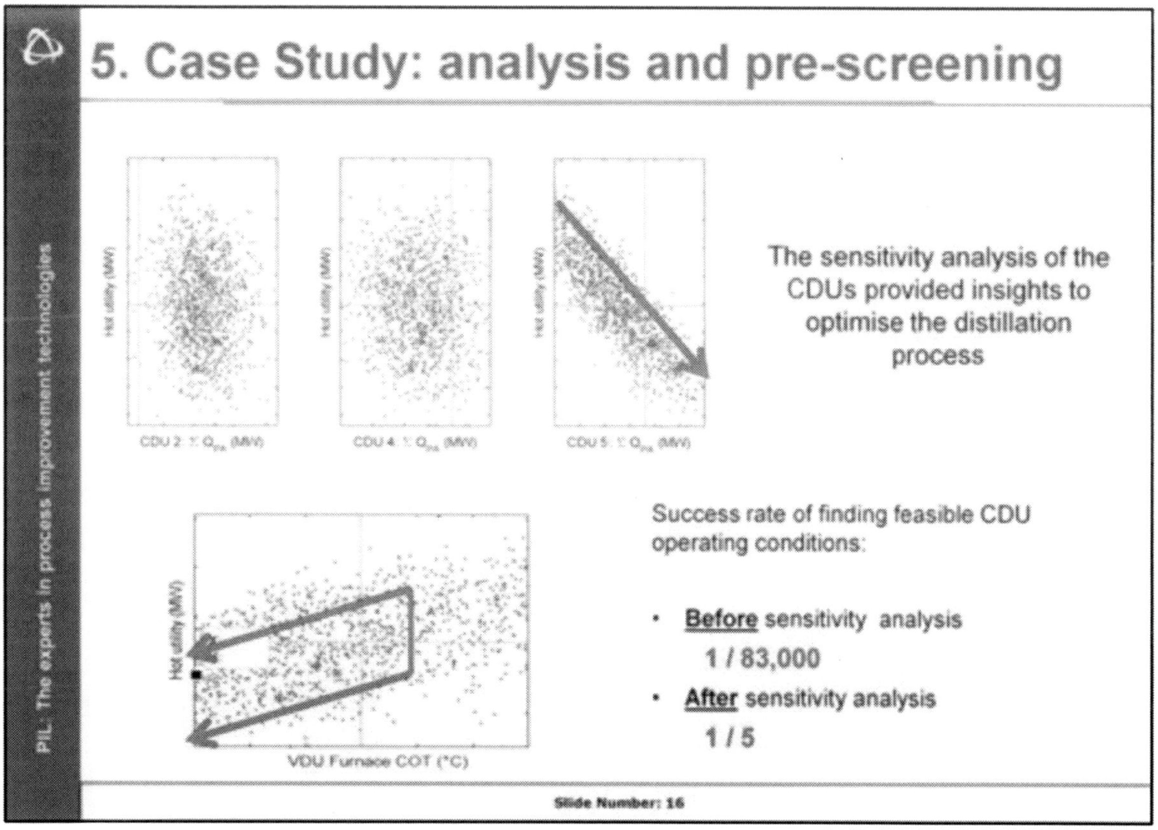

A sensitivity analysis of the CDUs is performed before developing the ANN models for the CDUs and before performing optimisation. This sensitivity analysis is very useful to gain insights of the distillation process, screen CDU operating conditions before optimisation, and identify the potential optimisation search space. This slide shows the analysis of some degrees of freedom of the CDUs and their effect on minimum hot utility requirements (calculated with the GCC). The horizontal and vertical lines (in green) indicate the base case values of these variables. The points are the samples obtained for the sensitivity analysis. The points form a "cloud" in each plot. The shape of this cloud is helpful to identify trends and operational limits. For example, in the first plot (top left), the cloud of points is far away from the lines that depict the base case values. This suggests that the feasible operating range in within the cloud rather than closer to the base case conditions. Thus, it is very likely that, during optimisation, the value for this variable lies inside the cloud.

The third (top right) and fourth (bottom) figures show trends (relationships) between hot utility requirements and the corresponding degree of freedom. The third figure indicates that increasing the total cooling duty of the VDU (i.e. CDU 5) reduces minimum hot utility requirements. On the other hand, the fourth figure suggests that decreasing the coil outlet temperature (COT) also reduces hot utility requirements. These trends can be applied during optimisation, for example, the upper bound of these degrees of freedom can be set to lower values to bias the optimisation to follow this trend.

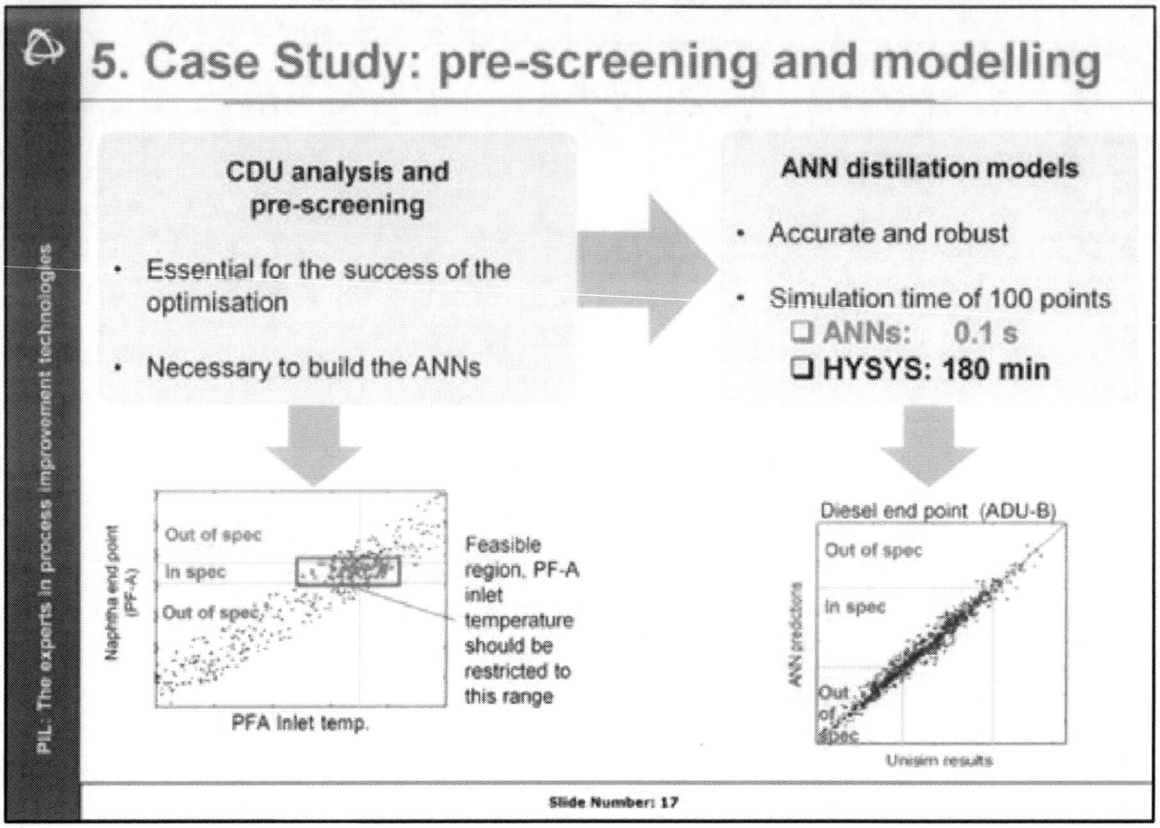

During the development of this project, it was found that the analysis and pre-screening of CDU operating conditions is essential for the success of the optimisation, especially for highly-constrained distillation systems. Analysis and pre-screening is also necessary to build accurate ANNs.

The figure in the left illustrates the sensitivity analysis of a CDU degree of freedom (PF-A inlet temperature) and one product quality specification (naphtha end point). Labels "in spec" and "out of spec" denote samples that meet and do not meet product quality specifications. It can be seen that the product meets constraints only in a limited range of PF-A inlet temperature values. This indicates that the PF-A inlet temperatures should be restricted to those values where product meets constraints.

The resulting ANN models are significantly faster in convergence and more robust than rigorous models built in Aspen HYSYS, with practically no detriment to ANN model accuracy. Once built, the ANN models can predict product quality and can also help to further screen operating conditions that do not meet product quality specifications, as shown in the figure in the right.

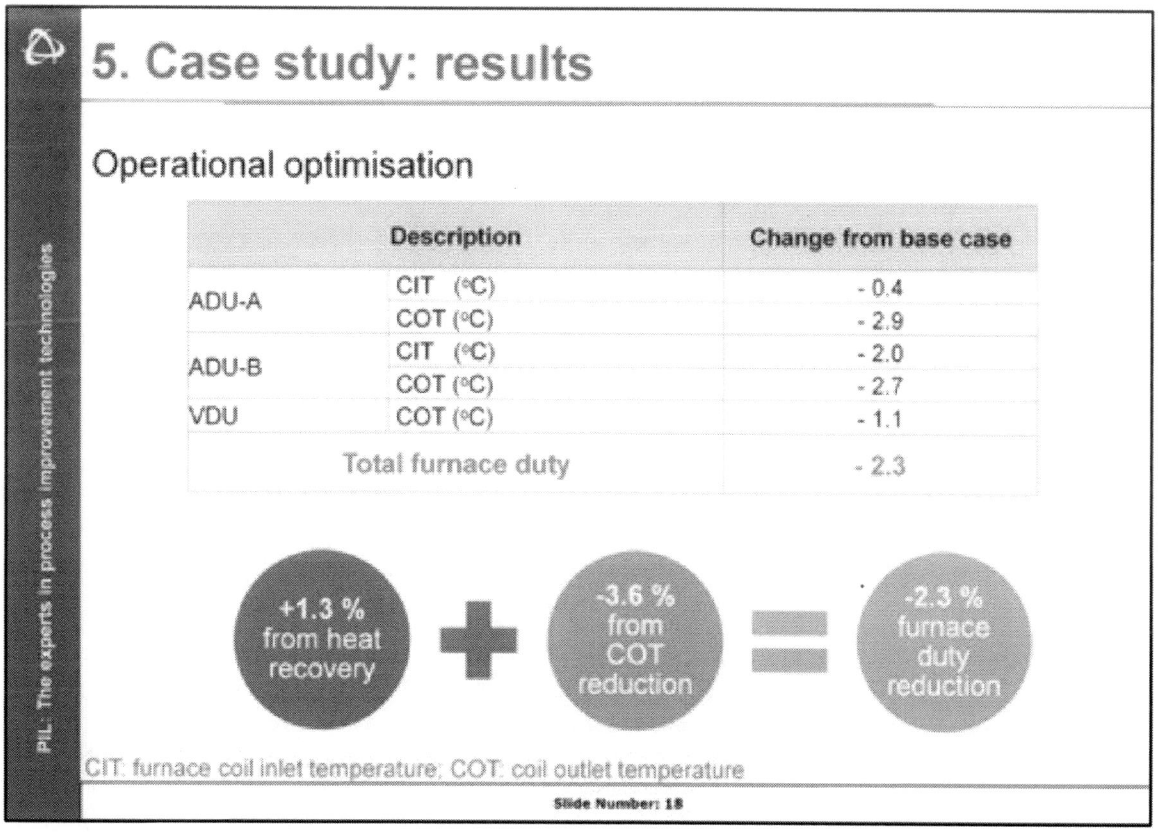

After the sensitivity analysis and pre-screening, the ANN models were implemented in the optimisation framework (using MATLAB and i-Heat at the same time). This slide shows some of the results for operational optimisation. For space saving and confidentiality, sensitive client information is removed from this presentation.

The furnace duty is reduced by 2.3 % for operational optimization. This reduction is indicated by a decrease in the coil outlet temperature (COT) that is considerably greater than the reduction in the coil inlet temperature (CIT).

The optimization framework is able to find a significant reduction in the furnace duty, which is directly related to fuel oil consumption and operating costs. For operational optimization, results can be readily implemented at no cost as no capital investment is needed.

6. Conclusions

PIL: The experts in process improvement technologies

- Recent developments in the optimisation of crude oil distillation systems include
  - New analysis tools to model CDUs and screen solutions
  - A robust, simple and accurate model to represent the complex CDUs
  - More realistic approach to represent the CDUs and HEN
  - An optimisation framework that considers practical constraints such as product quality, column flooding, installed heat transfer area, plant layout constraints, etc.

- New methodology is able to find reductions in energy consumption even for heavily constrained systems
  - Product quality kept during optimisation
  - Practicable solutions
  - Savings found for operational optimisation

Slide Number: 19

6. Future work

Scope	Development status	Software package
Off-line operational optimisation (CDU+HEN)	Current work	i-CDU
CDU operational optimisation + HEN retrofit	Current work	i-CDU
On-line operational optimisation	Future work	e-CDU
CDU retrofit + HEN retrofit	Future work	i-CDU
Grassroots design (CDU + HEN)	Future work	i-CDU
Use of plant data for operational optimisation	Future work	e-CDU

PIL: The experts in process improvement technologies

Slide Number: 20

This table shows current and future work to be developed by PIL

443712 Pilot Experiments of a Novel Catalytic Distillation Process for Sulfur Removal of FCC Gasoline

Abstract:

The environmental restriction regarding sulfur content in transportation fuels is currently one of the most important issues. Since major sulfur compounds in gasoline come from the fluid catalytic cracking (FCC) naphtha cut, sulfur removal of FCC naphtha is, therefore, essential for attaining the clean gasoline.

This paper will present pilot experiments of a novel catalytic distillation process for sulfur removal of FCC naphtha in a column packed with structural catalytic distillation cells containing Amberlyst35 resins. In comparison to the conventional hydrodesulfurization, this process can be conducted under mild conditions (Pressure of the catalytic distillation column is 100kPa and temperature of the bottom is lower than 150) and will consume lesser hydrogen.

In this process, thiophene and methyl-thiophenic compounds react with low carbon alkenes to form higher boiling point thiophenic derivatives, as well as mercaptans are etherified to higher boiling point alkyl-sulfides. Then, the higher boiling point sulfides are removed from light fractions by distillation and concentrated in the heavy fractions. Consequently, the sulfur free light cut is obtained from the top of distillation column. the fraction cut point can be raised from 75 to 120 , which will prevent olefins (with boiling point below 120) from saturation at the condition of hydrodesulfurization and avoid octane number loss with deeper desulfurization. This study would be of useful references to further industrial application of desulfurization of FCC gasoline.

Using a Large Data Bank to Develop a Robust and Simple Model to Estimate the Saturation Pressure of Oil Reservoirs

By

Hassan Alkandari[1], Adel Elsharkawy[1], Adel Malallah, Osamah Alomair

Petroleum Engineering Department, College of Engineering & Petroleum
Kuwait University
Po Box 5969 Safat 13060-Kuwait

Abstract

Measurements of saturation pressure are crucial for all hydrocarbon reservoir fluids. Below the crude oil saturation pressure gas reaches a critical saturation; then, a two phase flow occurs and results in decreasing oil production and recovery. To maximize oil production and recovery, the reservoir pressure has to be maintained closer to the original saturation pressure. This pressure is normally measured using bottom-hole samples or surface recombined samples of oil and gas. Occasionally, the samples become unavailable and the pressure needs to be estimated using computational methods.

In this study, a large data bank of 231 crude oil compositions and saturation pressure measurements including literature data, unpublished data, and newly measured data were used to develop two empirical models to predict the saturation pressure of a variety of crude oils if the oil sample is unavailable or the experimental measurements are unreliable. The first proposed model utilizes the extended compositions of hydrocarbons up to the heptane plus fraction in addition to non-hydrocarbons. The second model uses the lumped compositions of light, intermediate, and heavy components in addition to non-hydrocarbon components as the input variables. The lumped model has the advantage of using fewer input parameters while maintaining the thermodynamics. The accuracy and validity of both models to calculate the saturation pressure for volatile oils, black oils, and heavy oils are presented using several compositional data. The models performance is also compared to the Peng-Robinson equation-of-state (PR-EOS), and the Soave-Redlich-Kwong equation-of-state (SRK-EOS) as well as all published methods that use compositions as input variables. The comparison indicates that the proposed models are much simpler and more accurate than the other computational methods. The proposed models treat the heptane-plus fraction as a single component, thus eliminating the splitting and characterizing the plus fractions and the binary interaction parameters needed for the EOS's calculations.

Key words: Saturation pressure (Ps), Bubble-point pressure (BPP), Equation of state (EOS), Correlations, and PVT Data

[1] Corresponding Authors: Hassan Alkandari , Hassan@hkandari.com, Adel Elsharkawy, adel.elsharkawy@ku.edu.kw

Introduction:

Reservoir fluid properties are an essential source of information for all calculations on hydrocarbon reserves. Accurate estimation of the pressure-volume-temperature (PVT) measurement is crucial. Of all the PVT properties, saturation pressure (Ps) is considered the most important parameter. It is well known that gas starts to flow once the reservoir pressure drops below the bubble point pressure and the gas saturation reaches a critical value. To minimize the gas flow and maximize the oil recovery, it is very important to maintain the reservoir pressure close to the original bubble point pressure.

Although the PVT properties are experimentally measured in laboratories and provide reliable results, this process is expensive, time consuming, and sometimes unavailable; therefore, measured field data (separator gas oil ratio, separator pressure, stock-tank oil specific gravity, and reservoir temperature) or reservoir fluid compositions are used to estimate the bubble point pressure.

There are two methods in the petroleum industry to calculate the saturation pressure of crude oil when experimental is unavailable; either using production data or compositional data.

Methods Based on Production Data (Composition is unavailable)

Numerous attempts have been made since the 1940s to develop correlations to estimate the saturation pressure from production data, such as producing gas oil ratio, oil gravity, gas gravity, reservoir pressure and temperature. Table 1 presents summary of all the methods that use production data to estimate the saturation pressure.

Methods based on Compositional data

Many authors have reported that the application of any correlation in other regions results in serious errors. These conclusions arise from the fact that regional crude oils represent a physical-chemical trend; therefore, all attempts to develop a general correlation using field measured parameters to estimate the saturation pressure from a large data bank have failed. For this reason, Elsharkawy (2003) made the first attempt to correlate the saturation pressure to the reservoir fluid composition instead of using the production data. In recent years, there have been several attempts to estimate the saturation pressure of oil reservoirs. These methods are reported in Table 1.

Elsharkawy (2003) presented a correlation of the bubble point pressure for the oil reservoir using the same input data needed for the equation-of-state (EOS) calculations, i.e., the compositional data and the reservoir temperature. Elsharkawy's model is based on the saturation pressure measurements of 60 crude oil samples from the Middle East and 75 collected from the literature. He demonstrates that the calculation of the saturation pressure depends on the type of EOS, the number of pseudo components, the method used to characterize the pseudo-components and the binary interaction number. Elsharkawy compared the performance of his model to the SRK-EOS and PR-EOS for black oils, volatile oils, heavy oils, and synthetic oils. He indicated that his model eliminates splitting the heavy plus fraction, characterizing the pseudo-components and inclusion of the binary interaction parameters needed for the EOS calculations.

Later, Tale et al. (2009) made the second attempt to model the saturation pressure of crude oil using 132 data samples published by Elsharkawy (2003). Tale's model, which is based on the alternating conditional expectation (ACE), has a set of 14 equations with 42 constants. The average absolute relative error (AARE) reported by Tale et al. indicates that the model did not perform well with the selected data. The ACE model has an AARE of 12.3% compared with AARE of 6.7% by Elsharkawy (2003) and 8.1% AARE for the SRK-EOS.

AlQuraishi (2009) attempted to model the saturation pressure using Elsharkawy's (2003) data. The model was developed using linear genetic programming. He correlated the saturation pressure to the formation temperature, methane content, and heptanes plus molecular weight. Thus, AlQuraishi's model neglected the effect of the intermediate and heptane plus content on the saturation pressure. The error analysis reported by AlQuraishi indicates that his model has a 5.8% error compared to 7.7% by Elsharkawy, 9.9% using the SRK-EOS, and 10.1% using the PR-EOS.

Bandyopadhyay and Sharma (2011) presented a model to predict the bubble point pressure, using temperature and fluid composition. They used 129 experimental and literature data originally published by Elsharkawy (2003) to develop their model. The model introduced a temperature interaction parameter that depends on compositions. The proposed model has a set of 24 coefficients and the estimated bubble point pressure is calculated as the sum of the series calculations. They reported an average absolute error (AAR) of 7.83% using their model compared to 8.3% using Elsharkawy's simple model (2003), 10.15% using SRK-EOS, and 10.58% using the PR-EOS.

Khamehchi et al. (2011) used an intelligent proxy simulator (IPS), which is based on an adaptive neural fuzzy inference system (ANFIS) and a structured artificial neural network (ANN) to predict the bubble point pressure and asphaltene onset pressure. The model uses a reservoir fluid composition and temperature to predict the Ps. It is not clear how the Ps is calculated from the proposed method, and they did not present the accuracy of the proposed method compared to the other published methods.

Farasat et al. (2013) presented a new mathematical approach to calculate the crude oil saturation pressure as a function of the temperature and the reservoir fluid composition. The model developed by Farasat et al used 130 experimental data, which were originally published by Elsharkawy (2003). Farasat used a support vector machine (SVM) to model the saturation pressure. They did not clearly demonstrate how to calculate the saturation pressure from the SVM model. Moreover, they did not indicate the overall average absolute deviation (AAD) for their model compared to other published models and both the PR-EOS and SRK-EOS for the 130 data.

Ahmadi et al. (2014) also presented an ANN technique called gene expression programming (GEP) to determine the bubble point pressure of the oil samples. Similar to Farasat et al. (2013), Ahmadi's model used most of the data previously published by Elsharkawy (2003) to construct a non-linear equation using reservoir fluid compositions and temperatures as input parameters. The proposed model is highly complicated and consists of 8 equations that have several constants compared to the one simple equation proposed by Elsharkawy (2003). They compared the performance of their proposed model to the SRK-EOS and PR-EOS; however, their paper does not mention how the heptanes plus fraction were treated and characterized for the EOS calculations. They reported an AAD of 4% using the proposed model compared to 8% using Elsharkawy's model (2003), 10.9% by PR-EOS, 10.5% by SRK-EOS, and 15% by Bandyopadhyay and Sharma (2011).

Gholami et al. (2014) estimated the bubble point pressure using a support vector regression SVR and a supervised learning algorithm SLT based on the statistical learning theory, and then the ACE was used. The model used reservoir fluid compositions and temperature to estimate the bubble point pressure. Gholami et al. used the same data originally published by Elsharkawy (2003). They presented three attempts based on genetic programming to correlate the bubble point pressure: SVR, ACE, and power-law committee machine (PLCM). The ACE-based method resulted in a set of three equations with a matrix of [8 by14] constants. Gholami et al. compared the accuracy of their model to the SRK-EOS and PR-EOS predictions reported by Elsharkawy (2003). Based on their AARD analysis, the PLCM model has the lowest errors followed by the SVR model, Elsharkawy's model, ACE model, PR-EOS, and SRK-EOS.

Lately, Jarrahian et al (2015) presented two empirical models to estimate the saturation pressure of black oils. The first model is based on production information. The second model is based on compositional information mostly from Elsharkawy (2003) and Heidaryan and Moghadasi (2011). They have not presented comparison of the accuracy of their model to Farasat (2013) model, Ahmadi (2014) model, and Gholami (2014) model.

This paper is concerned with the estimation of saturation pressure using compositional data. Therefore, the objective is to develop a robust yet simple model to estimate the saturation pressure for various types of reservoir oils using a larger data bank than that used by other published studies. The accuracy of the proposed models is compared to previously published models as well as SRK-EOS and PR-EOS.

Data bank

This study uses a large data set of 231 crude oil compositions and saturation pressure measurements. The data set consists of 130 data sets originally reported by Elsharkawy (2003); in addition to others collected from literature (Jacopy, 1958; Vogel, 1980; Coats, 1986; Droham, 1980; Dansh, 1992; Moharam, 1996; Rosenegger, 1999); 94 new data sets that have not been utilized before; as well as 7 newly measured samples at the Petroleum Fluid Research Center (PFRC) in Kuwait University. These data represents a wide range of compositions, such as volatile oils, black oils, and heavy oils. The data cover a pressure range of 537 to 5065 psia and a temperature range of 58 to 319 °F, crude oil compositions from C_1 to C_{7+}, non-hydrocarbons, such as N_2, H_2S, and CO_2, and the molecular weight and the specific gravity of the heptane plus fraction.

Data screening and classification

The compositions and the saturation pressure measurements in the data bank are subjected to a meticulous screening process to check the validity of the data. Our screening criteria uses the following steps: the sum of the molar compositions of all the components must add up to one, otherwise, the sample is eliminated. Samples with a missing molecular weight or specific gravity of the plus fraction are also eliminated.

After filtering, the data were reduced to 214 and divided into 179 measurements used to develop the new models, and 35 data sets are used for the blind test of the robustness of the proposed models. Statistical analysis of the input variables of the data set used in training and testing the extended and the lumped models are illustrated in Tables 2.

Several of the reported compositional data have extended analysis up to (C_{20+}). The following techniques are used for lumping or regrouping the extended analysis into single pseudo-fractions (C_{7+}). The mole fraction (X_i) of the individual components of the extended analysis is equal to the plus fraction (Ahmad, 2011).

$$X_{C_{7+}} = \sum_{i=7}^{n} X_i \qquad i=7, 8 \ldots n \qquad (1)$$

The average molecular weight (Mw) of the individual component of the extended analysis is equal to the molecular weight of the plus fraction.

$$Mw_{C_{7+}} = \sum_{i=7}^{n} \frac{X_i Mw_i}{X_{C_{7+}}} \qquad i=7, 8 \ldots n \qquad (2)$$

The density of the total plus-fraction can be calculated as follows:

$$\gamma_{C_{7+}} = \frac{X_{C_{7+}} Mw_{C_{7+}}}{\sum_{i=7}^{n} \frac{X_i Mw_i}{\gamma_i}} \qquad i=7, 8 \ldots n \qquad (3)$$

It is important to note the significance of the density of plus fraction because its density determines its paraffinic naphthenic aromatic (PNA) content.

Equation-of-state calculations

The EOS calculation of the saturation pressure is compared with the newly developed models as well as the other methods. The two EOS calculations considered in this paper are the ones most considered and widely used in the petroleum industry to evaluate the volumetric and phase behavior of reservoir fluid properties: the Soave-Redlich-Kwong EOS (SRK-EOS) and the Peng-Robinson EOS (PR-EOS). It should be noted that the accuracy of any EOS calculation of the saturation pressure largely depends on the number of pseudo-components and the splitting schemes used to divide the heptane plus fraction; as well as the characterization and calculation of the critical properties of the pseudo-components and the binary interaction parameters. Therefore, the two EOS calculations of the saturation pressure are not the same, even though the number of pseudo-components, characterization of the pseudo-components, and binary interaction parameters were the same.

Model development

As previously stated, the objective of this work is to provide the petroleum reservoir engineer with a simple yet accurate tool to estimate the saturation pressure for a variety of crude oils using compositional data when experimental is unavailable. Two models are presented in this study to calculate the saturation pressure of the crude oil samples. The first model uses all input parameters, such as the molar compositions of the hydrocarbon and non-hydrocarbon components similar to the EOS models. This model is referred to as the extended model. The second model uses fewer input parameters in which some of the hydrocarbon components have been lumped into a single group; this is designated as the lumped model. Both models are simpler than all published correlations, physically correct, and have an acceptable accuracy compared with the well-known EOS.

Figure 1 shows a correlation matrix of all the data available under consideration. This figure also indicates the relative importance of each of the input variables into the first proposed model. It agrees with most published correlations and the EOS that the methane content, heptane plus content and temperatures have the greatest impact on the saturation pressure followed by the intermediate content. The data understudy indicates that the non-hydrocarbon has the smallest impact. This result is simply a reflection of the molar contribution of each component comprising the crude oil composition.

Model I - Extended Model

The extended model uses 13 variables comprising the molar compositions of the pure components, heptane plus properties, and reservoir temperatures, which are similar to the input parameters for the EOS. Figure 2 shows the correlation coefficient graph of all the input parameters in the extended model. Thus, the elimination of the intermediate effect has a negative impact on the accuracy of the predictive model. Our proposed extended model has the following form:

$$Ps = 124.72N_2 + 17.57CO_2 + 22.55H_2S + 64.22C_1 - 9.80C_2 - 52.49C_3 + 6.16C_4 - 19.22C_5 - 23.63C_6$$

$$- 21.43C_{7+} + 435.31\gamma_{C7+} + 1.14Mw_{C7+} + 4.29T \qquad (4)$$

Where Ps is the saturation pressure in psia, T is the temperature in °F, and the hydrocarbon and non-hydrocarbon compositions are expressed in mole.

Model II - Lumped Model

The second model is much simpler and uses only 7 -variables (the lumped model). Several lumping schemes are adapted in the petroleum industry for phase behavior modeling and reservoir simulation of the petroleum reservoir fluids (Ahmed, 2011). Hydrocarbon components are grouped into a pseudo fraction based on their molecular weight (Whitson, 1980), the similarity of their physical-chemical properties (Lee, 1979), their carbon number, their true boiling point or their number of families based on the chemical structure (Becker et al., 2015). (Emera and Samara, 2005) proposed lumping the hydrocarbon components into volatile (C_1 & N_2), intermediate (CO_2, H_2S, C_2 to C_4) and heavy categories (C_{5+}) to calculate the minimum miscibility pressure. (Duan et al., 2013) proposed lumping several components into a single carbon number (SCN) with predetermined values for properties, such as critical pressure, temperature, and eccentric factor. Others have proposed optimal lumping schemes based on the objectives (Lin et al., 2008). In lumped model, $C_{1+}N_2$, C_2, CO_2, and H_2S, the intermediate (C_3 to C_6), heptanes plus (C_{7+}) and its properties, and temperature are the input variable. The correlation matrixes of the lumped components are provided in Figure 3. Even though the non-hydrocarbon has the least effect their contribution is important, especially the H_2S in the case of sour crudes. Both models presented by AlQuraishi (2009) and Bandyopadhyay and Sharma (2011) have eliminated the effect of H_2S.

The proposed model has the following form:

$$Ps= 82.115*(C_1+N2)-11.635*C_2+39.158*CO_2+ 38.244*H_2S -1.217*(C_3—C_6) - 890.701*(C_{7+}/MW_{7+}*\gamma_{7+}) + 4.217 *T-1042 \quad (5)$$

Where Ps is the saturation pressure in psia; and T is the temperature in °F.

Result and discussion

Model accuracy

The accuracy of the two proposed models used to calculate the saturation pressures of the previously described crude oil samples is shown in Table 3. This table shows reports the average relative error, average absolute relative error standard deviation and the correlation coefficients for all the methods considered in this study. Table 3 also reports the accuracy of the PR-EOS, SRK-EOS, Elsharkawy's model, AlQuraishi's model, and Bandoypandyay and Sharma's model and Jarrahian model. The methods proposed by Gholami et al., Ahmadi et al., and Jarrahian et al are included in Table 3, even though they produce a unreasonable estimate of the saturation pressures, as explained in the appendix.

This table indicates that the extended and the lumped models have an (AARE) in the order of 8% and 8.5%, respectively. It should be noted that the complications in the Bandoypandyay and Sharma model did not improve the accuracy of the saturation pressure, because it had a large error. It is also noted that Jarrahian et al model produces an average error of 35%. Compared with the EOS, the lumped model has a few constants and eliminates the splitting, characterizing the pseudo-components and binary interaction parameters necessary for the EOS calculations. The lumped model as well as the extended model are very simple and require one-step calculations compared with all the other methods reported in Table 3.

A cross-plot of the calculated saturation pressure using the extended model based on the training data versus the measured data is illustrated in Figure 4A. This figure indicates that most of the data falls on the unit slope line, and there is no over-fit. This trend is supported by the high correlation of regression of the model (97.6%) presented in Table 3. Figure 4B shows a cross-plot of the testing data (28 from literature and 7 new measured in PFRC at Kuwait University). The extended model indicates an excellent agreement with all of the 28 data sets chosen at random, with the exception of a few data points at higher pressures. Additionally, the model has a perfect match for the 7 newly measured samples.

Figure 5A shows the performance of the lumped training-model for capturing the measured saturation pressure for most of the 179 samples. A cross-plot of the calculated saturation pressure using the lumped model based on the training data versus measured data is depicted in Figure 5B. This figure indicates that most of the data scatters falls to the unit slope line. The lumped model also has a high coefficient of regression of 97.4%, which is indicated in Table 3.

Models Validity

The accuracy and validity of the models for estimating the saturation pressure of various types of crude oil are reported in Table 4. This table illustrates the capability of the newly proposed models to predict the saturation pressure for volatile oils, black oils, and heavy oils. Table 4; provides the compositions, properties of the heptane plus fraction, temperature, and measured saturation pressures of the samples. Additionally, the calculated saturation pressure using the PR-EOS, SRK-EOS, Elsharkawy's model (2003), AlQuraishi's model (2009), and Bandoypandhyay and Sharma's model (2011), Jarrahian et al (2015); as well as the extended and lumped models are provided in Table 4. It is clear from this table that the proposed models have a wide range of applications regardless of crude oil types. We have also included information regarding the 7 newly measured data; at the PFRC, compositional, and saturation pressure. At the bottom of Table 4, we have provided the unpublished data of the compositions; and the saturation pressures at low temperatures. It is well known that the saturation pressure of a crude sample is measured at low temperatures in addition to the reservoir temperature as a method to check the quality of the sample.

Figure 6A reports the error of all the methods considered in this paper to predict the saturation pressure for volatile oils (heptane plus content below 30%). Both the extended and the lumped models have an AARE of nearly 4% followed by Bandoypandhyay and Sharma's model (2011); and Elsharkawy's model (2003); the SRK-EOS has nearly twice the AARE of the newly proposed model. AlQuraishi's model (2009) has an AARE of 12% because the model is based on methane and neglects the effect of heavy and intermediate components, whose contribution is highly significant in the reservoir and production performance of volatile crudes. Jarrhaine et al (2015) model is not suitable for volatile oils. It shows a 35% error in estimated saturation pressure.

The performance of the extended and lumped models compared to the various methods for the black oil samples (35< C_{7+} <55%) is provided in Figure 6B. The newly proposed models have an AARE close to 10%; followed by Bandoypandyay and Sharma's model 14%, the SRK-EOS and PR-EOS 16%, and Elsharkawy 17%. AlQuraishi's model has an AARE of 24%. Jarrhain et al (2015) model shows a 33% error in estimated saturation pressure even though the authors claim that the model is suitable for black oils.

Figure 6C shows the error level of all the methods as well as the proposed models for the newly measured crude oil samples at Kuwait University using the PFRC. Again, Jarrhain et al (2015) model show a 40% error in estimated saturation pressure.

The performance of all published models at low temperature is illustrated in Figure 6D. The low-temperature calculation indicates that Bandoypandhyay and Sharma's model (2011) has accuracy problems and is not valid at low temperatures because the authors proposed a temperature-composition related coefficient. Also, Jarrhain et al (2015) model is not suitable at low temperature. It shows a 40% error in estimated saturation pressure.

The ability of the proposed models to capture the thermodynamic is thoroughly checked against its independent variables. The validity of the model is compared to the experimentally measured data as well as the predictions by other methods used to capture the physics of increasing the saturation pressure as a function of temperature is illustrated in Figure 7. Two main points are observed here; first, the model matches the measured saturation pressures much better than the other methods. Second, it follows the physical trend predicted by the EOS's. Figure 8 shows the model performance for capturing the effect of changing the saturation pressure as a function of the molecular weight of the heptane plus fraction compared with the SRK-EOS and PR-EOS. The extended model follows the trend of the PR-EOS in increasing the saturation pressure. Elsharkawy's model (2003) was unable to predict the correct trend as predicted by the EOS's. The model sensitivity in capturing the change in the saturation pressure as a function of increasing the methane content is shown in Figure 9. The proposed model as well predicts the correct trend of increasing the saturation pressure when increasing the methane content.

Conclusion

In this paper, two models are presented to estimate the saturation pressure of a variety of crude oils when measures are unreliable or samples are unavailable. The two models are developed using compositional data and saturation pressure measurements from a large data bank comprising literature data, unpublished data, and newly measured and reported data of 231 crude oil samples. The input parameters in the first model are the extended hydrocarbon and non-hydrocarbon molar percentages and reservoir temperatures. The second model uses lumped hydrocarbon fractions of intermediate, and heptane plus fractions in addition to its properties; and the reservoir temperature as input parameters. The accuracy of the two models has been tested and compared to that of the SRK-EOS and the PR-EOS; as well as published correlations that use compositional data as the input parameters. The comparison indicates that the two models are much more accurate than all the published methods. The validity of both models has also been tested using various types of crude oils samples and compared to experimentally measured saturation pressures and estimates using the EOS as well as the published correlations. The proposed models are physically correct and much simpler than all the published methods. The models do not require tedious computations and eliminate splitting of the plus fraction, characterization of the pseudo-components; as well as the binary interaction parameters that are required for the EOS calculations.

Acknowledgement

The authors thank the General Facility Lab at Kuwait University for their cooperation and conducting the compositional and saturation pressure measurements of the 7 new samples described in this paper using Petroleum Fluid Research Center (PFRC) project no. GE 01107.

References

Agarwal, R.K., Li, Y.K., Nghiem, L., 1990. A regression technique with dynamic parameter selection for phase – behavior matching. SPE Res. Eng. February, 115 – 120.

Ahmadi, M. A., Zendehboudi, S, James, L., Elkamel, A., Dusseault, M., Chatsiz, I., Lohi, A., 2014, New Tools to determine Bubble Point Pressure of Crude Oils: Experimental and Modeling Study, *Journal of Petroleum Science and Engineering, v.123, 204-216.*

Ahmed, T., 2011, Splitting and lumping schemes of the plus fraction, working guide to vapour-liquid phase equilibria calculations, Chapter 9, Hydrocarbon Phase Behaviour, Gulf Publishing Co. 111-126.

Al-Gathe, A., Abd-el Fattah, Kh., El-Mtwally, K., 2014, New artificial intelligent approach for bubble point pressure, Int. J innovation and Sci. Research, Vol.4, No.2, July, 121-135.

Al-Marhoun, M.A. 1985. Pressure-Volume-Temperature Correlations for Saudi Crude Oils. Presented at the SPE Middle East Oil Technical Conference and Exhibition, Bahrain, 11–14 March. SPE-13718.

Almehaideb, R.A. 1992. Improved PVT Correlations for UAE Crude Oils. Presented at the Middle East Oil Show and Conference, Bahrain, 15-18 March. SPE-37691.

Al-Najjar, H.S., Al-Soof, N.B.A., and Al-Khalisy, K.M. 1988. Correlations for Bubble-Point Pressures, Gas Oil Ratios and Formation Volume Factors for Iraqi Crude Oils. *Journal of Petroleum Research* (June 1988): 13.

AlQuraishi, A. A., 2009, Determination of crude oil saturation pressure using linear genetic programing, Energy & Fuel, 23, 884-887.

Al-Shammasi, A.A. 2001. A Review of Bubblepoint Pressure and Oil Formation Volume Factor Correlations. *SPE Res. Eval. & Eng.* 4 (2): 146-160. SPE71302.

Asgarpour, S., McLauchlin, L., Wong, D. et al. 1989. Pressure-Volume-Temperature Correlations for Western Canadian Gases and Oils. *J Can Pet Tech.* 28 (4): 103.

Asoodeh, M., Bagheripour, P.,2012, Estimation of bubble point pressure from *PVT* data using a power-law committee with intelligent systems, *J. Pet. Sci. and Eng., Vol. 90–91, July 2012, Pages 1-11.*

Bandyopadhyay, P., Sharma, A., 2011, Development of a new semi analytical model for prediction of bubble point pressure of crude oils, *Journal of Petroleum Science and Engineering, Volume 78, Issues 3–4, September 2011, Pages 719-731.*

Becker, P. J., Celse, B., Guillaume, D., Dulot, H., Costa, V., 2015, Hydrotreatment modelling for variety of VGD feedstocks: a continuous lumping approach, Fuel, 139, 133-143.

Coats, K.H., and Smart, G.T., 1986, Application of regression base EOS PVT Program to laboratory data, SPERE, May 277-299.

Dansh, A. Xu, D. H., Todd, A. C., 1992 A grouping method to optimize reservoir description for compositional simulation of gas injection process, SPERE, Aug. 343-348.

De Ghetto, G. and Villa, M. 1994. Reliability Analysis on PVT Correlations. SPE-28904, Presented at the European Petroleum Conference, London, United Kingdom, 25-27 October.

Dindoruk, B. and Christman, P.G. 2001. PVT Properties and Viscosity Correlations for Gulf of Mexico Oils. Presented at the SPE Annual Technical Conference and Exhibition, New Orleans, 30 September-3 October. SPE-71633

Dokla, M.E. and Osman, M.E. 1992. Correlation of PVT Properties for UAE (United Arab Emirates) Crudes. *SPE Form Eval.* (March 1992): 41.

Droham, J.K., 1988, Enhancing the evaluation of PVT data, OSEA paper presented the 7th Offshore South East Asia, Feb. 2-5, Singapore.

Duan, J., Wang, W., Liu, H., Cong, J., 2013, modelling the characterization of the plus fraction by using continuous distribution function, Fluid Phase Equilibria, 345, 1-10.

Elam, F.M. 1957. *Prediction of Bubble Point Pressures and Formation Volume Factors from Field Data.* MS thesis, University of Texas at Austin, Austin, Texas.

Elmabrouk, S., Zekri, A., Shirif E., 2014, The prediction of bubble point pressure and bubble point oil formation volume factor in absence of PVT analysis, PET. Sci. & Tech., 32, 1168-1174.

Elsharkawy, A. M., 2003, An empirical model for estimating the saturation pressures of crude oils, Journal of Petroleum Science and Engineering 38 (2003) 55-77.

Elsharkawy, A.M. and Alikhan, A.A. 1997. Correlations for predicting solution gas/oil ratio, oil formation volume factor, and undersaturated oil compressibility. *J. Pet. Sci. Eng.* 17 (3–4): 291-302.

Emera, M., Samara, H., 2005, Use of genetic algorithm to estimate CO2-oil minimum miscibility pressure- a key parameter in design of CO2 miscible flood, J. Pet. Sci. Eng., Vol. 46, issue 1-2, 37-52.

Farasat, A., Shokrollahi, A., Arabloo, M., Gharagheizi, F., Mohammadi, A., 2013, toward an intelligent approach for determination of saturation pressure of crude oil, Fuel Processing Technology 115 (2013) 201-214.

Frashad, F., LeBlanc, J.L., Garber, J.D. et al. 1996. Empirical PVT Correlations for Colombian Crude Oils. SPE-36105 Presented at the SPE Latin American and Caribbean Petroleum Engineering Conference, Port of Spain, Trinidad and Tobago, 23–26 April. -MS.

Gharbi, R.B.C., Elsharkawy, A.M., 1999, neural network model for estimating the PVT properties of Meddle East crude oils, SPE Reservoir Eval. Eng. 2 (3), 255-265.

Gholami, A., Asoodeh, M., Bagheripour, P., 2014 How committee machine with SVR and ACE estimates bubble point pressure of crudes, *Fluid Phase Equilibria, Volume 382, 25 November 2014, Pages 139-149.*

Glasø, Ø. 1980, Generalized Pressure-Volume-Temperature Correlations. *J Pet Tech.* 32 (5): 785-795.

Heidaryan, E., Moghadasi, J., 2011, Crude saturation pressure through imperial correlation, Thermodynamics 2011, Athens, 31 August- 3 September 2011.

Hasan, M., 1993, PVT Correlations for Indonesian Crude Oils. *Proc.* 22nd Annual Convention of the Indonesian Petroleum Association, Indonesia, Vol.1, IPA93-2.1-070.

Jarrhian, A., Moghadsi, J., Heidaryan, E., 2015, Empirical estimating of black oil bubblepoint (saturation) pressure, J. Petroleum Science and engineering, (126), 69-77.

Jacopy R.H., Berry V.J.,1958, A method for predicting the pressure maintenance for reservoir producing volatile oil, AIME 213, 59-65.

Kartoatmodjo, T., Schmidt, Z. 1994, Large Data Bank Improves Crude Physical Property Correlations. *Oil Gas J.* 92 (27): 51–55.

Kartoatmodjo, T.R.S. and Schmidt, Z. 1991, *New Correlations for Crude Oil Physical Properties*, SPE 23556.

Khairy, M., El-Tayeb, S., Hamdallah, M., 1998. PVT Correlations Developed for Egyptian Crudes. *Oil Gas J.* 96 (18): 114.

Khamekhcki, E., Behvandi, R., Rashid, F., 2011, Prediction of bubble point pressure & asphaltene onset pressure during CO2 injection using ANN & ANFIS models, Pet. Sci. & Tech. Vol. 1 No.2, 35-45.

Labedi, R.M. 1982, *PVT Correlations of the African Crudes*, PhD thesis, CSM.

Labedi, R.M. 1990, Use of Production Data to Estimate the Saturation Pressure, Solution GOR, and Chemical Composition of Reservoir Fluids, SPE-21164-MS. Presented at the SPE Latin America Petroleum Engineering Conference, Rio de Janeiro, Brazil, 14-19 October,

Lasater, J.A. 1958, Bubble Point Pressure Correlations. *J Pet. Tech.*10 (5): 65–67. SPE-957-G.

Levitan, L.L. and Murtha, M. 1999. New Correlations Estimate Pb, FVF. *Oil Gas J.* 97 (10): 70.

Lee, S., 1979, Experimental and theoretical studies on the fluid properties required for simulation of thermal process, Paper SPE 8393.

Li, Y.K., Nghiem, L.X., Siu, A., 1985. Phase behavior computations for reservoir fluids: effect of pseudo-components on the phase diagrams and simulation results. J. Can. Pet. Technol. Nov. – Dec., 29 – 36

Lin, B., Leibovici, C. F., Jorgensen, S. B., 2008, optimal component lumping: problem formulation and solution techniques, Computer & Chem. Eng., 32, 1167-2272.

Macary, S.M. and El-Batanoney, M.H. 1992, Derivation of PVT Correlations for the Gulf of Suez Crude Oils. *Proc. of the* 11th EGPC Pet. Exp. & Prod. Conference, Vol. 1, 374, Cairo, Egypt,

Moharam, H.M., 1995. Prediction of viscosity of heavy petroleum fractions and crude oil using a corresponding states method. Ind. Eng. Chem. Res. 34, 4140 – 4144.

Moradi, B., Malekzadeh, E., Amani, M., Boukadi, F.H., Kharat, R., 2101, Bubblepoint pressure imperial correlation, SPE 132756.

Obomanu, D.A. and Okpobiri, G.A. 1987, Correlating the PVT Properties of Nigerian Crudes. *J. Energy Res. Tech.* 109 (4): 214-217.

Okoduwa, I. G., Ikiensikimama, S.S., 2010, Bubble point pressure correlation for Niger Delta Crude Oils, SPE 136968.

Omar, M.I. and Todd, A.C. 1993. Development of New Modified Black Oil Correlations for Malaysian Crudes. SPE-25338 Presented at the SPE Asia Pacific Oil and Gas Conference, Singapore, 8-10 February.

Owolabi, O.O. 1984. *Reservoir Fluid Properties of Alaskan Crudes*. MS thesis, University of Alaska, Fairbanks, Alaska (May 1984).

Pedersen, K.S., Thomassen, P., Fredenslund, Aa., 1988a. Characterization of gas condensate mixtures. AIChE Spring National Meeting, March 6 – 10, New Orleans, LA.

Peng, D. y., Robinson. D. B., 1976. A new two-constant equation of state , Ind. Eng. Chem. Fund. 15, 59-63.

Petrosky, G.E. Jr. 1990. *PVT Correlations for Gulf of Mexico Crude Oils*. MS thesis, University of Southwestern Louisiana, Lafayette, Louisiana.

Petrosky, G.E. Jr. and Farshad, F. 1998. Pressure-Volume-Temperature Correlations for Gulf of Mexico Crude Oils. *SPE Res Eval & Eng.*1 (5): 416-420. SPE-51395.

Riemens, W.G., Schutle, A.M., Jong, L.N.J., 1988. Birba field PVT variations along the hydrocarbon column and confirmatory field tests. J. Pet. Technol. Jan., 83 – 88

Rosengger, L., Wu., R.1990, Integrated PVT data characterization- lessons from four case histories, J. Can. Per. Tech., 38, (13), (paper 97-05)

Shojaei, M., Bahrami, F., Barati, P., Riahi, S., 2014, Adaptive neuro-fuzzy approach for reservoir oil bubble point pressure estimation, *J. of Natural Gas Sci. and Eng., Volume 20, September 2014, Pages 214-220.*

Soave, G. 1972. Equilibrium constants from a modified Redlich-Kwong equation of state, Chem. Eng. Sci. 27, 1197-1203.

Standing, M.B. 1947. A Pressure-Volume-Temperature Correlation for Mixtures of California Oils and Gases. *API Drilling and Production Practice* (1947): 275-287.

Tale, M., Alquraishi, A., Al Mushaigeh, E. 2009, New model for crude oil saturation pressure prediction using alternation conditional expectation, Offshore Mediterranean Conf. & Exhibit. Italy 25-27, 2009.

Talebi, R., Ghiasi, M. Talebi, H., Mohammadyian, M., Zendehboudi,S., Arabloo, M., Bahadori, A., 2014, Application of soft computing approaches for modelling saturation pressure of reservoir oil, *J. Nat. Gas Sci. and Eng., Vol. 20, Sep. 2014, 8-15.*

Vazquez, M., Beggs, H. D. 1980, Correlations for Fluid Physical Property Prediction. *J Pet. Tech.* 32 (6): 968-970. Vazquez, M.E. 1976. *Correlations for Fluid Physical Property Prediction,* MS thesis, University of Tulsa.

Velarde, J., Blasingame, T.A., and McCain Jr., W.D. 1997. Correlation of Black Oil Properties at Pressures below Bubble Point Pressure - A New Approach. Presented at the Annual Technical Meeting of CIM, Calgary, Alberta, 8–11 June. PETSOC 97-93.

Vogel, J.R., Yarborough, L. 1980, the effect of nitrogen on phase behavior and phase properties of reservoir fluids, SPE 8815.

Whitson, C. H., 1980, Characterizing hydrocarbon plus fraction, SPEJ, 24 (4), 683-694.

Williams, C.A., Zana, E.N., 1980. Use of the Peng-Robinson equation of state to predict hydrocarbon phase behavior and miscibility for fluid displacement. SPE 8817, SPE/DOE on Enhanced Oil Recovery, April 20–23, Tulsa, OK.

Appendix

In this appendix, sample calculations are presented to show the deficiency of some of the published methods.

Input parameters:

N_2	CO_2	H_2S	C_1	C_2	C_3	C_4	C_5	C_6	C_{7+}	GR.C7+	MWC7+	T, °F	Ps, psia
0.65	0.02	0	45.02	12.45	8.93	6.03	3.02	1.44	22.44	0.81	184	140	3002

Ahmadi et al (2014)

Ahmadi et al. proposed an 8 steps calculation (copied from their paper) to estimate the BPP as follows:

$$\text{BPP} = A \times \text{Mw}_{C7+} - \frac{B}{C_{7+}} + C \times \text{Vol./Inter.} + D \times \text{SG}_{C7+} \frac{E}{\text{SG}_{C7+}} + \frac{F}{\text{Mw}_{C7+}} + T \times 505.967 + \frac{2.45789 \times 10^{6}}{T} + \frac{G}{\text{Vol./Inter.}} \quad (5)$$

$$A = 1488.46 - \text{SG}_{C7+} \times 1087.64 - \frac{601.512}{\text{SG}_{C7+}} + 1670.58 + T \times 0.40771 + \frac{6345.73}{T} + \frac{8.72863}{\text{Vol./Inter.}} \quad (6)$$

$$B = \frac{4132.53}{\text{Vol./Inter.}} + \frac{6.1977 \times 10^{7}}{\text{Mw}_{C7+}} + T \times 3181.39 + \frac{1.06393 \times 10^{8}}{T} \quad (7)$$

$$C = \frac{20964.7}{C_{7+}} - \frac{1.76162 \times 10^{6}}{\text{Mw}_{C7+}} - T \times 19.1963 - \text{Mw}_{C7+} \times 21.5427 + \frac{12851.5}{\text{SG}_{C7+}} - \frac{712613}{T} - 3477.49 + \text{SG}_{C7+} \times 9532.73 \quad (8)$$

$$D = 1.05345 \times 10^{+06} - \frac{1.50748 \times 10^{8}}{\text{Mw}_{C7+}} - T \times 676.048 - \frac{1.20252 \times 10^{7}}{T} - \frac{12178.2}{\text{Vol./Inter.}} \quad (9)$$

$$E = 641357 - \frac{1.02891 \times 10^{8}}{\text{Mw}_{C7+}} - T \times 301.034 - \frac{1.91446 \times 10^{6}}{T} - \frac{3873.62}{\text{Vol./Inter.}} \quad (10)$$

$$F = 2.51471 \times 10^{+08} + T \times 25332.5 + \frac{3.47893 \times 10^{8}}{T} - \frac{244573}{\text{Vol./Inter.}} \quad (11)$$

$$G = 20125.6 - T \times 23.9271 - \frac{703193}{T} \quad (12)$$

Ahmadi et al. define the Vol. /Inter. as the mole percentage ratio of the volatile (C_1& N_2) to the intermediate components (C_2 to C_6, CO_2 and H_2S). Application of the above equations from their paper using the input parameters results in the following:

Vol./Inter.	A	B	C	D	E	F	G	BPP
1.43211	1643.9	1545062	-271.127	45122.71	23642.47	2.57E+08	11753	1,794,123

The calculated saturation pressure (BBP) from Eq.5 is 1,794,123 psia which is unreasonable. This indicates a deficiency in the model.

Gholami et al (2014)

Gholami et al. presented 3 steps calculation (copied from their paper) to estimate the BPP as follows:

$$Tr(M) = \sum_{i=0}^{5} B_i M^i \quad (15)$$

Application of Eq. 15 resulted in the following

N_2	0.00584
CO_2	0.060917
H_2S	0.050636
C_1	-0.47658
C_2	-0.41547
C_3	-0.45083
C_4	-0.02003
C_5	0.068769
C_6	0.273921
C_{7+}	1.814523
SG.C7+	-0.0197
MWC7+	37.03525
T., F	-0.11139

Application of Eq. 16

$$Tr(P_b) = Tr(N_2) + Tr(CO_2) + Tr(H_2S) + Tr(C_1) + Tr(C_2) + Tr(C_3) + Tr(C_4) + Tr(C_5) + Tr(C_6) + Tr(C_{7+}) + Tr(SG_{C7+}) + Tr(MW_{C7+}) + Tr(T) \quad (16)$$

Resulted that Tr(Pb) = 37.8159

$$P_b = (57.08403877741 \times (Tr(P_b))^2) + (1166.9102304199 \times (Tr(P_b))) + 2207.3573204358 \quad (17)$$

Application of Eq. 17 resulted in saturation pressure (P_b) = 127,967 psia, which is unreasonable, compared to the measured one is 3002 psia. This indicates that there is a deficiency in the model.

Jarrahian et al (2015)

Jarrahian et al. proposed the following Eq. (copied from their paper) to estimate the saturation pressure:

$$P_{sat} = A_0 \exp\left(A_1 z_{N_2} + A_2 z_{CO_2} + A_3 z_{H_2S} + A_4 z_{C_1}^{A_5} + A_6 \sum_2^6 z_{C_i} + A_7 z_{C_{7+}}^{A_8} \gamma_{C_{7+}}^{A_9} Mw_{C_{7+}}^{A_{10}} + A_{11}(T - 459.67)\right) \quad (2)$$

Application of Eq. 2 resulted in saturation pressure (Ps) = 1812 psia, compared to the measured one is 3002 psia. This indicates that model has inaccuracy problem. This has been the pattern in all the crude oil samples shown in Table 4 and figures 6A through 6D.

Table 1- Summary of attempts to calculate the saturation pressure of crude oils

Method based on production information	Method based on compositional data
Standing, 1947;	Elsharkawy , 2003
Elam, 1957;	Tale et al., 2009
Lasater, 1958;	AlQuraishi, 2009
Vasquez, 1976;	Bandyopadhyay and Sharma, 2011
Vasquez and Beggs, 1980;	Khamehchi et al., 2011
GlasØ, 1980;	Farasat et al., 2013
Labedi, 1982&1992;	Ahmadi et al., 2014
Owolabi, 1984;	Gholami et al. 2014
Al-Marhoun, 1985;	Jarrahian et al 2015
Obomanu and Okpobiri, 1987;	
Asgarpour et al., 1989;	
Al-Najar et al., 1988;	
Dokla and Osman, 1990;	
Petrosky, 1990;	
Kartoatmodjo and Schmidt, 1991&1994;	
Macary and El-Batanoney, 1992;	
Omar and Todd, 1993;	
Hasan, 1993;	
De Ghetto and Villa, 1994;	
Farashad et al., 1996;	
Almehaidaeb, 1997;	
Elsharkawy, 1997;	
Velarde et al., 1997;	
Kairy et al., 1998;	
Petrosky and Farshad, 1988;	
Al-shammasi, 2001;	
Gharbi & Elsharkawy 1999	
Levitan and Murtha, 1999;	
Dindrouk and Christman, 2001	
Moradi et al., 2010	
Okoduwa and Ikiensikimama 2010	
Asoodeh and Bagheripour, 2012	
Shojaei et al., 2014	
Talebi et al., 2014	
Elmabrouk et al., 2014	
Kanu and Ikiensikimama, 2014	
Jarrahian et al 2015	

Table 2- Description of the data used to develop and test the extended model

Input	Training model (179 Data)			Testing (28 Data)			Testing (7 new experimental)		
	Min	Ave	Max	Min	Ave	Max	Min	Ave	Max
N_2	0.00	0.35	1.67	0.00	0.27	1.67	0.10	0.26	0.35
CO_2	0.00	1.27	11.37	0.00	1.17	11.37	0.14	0.41	0.83
H_2S	0.00	0.15	3.68	0.00	0.15	3.22	0.00	0.00	0.00
C_1	13.16	33.86	74.18	19.50	34.79	73.36	20.64	28.25	42.79
C_2	3.36	8.07	13.71	4.64	8.26	11.63	6.69	7.88	10.82
C_3	0.89	6.59	11.87	2.46	6.76	12.03	6.05	6.59	7.10
C_4	0.95	4.46	8.4	1.66	4.41	6.58	4.03	5.05	5.93
C_5	0.40	3.21	5.95	1.00	3.11	5.23	3.36	4.33	5.52
C_6	0.00	3.14	6.37	0.99	3.15	5.46	2.88	3.74	4.33
C_{7+}	10.72	38.89	57.73	11.18	37.88	51.22	27.61	43.49	51.31
$SG.C_{7+}$	0.74	0.87	0.959	0.70	0.86	0.93	0.85	0.89	0.92
MWC_{7+}	134	240	368.9	105	223	292	216	262	298
T, °F	58	156	319	58	165	261	121	157	230
Ps, psia	537	2198	5065	1025	2307	5000	1227	1779	3344

Table 3: Error analysis of the proposed models in comparison with all the investigated methods

Method	ARE	AARE	SD	R^2%
PR-EOS	6.24%	10.19%	1028	97.9
SRK-EOS	-3.12%	10.36%	1024	96.7
Elsharkawy (2003)	0.85%	8.57%	1015	96.4
Alquraishi (2009)	-6.06%	10.16%	1088	96.7
Bandyopadhyay (2011)	8.05%	12.28%	1054	96.8
Farasat (2013)	NA	NA	NA	NA
Gholami et al. (2014)	?	?	?	?
Ahmadi et al. (2014)	?	?	?	?
Jarrahian et al (2015)	35.52	35.56	983	95.2
New Extended Model	-0.37%	8.20%	1030	97.6
New Lumped Model	-0.50%	8.46%	1036	97.4

ARE = Average Relative Error
AARE = Average Absolute Relative Error
SD = Standard Deviation
R^2 = Correlation Coefficient
NA = Calculation was not possible, because the weight and bias were not provided.
? = The method produces unreasonable results as indicated in the appendix.

Table 4- Comparing model accuracy for calculations of saturation pressures for various crude oils

Volatile-Oil-Samples (C7+<30)																						
Fluid Source	index	N2	CO2	H2S	C1	C2	C3	C4	C5	C6	C7+	GR.C7+	MWC7+	T, °F	Ps, psi	SRK	PR	Elsh	Alq	Bandy	Jarr	Extd	Lump
Rosenegger	7	0.03	8.39	0	47.4	10.3	6.12	4.2	2.89	2.05	18.6	0.83	180	295	4000	3933	3879	4460	4420	4165	3317	4139	4178
Vogel-80	13	0.65	0.02	0	45	12.5	8.93	6.03	3.02	1.44	22.4	0.81	184	140	3002	2624	2548	2742	2895	2643	1812	3019	2996
Dansh-92	18	0	0	0	46.8	8.77	7.44	4.01	2.56	4.02	26.4	0.77	158	212	2941	2550	2594	3248	3617	3269	1944	3244	3374
	31	0.44	0.38	0	49.1	7.6	6.13	3.84	2.52	2	28	0.84	231	199	3739	3522	3507	3418	3951	3508	2309	3625	3644
Middle East	45	0.03	0.97	0	41.6	10.8	7.52	4.5	3	2.43	29.1	0.85	208	241	3069	3109	2979	3080	3513	3138	1962	3127	3140
Coats-86	51	0.55	1.03	0	36.5	9.93	8.85	6	3.78	3.56	30.4	0.84	200	234	2746	2537	2472	2714	2945	2643	1682	2777	2720

Black-Oil -Samples (35<C7+<55)																							
Fluid Source	index	N2	CO2	H2S	C1	C2	C3	C4	C5	C6	C7+	GR.C7+	MWC7+	T, °F	Ps, psi	SRK	PR	Elsh	Alq	Bandy	Jarr	Extd	Lump
Coats-86	73	1.64	0.08	0	28.4	7.16	10.5	8.4	3.82	4.05	36	0.84	252	131	1694	1674	1618	1615	1738	1644	1084	1713	1713
Middle East	88	0.08	1.37	0.69	36	8.67	5.94	2.97	1.5	3.1	39.7	0.9	274	168	2505	2710	2318	2247	2608	2321	1476	2441	2445
Middle East	114	0.29	0.48	0	28.4	8.29	7.38	5.06	3.42	4.41	42.3	0.88	252	133	1632	1672	1477	1544	1748	1517	984	1587	1598
Middle East	122	0	1.25	0	33.4	9.24	6.07	2.4	1.62	2.78	43.3	0.85	252	230	2110	2313	2278	2424	2737	2438	1410	2388	2412
Middle East	134	0	0.17	0	28.6	8.63	7.43	3.81	2.61	4.11	44.7	0.86	249	134	1548	1551	1429	1477	1764	1433	937	1503	1567
Middle East	146	0.44	0.83	0	27.8	7.54	6.9	4.61	3.1	3.11	45.7	0.9	272	133	1565	1799	1505	1558	1730	1528	1002	1602	1590
moharam-96	149	0.21	0.34	0	20	7.93	8	6.6	5.87	5.08	45.9	0.86	230	235	900	1353	1290	1531	1514	1225	773	1303	1294
Middle East	162	0.35	0.47	0	26.5	7.71	6.05	4.03	4.08	4.22	46.6	0.89	253	135	1500	1613	1380	1427	1631	1391	923	1478	1456
Middle East	176	0.12	0.51	0	28.8	7.91	6.46	3.17	2.07	3.25	47.7	0.88	250	134	1615	1665	1476	1574	1783	1502	981	1571	1614
Middle East	194	0.21	0.15	0	27.8	7.68	6.19	3.41	1.81	2.84	49.9	0.86	228	134	1590	1532	1377	1545	1680	1426	923	1466	1493
Rosenegger	207	0.37	0.02	0	17.3	6.11	8.23	5.47	3.76	3.96	54.8	0.85	242	156	537	938	888	914	1064	836	558	661	731

Newly-Measured Kuwait University Laboratory																						
Fluid Source	index	N2	CO2	H2S	C1	C2	C3	C4	C5	C6	C7+	GR.C7+	MWC7+	T, °F	Ps, psi	SRK	PR	Elsh	Alq	Bandy	Jarr	Extd	Lump
KU-1	209	0.33	0.22	0	25.6	6.87	6.39	5.61	4.68	4.33	46	0.879	222	133	1595	1360	1530	1442	1514	1394	860	1357	1337
KU-2	210	0.35	0.47	0	26.5	7.71	6.05	4.03	4.08	4.22	46.6	0.893	253	135	1500	1400	1620	1427	1631	1391	924	1478	1456
KU-3	211	0.27	0.74	0	23.1	7.2	6.45	5.93	4.66	4.16	47.5	0.906	290	138	1283	1140	1360	1284	1448	1296	800	1287	1219
KU-4	212	0.22	0.26	0	20.6	6.69	6.65	5.11	5.52	3.6	51.3	0.919	298	171	1227	1120	1300	1188	1415	1170	742	1176	1131
KU-5	213	0.19	0.14	0	27.8	8.37	6.88	5.28	4.33	3.85	43.2	0.89	278	121	1524	1300	1560	1402	1659	1409	928	1534	1489
KU-6	214	0.33	0.24	0	31.3	7.48	6.58	4.86	3.7	3.17	42.3	0.9	274	171	1978	1800	2200	1975	2242	1985	1224	2060	2026
KU-7	215	0.1	0.83	0	42.8	10.8	7.1	4.51	3.36	2.88	27.6	0.85	216	230	3344	3070	3400	3083	3568	3139	2003	3213	3200

Low-Temperature Oil Samples (59°F<T<71°F)																						
Fluid Source	index	N2	CO2	H2S	C1	C2	C3	C4	C5	C6	C7+	GR.C7+	MWC7+	T, °F	Ps, psi	SRK	PR	Elsh	Alq	Bandy	Jarr	Extd	Lump
Middle East	101	0	1.12	0.21	27	10.8	9.23	4.97	2.62	3.33	40.8	0.88	249	71	1247	1090	1033	1040	1235	564	808	1148	1205
Middle East	117	0.36	0.17	0	27.2	8.93	8.6	6.07	3.71	2.55	42.4	0.88	271	68	1130	1028	1245	1094	1248	766	820	1239	1228
Middle East	135	0	0.17	0	28.6	8.63	7.43	3.81	2.61	4.11	44.7	0.86	249	70	1270	1055	1255	1130	1311	783	826	1229	1297
Middle East	138	0.14	0.54	0	29.4	7.96	6.19	3.96	3.25	3.84	44.7	0.9	290	59	1265	1065	1350	1191	1301	880	894	1408	1390
Middle East	147	0.44	0.83	0	27.8	7.54	6.9	4.61	3.1	3.11	45.7	0.9	272	67	1320	1125	1400	1201	1268	921	879	1319	1312
Middle East	173	0.06	0.81	0	31.3	7.42	5.68	2.82	1.63	2.64	47.6	0.89	270	68	1460	1215	1535	1441	1463	1026	975	1548	1576
Middle East	193	0.16	0.3	0	24.7	7.51	6.92	4.76	2.65	3.43	49.6	0.88	239	71	1050	975	1156	1008	1107	701	714	958	986

Elsh: Elsharkawy (2003)
Extd: Extended model

Alq: AlQuraishi (2009)
Lump: Lumped model

Band: Bandyopadhyay and Sharma (2011)

Jarr: Jarrahian et. al (2015)

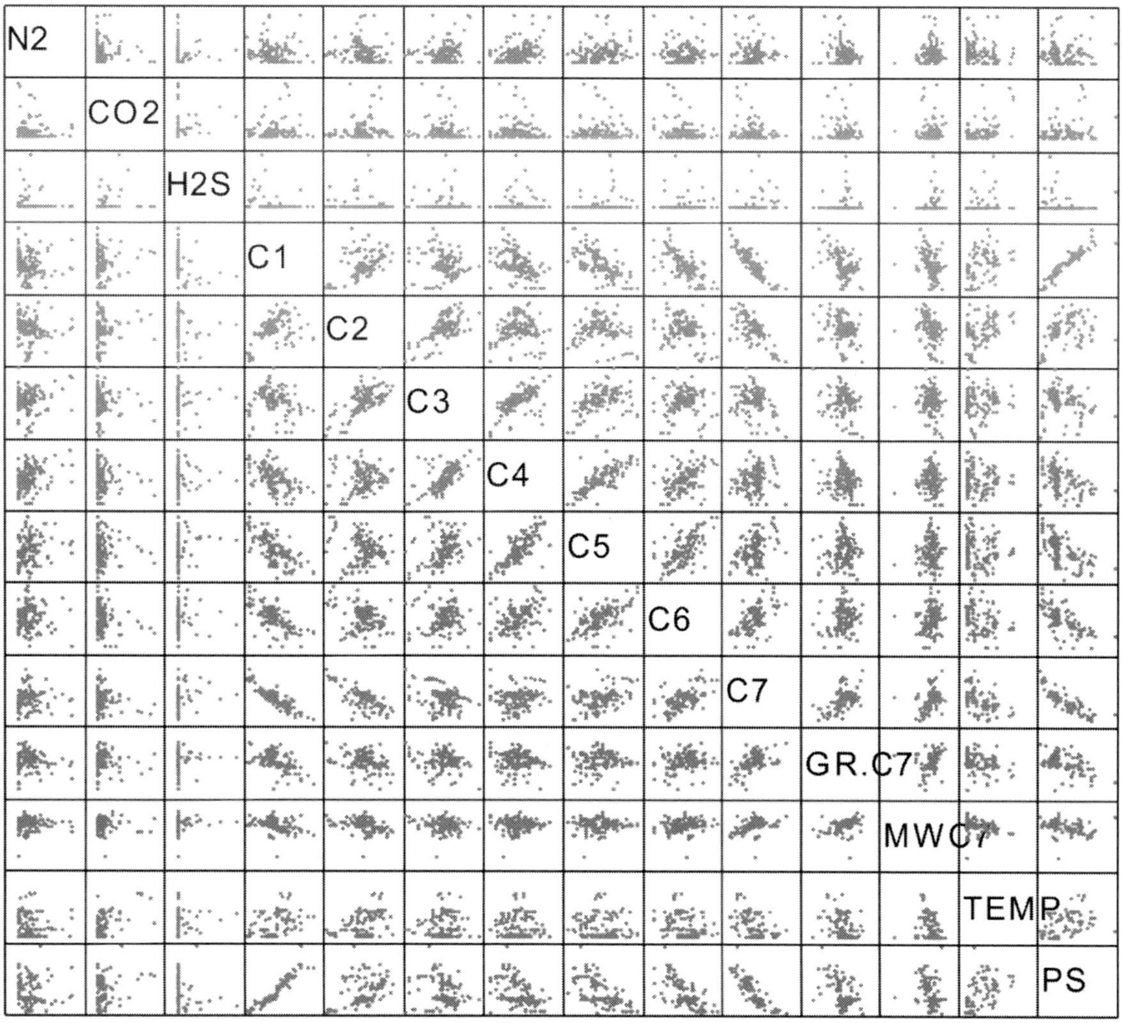

Figure 1: Correlation matrix for all the variables in the oil data bank

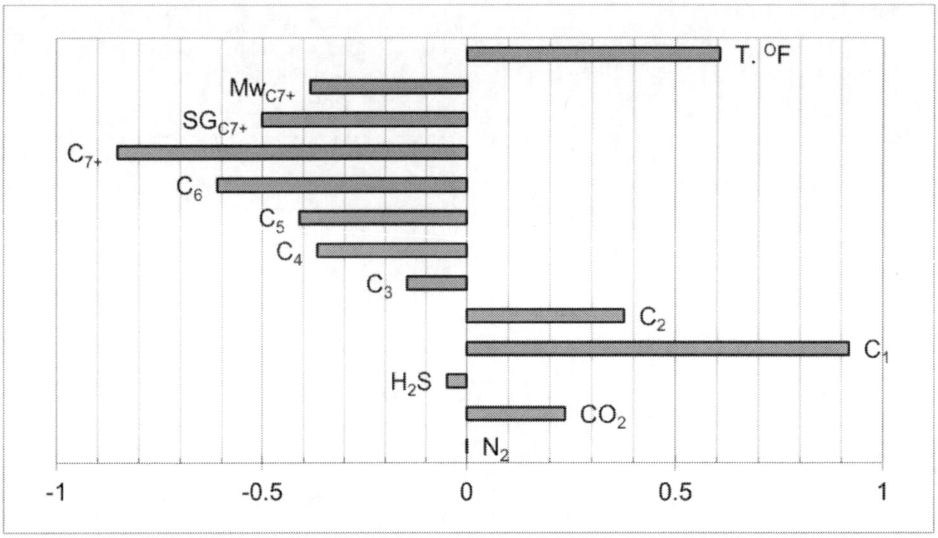

Figure 2: Correlation coefficient for all the variables in the training model (179 data)

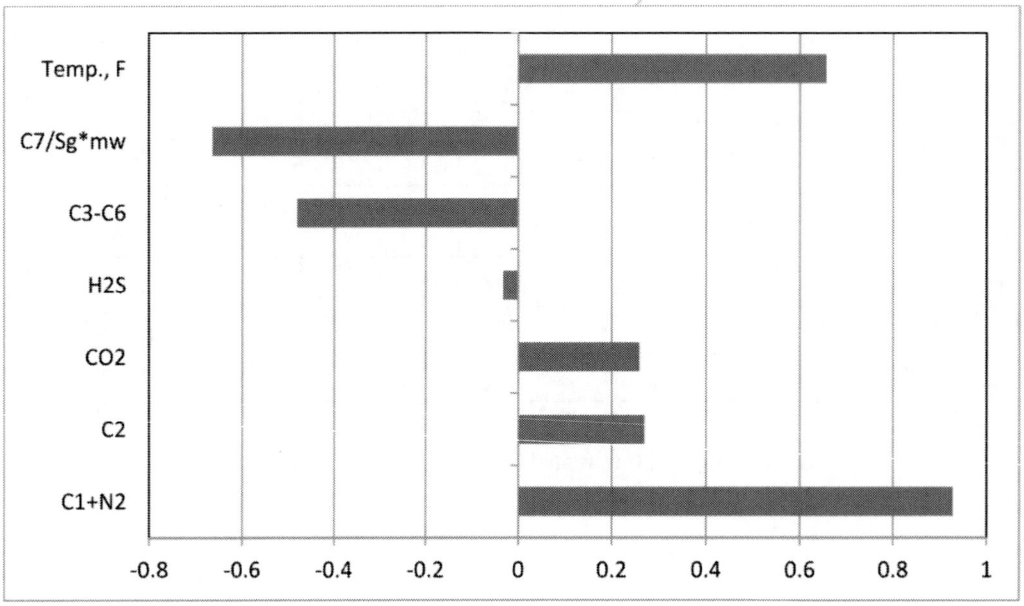

Figure 3: Correlation matrix for the lumped model (179 data)

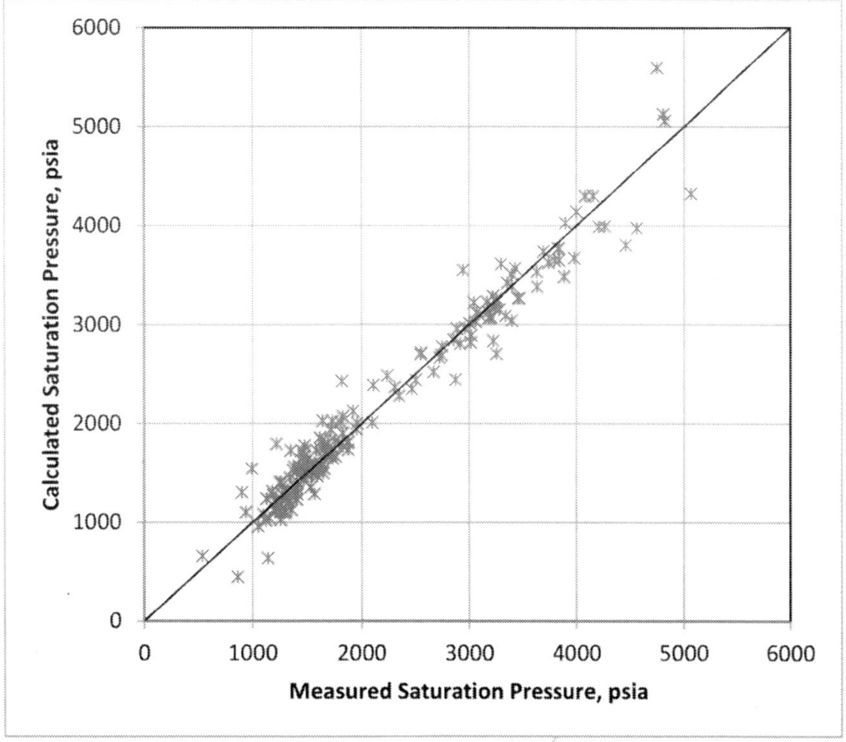

Figure 4A: Cross plot of the extended model using 179 training data

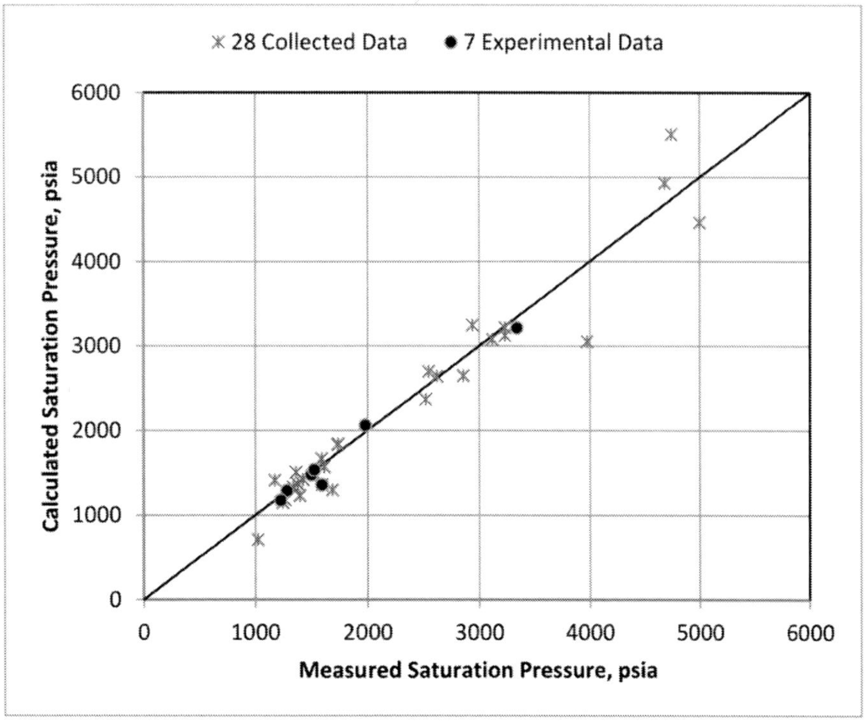

Figure 4B: Cross-plot of the extended model using 35-testing data

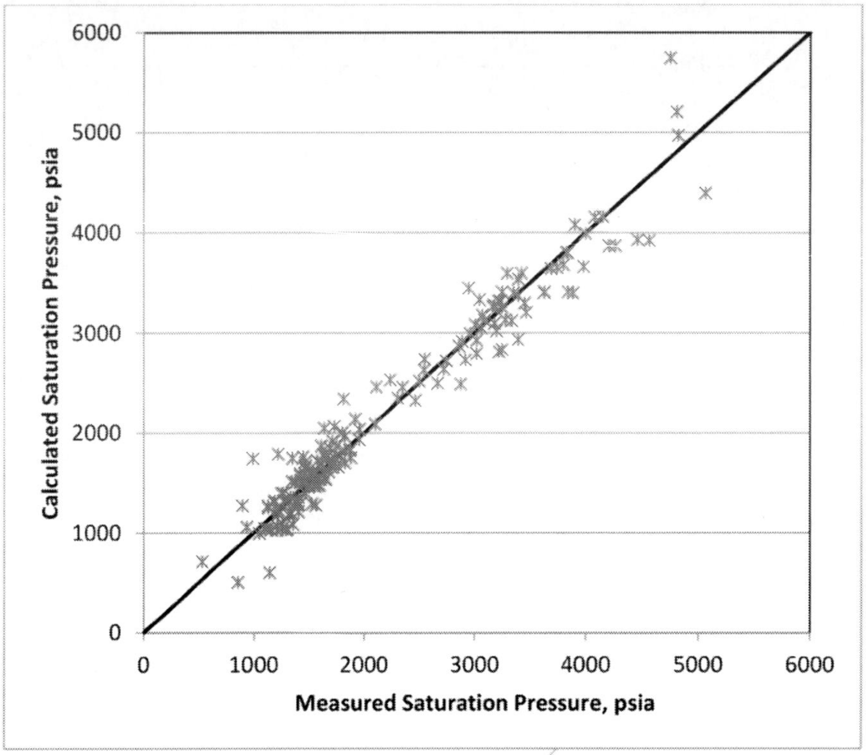

Figure 5A: Cross-plot of the lumped model using 179 training data

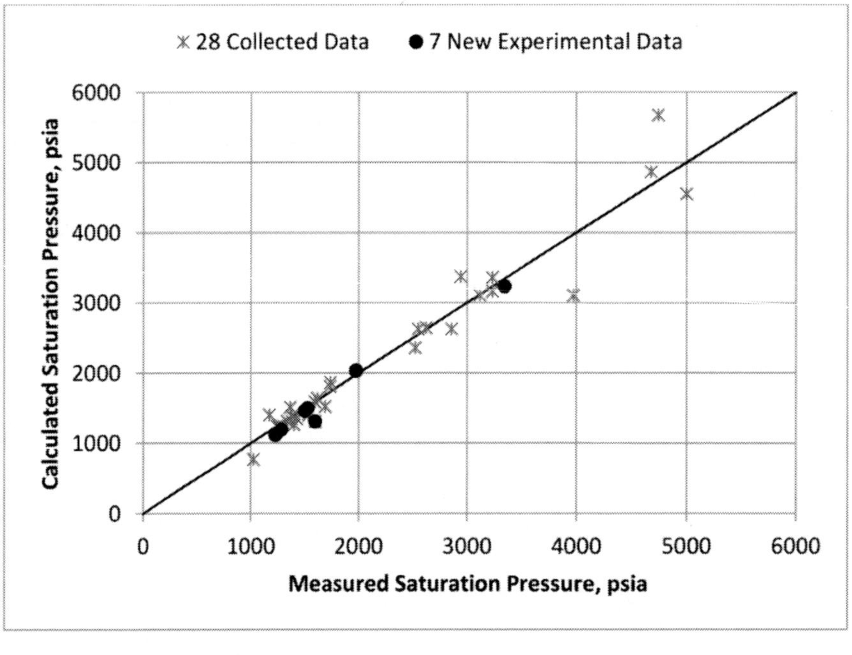

Figure 5B: Cross-plot of testing data for the lumped model

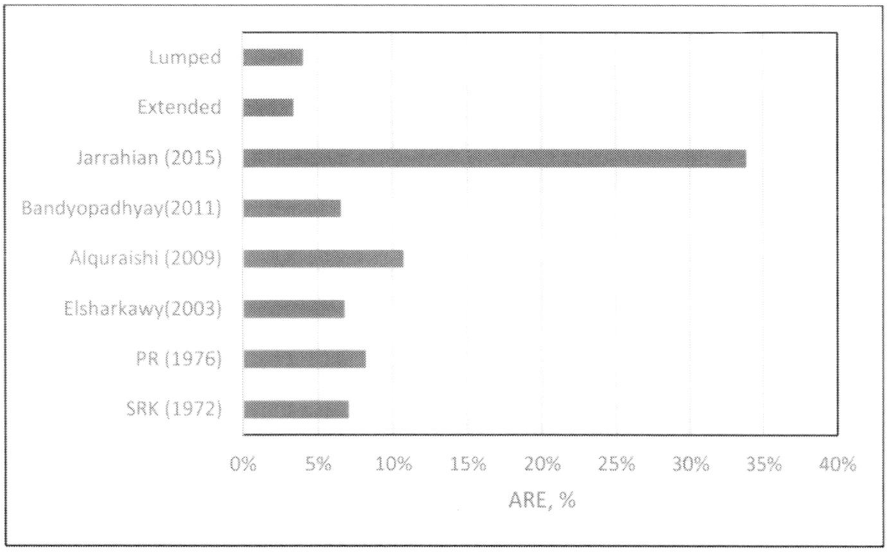

Figure 6A: Comparison of all the models' accuracy for the volatile oil samples

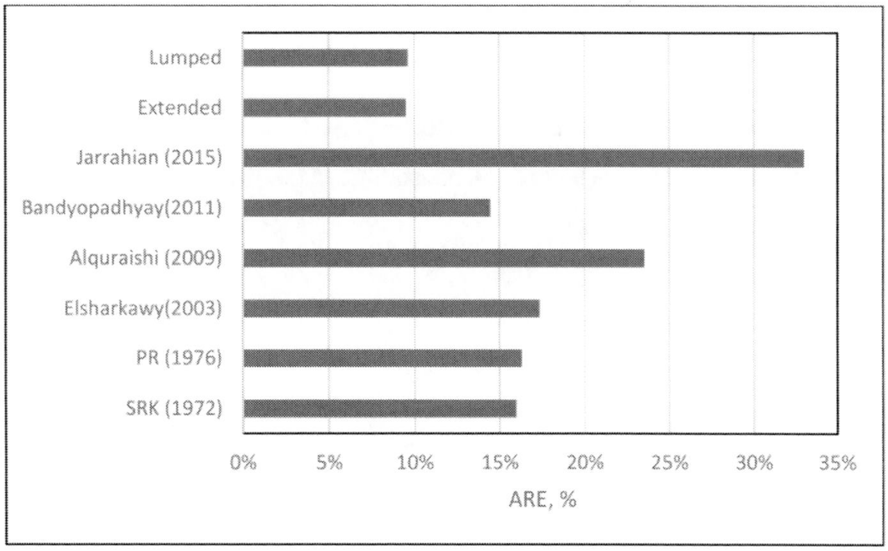

Figure 6B: Comparison of all the models' accuracy for the black oil samples

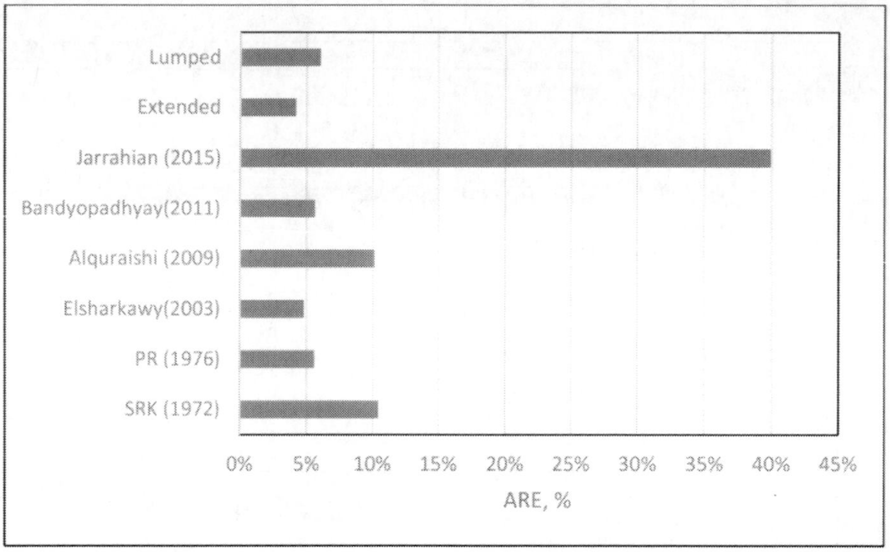

Figure 6C: Comparison of all the models' accuracy for the newly measured samples at KU

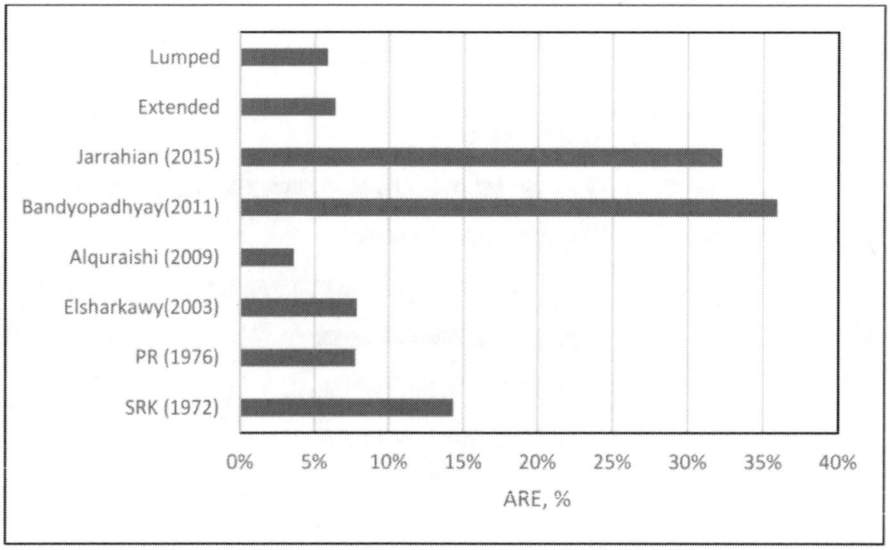

Figure 6D: Comparison of all the models' accuracy for low-temperature oil samples

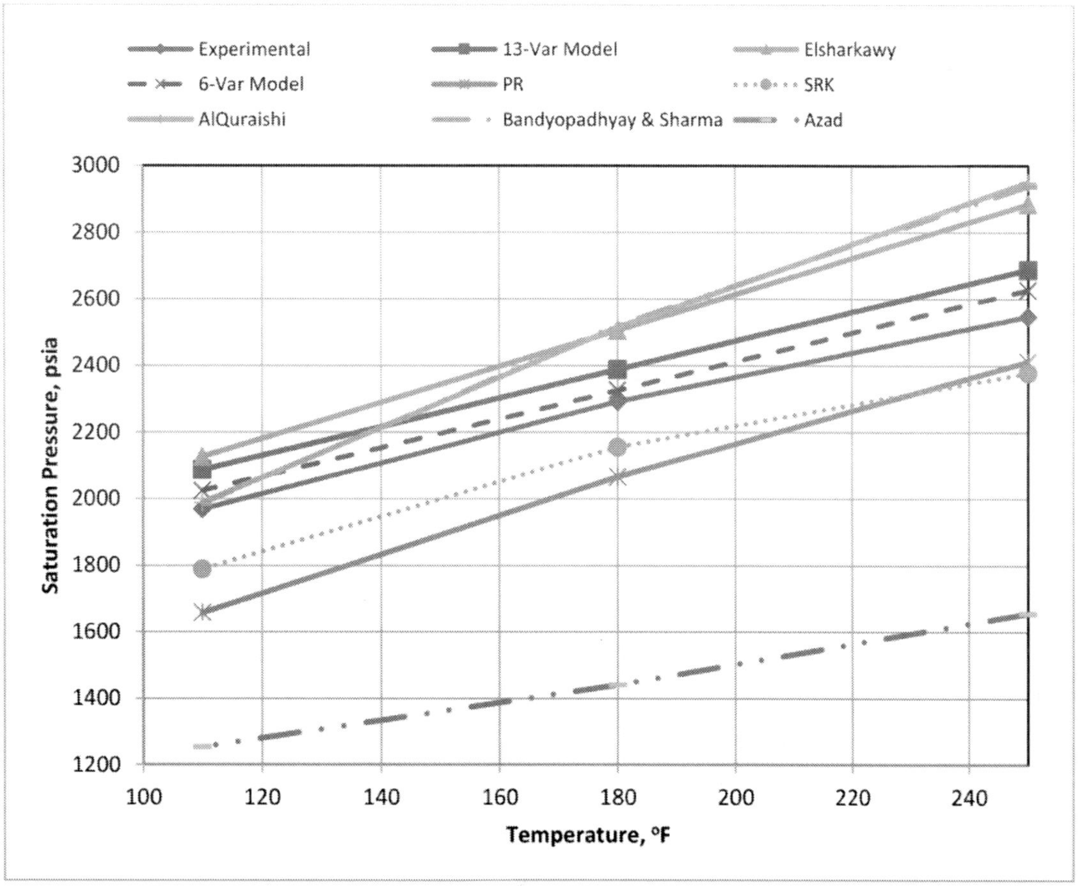

Figure 7: Changes in saturation pressure as a function of temperature for the different models (Sample #70)

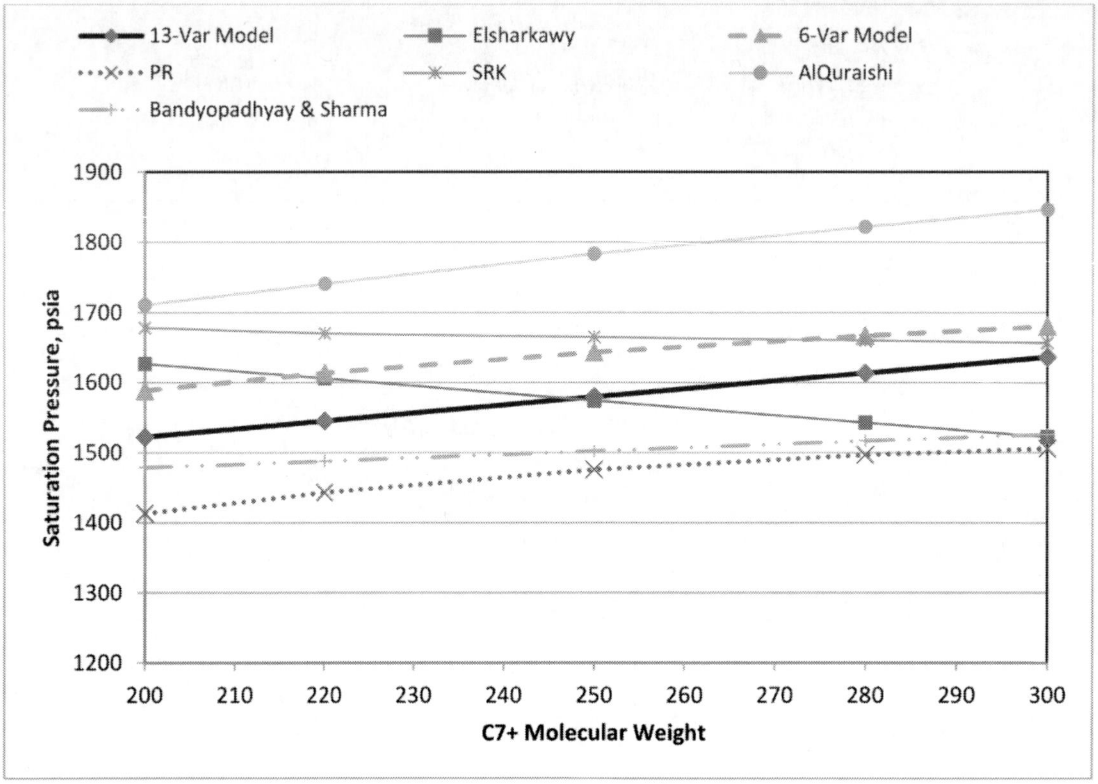

Figure 8: Comparing model performances: Changes in BPP as a function of C_{7+} molecular weight at 134°F (Sample #176)

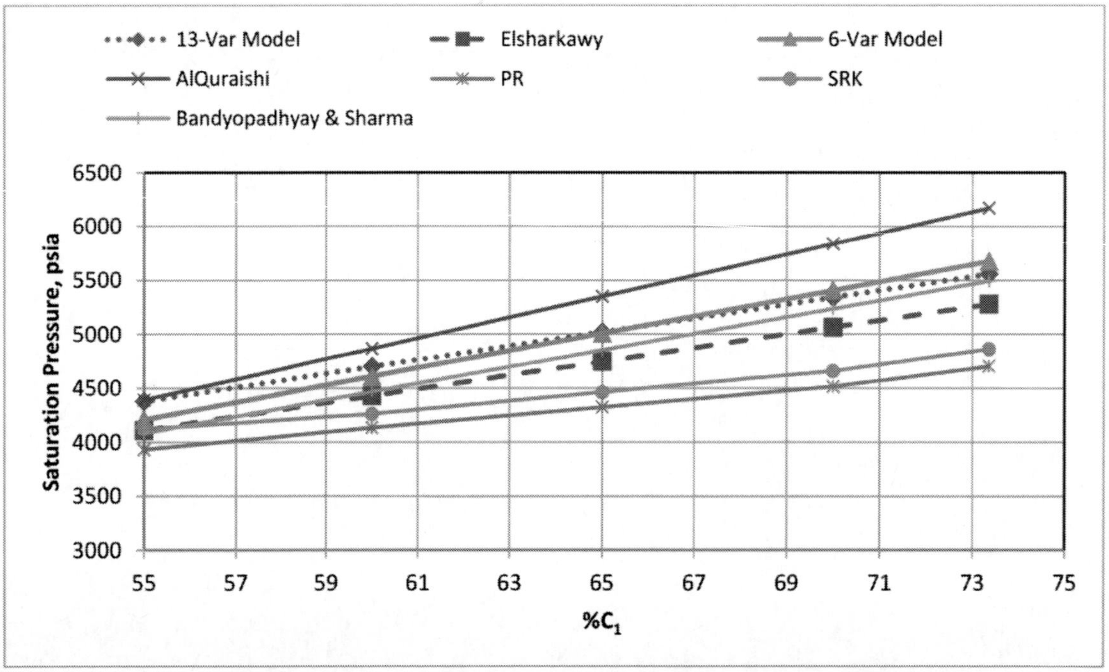

Figure 9: Sensitivity of saturation pressure to methane content at 212°F (sample #2)

439811 Purification of Ibuprofen Enantiomers By a New Technique Combining Distillation and Crystallization

Abstract:

Stripping crystallization (SC) is a new separation technique which combines distillation and crystallization. Basically, SC is operated at a triple-point condition, in which the liquid is simultaneously vaporized and crystallized due to the three-phase equilibrium. In essence, the process is continued until the liquid phase is completely eliminated and only the pure solid crystals remain in the feed. Thus, crystal washing is not required since only pure solid crystals remain in the feed and no impurities are adhered on the crystal surfaces at the end of SC. In the previous research, SC has been successfully applied to separate p-xylene from the mixed xylenes by Shiau et al. (2005, 2008, 2013).

Ibuprofen, also called 2-(isobutylphenyl)-propionic acid, is a non-steroidal anti-inflammatory drug widely used to treat headaches and minor pains. Although it is known that S-ibuprofen is responsible for the anti-inflammatory effects while R-ibuprofen might cause some long-term side effects, ibuprofen is still sold as a racemic mixture in the market due to the complexity involved in the separation of enantiomers. A thermodynamic model is developed in this work to simulate the three-phase equilibrium conditions during the SC operation. The experiments show that seeding with ultrasound mixing or magnetic stirring can be efficiently employed to purify S-ibuprofen from ibuprofen enantiomers. The experimental results, including the final enantiomeric purity and recovery ratio of S-ibuprofen, are consistent with the simulation results predicted by the model.

References:

1. Shiau, L. D.*, Wen, C. C. and Lin, B. S. "Separation and purification of p-xylene from the mixture of m-xylene and p-xylene by distillative freezing", *Ind. Eng. Chem. Res*, **44(7)**, p2258-2265, 2005.

2. Shiau, L. D.*, Wen, C. C. and Lin, B. S. "Separation of p-xylene from the multicomponent xylene system by stripping crystallization", *AIChE Journal*, **54(1)**, p337-342, 2008.

3. Shiau, L. D.*; Liu, K. F. "Investigations into the effects of the cooling rate on stripping crystallization", *Ind. Eng. Chem. Res*, **52**, p1716-1722, 2013.

441535 Hydrogen Disposal Flare System Hazard Identification and Mitigation Analysis

Abstract:

A hazard assessment on a hydrogen disposal flare system was conducted in preparation for a hydrogen turbopump test at the Air Force Research Laboratory (AFRL), Aerospace Systems Directorate at Edwards AFB, CA. The test operations required the disposal of large quantities of liquid and gaseous hydrogen through a commercial flare system to prevent the formation of an explosive environment. As part of the assessment, the primary steps in the test operation process were evaluated for the formation of a potential hazard by looking at each associated failure mode and the resulting affect on personnel and facilities. Evaluation of each operational step was conducted in a Process Failure Modes and Effects Analysis (PFMEA) that identified the probability and severity of the identified effects and associated hazards. If the probability and severity were unacceptable, mitigating steps were applied until the overall hazard risks were reduced to acceptable levels. The recommendations of this assessment were incorporated into both the flare system design and the operational procedures. This paper outlines the results of the PFMEA which include identification of the effects/hazards, available risk-mitigating techniques, and implementation of these techniques used for the mitigation of the identified hazards. Based upon the findings significant design modification to the vender's commercial equipment was needed to comply with recent changes in the code. Most of the identified hydrogen flare system hazards/effects were categorized into four groups: internal explosions, obstructions, external explosions, and maintenance degradation. Many corrective actions were implemented to mitigate both the probability and severity of the identified hazards. In addition, this paper discusses the installation of oxygen monitoring systems on the flare system, determination of appropriate system purge requirements, installation of purge system status alarms, incorporation of a triple redundant flare ignition system with status indicators, and a number of other hazard risk mitigation techniques. This comprehensive description of the hazards and mitigation of a hydrogen disposal system is a plant is vital part to safe operation of both hydrogen commercial and aerospace facilities. <align="center">

442068 Enhanced Furfural Production By an Innovative Bottom Dividing Wall Column with a Decanter Configuration

Abstract:

Enhanced Furfural Production by an Innovative Bottom Dividing Wall Column with a Decanter Configuration

Nguyen Van Duc Long, Le Cao Nhien, Le Quang Minh and Moonyong Lee[†]

School of Chemical Engineering, Yeungnam University, Gyeongsan 712-749, South Korea

Furfural, which are used to derive other industrial chemicals, has been identified as one of the major bio-based platform chemicals that can compete with petroleum-based chemicals. However, the current commercial furfural process has low yield and requires intensive energy. This paper reports the simulation, detailed design and optimization of an industrial process recovering furfural from lignocellulosic biomass, a renewable and inexpensive feedstock. The aqueous feed mixture from the acidic hydrolysis of biomass was fed to the separation process to recover furfural as the main product, along with methanol as valuable byproducts. The purification process was intensified by applying an innovative bottom dividing wall column with a decanter configuration (BDWC-D). The response surface methodology (RSM), which allows the interactions between variables to be identified and quantified, was used to optimize the BDWC-D structure. The predictions by the RSM showed satisfactory agreement with the rigorous simulation results. The heat pump technology and heat integration were then employed to improve the purification process. The results showed that the proposed sequence could save the overall energy requirements and total annual cost significantly as compared to the conventional sequence. This sequence can be employed for both grass-root and retrofit designs.

Keywords: Lignocellulosic Biomass; Furfural; Energy Efficiency Improvement; Dividing Wall Column; Response Surface Methodology

Acknowledgments

The authors wish to acknowledge the financial support from the R&D Convergence Program of NST (National Research Council of Science & Technology) of Republic of Korea and KITECH (Korea Institute of Industrial Technology) (ES150001).

442557 Flow Field Simulation and Distribution Performance Analysis of a Gas Pipe Distributor in Fccu

Abstract:
A fluidized catalytic cracking (FCC) is an important petroleum refinery process, which, aided by highly selective zeolite catalysts, cracks large-molecule low-value heavy oil feedstock into small-molecule highvalue
products. When spend catalyst is fed into the regenerator for reactivating, it is fluidized by the gas distributor at the bottom of regenerator. The gas distributor is designed to uniformly distribute gas to maintain the fluidized state of the fluidized bed and to provide the reaction gas to guarantee the burning coke deposited on the spend catalyst. Therefore, the gas distribution performance of the gas distributor influences the quality of initial gas distribution directly. So the pipe distributor used in a industrial FCC unit, which is composed by main pipes and a number of branch pipes with many nozzle tubes, was investigated by means of flow field calculation. The gas distribution performance of the pipe distributor was analyzed based on the simulated flow field. The simulation result shows that there are a obvious nununiform
gas distribution from the different nozzle gas flow rate. The static pressure increases gradually along the flow direction in the nozzle tube while the axial flow rate decreases, which leads to the nozzle gas flow rate increasing gradually. Furthermore, an obvious bias flow occurs at the inlet of the nozzle tube, which influences the gas flowing into the nozzle. The pressure drops performance of nozzle and pressure varying in the branch pipe are analyzed comprehensively. Due to the nonuniform
gas distribution by the gas pipe distributor in the practical usage, the problems such as overlarge temperature difference among the fluidized bed layers in the regenerator have often happened, which will not only reduce burning coke efficiency and but also decrease activity of the regenerated catalyst. Meanwhile, the erosion of the distributor also occurs, and in consequence the function of the distributor fails, which even causes the nonscheduled shutdown of the FCC unit and leads to tremendous economic loss. The support is provided for optimizations of both structural parameters and operation parameters of the gas pipe distributor and improvement of the gas distribution performance. Furthermore, improvement suggestions were provided based on the above reasoning. The results are expected to be useful for the distributor design.

442925 Identifying Gas Liquid Contactor Performances for Optimized Natural Gas Desacidification Amines Units - from Process Calculations to Experimental Strategy

Abstract:

Amine units are widely used for gas sweetening. The absorber performances of these reactive absorption processes are typically mass transfer limited. The proper selection of gas liquid contactor is thus critical to optimize the size of towers, the solvent flowrate and the duty of the reboiler. This work proposes an approach to identify the most suitable packings for a given application.

The case-study selected here to illustrate the approach deals specifically with a CO_2removal application where the sour gas is at 7 % vol, and the sweet gas specification is 2% vol. The gas flowrate is fixed at 0.18 MMSCFD with a solvent flowrate fixed at the minimum of the thermodynamic constraints (via equilibrium curves). An in-house rate-based simulator with accessible forms of mass transfer correlations, based on a double film model, is used to calculate the absorber height and diameter which are needed to reach the sweet gas specifications. A reference structured packing is compared with standard valve trays.

The calculations show that the bed height required with the first contactor is considerably higher than with the second one while the column diameter is reduced. In this case the use of standard valve trays is more attractive for cost optimization. Interestingly, because mass transfer is accelerated, the absorption rate with the structured packing is found to be thermodynamically limited due to a more severe temperature increase which reduces the driving force in the bottom section of the absorber.

The study determined that the optimum design for the investigated case is a compromise between a full trayed tower, that generally imposes a larger diameter, and a packed tower where the higher mass transfer forces result in a temperature rise in the bottom of the tower which in turn imposes a height increase and a much higher solvent flowrate. . To determine the mass transfer characteristics of a gas-liquid contactor providing this optimum, a sensitivity analysis has been conducted. Mass transfer parameters, i.e., liquid and gas side mass transfer coefficients and interfacial area, were varied by a factor 2 relative to the ones obtained with the reference structured packing. Calculations show that one should use a contactor that reduces liquid phase resistance, specifically, the mass transfer conductance $k_L a_e$ should be increased by at least 50%. This target could be reduced to +30% if one could simultaneously reduce the gas phase conductance $k_G a_e$ by 20%. This result can be explained by the link between mass and heat transfer (Chilton-Colburn analogy) whereby reducing $k_G a_e$decreases the temperature level which in turn thermodynamically increases the mass transfer driving force.

The next step will be to review the literature more extensively to make a short list of pre-selected random and structured packings that best meet the targets. Experimental tests will then be run on IFPEN facilities to measure flooding limits and mass transfer parameters in a small 150 mm diameter column. This step is required to confirm packing's behavior, considering the disparity of the values reported by different authors in the literature. The last step is to consolidate the existing correlations for the most promising packing, and mitigate differences between various authors by performing characterization tests on larger diameter test columns. The final correlations will be implemented in our in-house rate based simulator for industrial use.

The present approach can be applied to determine the required mass transfer characteristics for optimizing other gas treatment applications. However, the conclusions reached here as regards to mass transfer targets and the problematic temperature levels, are valid only to the present case -study, and are not directly applicable to other gas sweetening applications (selectivity, LNG, ...). For example, in a case where H_2S and CO_2 are treated simultaneously the mass transfer limitations could be very different.

443365 Technical Reconstruction of Double Effect Distillation for Methanol Process

Abstract:

Abstract: Methanol is one of important basis raw materials of chemical industry and also an excellent environment-friendly fuel, which make it as a significant commodity in the global market. Distillation is an essential unit operation for purified production of methanol. Product quality, production cost, environmental impact and energy consumption are largely related to distillation process.

Our research effort, under this circumstance, has been devoted since 2002 to energetically optimize 4-column methanol distillation system that is widely adopted in methanol units from coal, with reduced cycling water requirement. This patented process technology saw its industrial application in Chinese largest single unit of 1800Kt/a of Baofeng Group located at North China, with the lowest steam consumption of 0.9/1 t(steam)/t(methanol) compared with normally 25% higher for both simulative case study and on-site operation. Methanol product is in accordance with O-M-232G and the recycling water is reduced significantly from 67 t(CW)/t(methanol) to 52 t(CW)/t(methanol).

The flow diagram of this advanced double effect distillation process is showed in Fig.1. Its advantages arise for limitations found in current units: (1) heat exchange network failed to fully recover energy causing large CW flow to remove extra heat; (2) for pressured column, bubble point reflux need larger reflux ratio than the sub-cooled; (3) the heat of distillate from pressured column was normally ignored; (4) the low concentration of methanol of feed inlet to recovery column leads to higher reflux radio, responsible for unsatisfied unit energy efficiency.

Fig.1 The double effect distillation process.

(Column1.Light ends column; Column 2.Pressured column;

Column 3.Atmospheric column; Column4.Recovery column)

As a solution, a new flow diagram is built up and showed in Fig.2. (1) HEN totally reorganized (not completely show); (2) sub-cooled reflux to pressured column caused 10% reduction to the reflux ratio, economizing steam consumption; (3) top product of pressured column give heat to crude methanol before feeding light ends column; (4) change product and by-product position of recovery column to give efficient unit energy consumption.

Fig.2 The reconstruction of distillation process.

(Column1.Light ends column; Column 2.Pressured column;

Column 3.Atmospheric column; Column4.Recovery column)

It is estimated that the production of methanol will be over 60 million tons in 2014. If this reconstruction could be applied in the whole methanol industry, 1.3 million tons of standard coal would have been saved. This is equivalent to reduce 5 million tons emission of carbon dioxide, so the environmental impact can be largely decreased. Field operation shows reconstruction perfect substitution worth to spread to the whole methanol industry, energetically efficient and low environmentally impacting.

443095 Experimental and Numerical Study of Transport Phenomenon of Falling Film Flow with Different Heating Methods

Abstract:

Falling film flow is a common phenomenon in chemical industry. It is widely used in chemical instruments, such as structured packed columns and falling film evaporators. An experiment was implemented to study the mass transfer phenomenon of binary-mixture (ethanol-water) falling film flows with different heating methods (uniform or concentrated heating). In order to study the detailed flow and mass transfer behaviors, a 2-D multiphase computational fluid dynamics (CFD) model based on volume of fluid (VOF) was developed. This model divides mass transfer into two parts: one is caused by the vapor-liquid equilibrium (VLE) and the other is caused by the heat flux. Results from simulation are inconsistent with the experimental data. Both experiments and simulations show that the ethanol concentration of the vapor with concentrated heating is 1.5% higher than the one with uniform heating.

443101 Thermodynamic Modeling of HNO_3-H_2so_4-H_2o Ternary System with Symmetric Electrolyte NRTL Model

Abstract:

Nitric acid concentration/sulfuric acid concentration (NAC/SAC) process has been widely used for concentrating dilute aqueous nitric acid and recovering spent sulfuric acid. Nitric acid is concentrated by employing sulfuric acid to bind water and break nitric acid-water azeotrope at approximately 68 wt. % nitric acid. To support heat and mass balance calculations and process simulation for NAC/SAC processes, we develop a comprehensive thermodynamic model for nitric acid-sulfuric acid-water ternary system based on previously published thermodynamic models of nitric acid-water [Wang et al., Fluid Phase Equilibria, (2015) (DOI: 10.1016/j.fluid.2015.04.015)] binary and sulfuric acid-water [Que et al., Journal of Chemical & Engineering Data, 56 (2011) 963-9772] binary systems developed with symmetric electrolyte NRTL (eNRTL) activity coefficient model. The present work takes into account one-step hydration of nitric acid molecules, two-step hydration of sulfuric acid molecules, and further hydration of hydronium ions. With up to three temperature coefficients for each eNRTL binary interaction parameter and up to two temperature coefficients for each chemical equilibrium constant, the ternary model is validated with boiling point data from 383.15 to 471.45 K and activities at 273 K. Furthermore, contour plots of boiling points, vapor phase composition, and specific heat capacity of the ternary system are generated for engineering use.

Keywords:

Nitric acid, Sulfuric acid, Electrolyte NRTL model, Vapor-liquid equilibrium, Boiling point, Heat capacity, Nitric acid concentration/sulfuric acid concentration (NAC/SAC)

443503 A Computational Analysis of Fire Scene for Safety in a Compartment

Abstract:

This article details numerical prediction of fire spread carried out to study the distribution of carbon monoxide concentration, smoke dispersion, soot concentration and temperature in laboratory compartment, using a commercial fire modeling software tool. The heat release rate curve predicted from the simulation was compared with literature values carried out in a road tunnel. The time variations of carbon monoxide transportation, smoke concentration and temperature spread in compartment room are reported to evaluate the effect on humans. The soot concentration was described and used throughout this study as an indicator of safety in the compartment room. The effects of vent dimension and ventilation capacity were analyzed by smoke detector response times. This work provides improvements derived from reinterpretation of actual fire scene considering the ability of mechanical exhaust system and ventilation in compartment room to mitigate the hazard of fire.

443640 Preparation and Characterization of (*FURCRAEA spp*) as Green Support of Iron Nanostructured Catalyst

Abstract:

PREPARATION AND CHARACTERIZATION OF (*FURCRAEA spp*) AS "GREEN" SUPPORT OF IRON NANOSTRUCTURED CATALYST

Karen G. Bastidas G[1], Anamaria Barrera B[1], Cesar A. Sierra A[2], Hugo R. Zea R[3]

[1] Department of Chemical and Environmental Engineering. National University of Colombia (Bogota).

[2] Associate professor. Director of the Research Group of Macromolecules. National University of Colombia (Bogota).

[3] Associate professor. Director of the Research Group of Materials, Catalysis and Environment. National University of Colombia (Bogotá).

Abstract.

The availability of natural water resources has declined over time, becoming one of the biggest problems today. Environmental risk in Colombia's rivers has increase due to the widespread use of harmful chemicals in a variety of processes, ranging from illegal mineral extraction to industry wastewater, e.g. Medellin's river, due to its strategic location along a textile industrial corridor might be subject to dye contamination; the gold and platinum mining zones in Choco and Antioquia regions are being heavily affected by the use of mercury in the extraction processes. In 2009, 520 tons of mercury were imported by miners working in this area, amount that exceeds several times the mercury used by regulated industry in previous years (Ministry of Mines and Energy of Colombia, 2007). Not only the environmental risk is a concern but also the economic conditions of many low-income families whose livelihood in many cases depends exclusively of the of the mining.

Due to the issues raised above, the main goal of this study is to research, develop and implement innovative technologies in the synthesis of highly efficient, low cost and environmentally friendly iron based catalysts capable of degradate pollutants, ensuring operational stability. A native fiber (Fique, FURCRAEA spp) has been selected as a solid support for these catalysts due to its abundance, low cost and special physical and chemical characteristics.

The combination of cellulosic fibers with iron oxide nanoparticles could provide exceptional biodegradable composite materials for the treatment of organic dyes and heavy metals present in wastewater. Likewise, the use of iron particles immobilized on fique fibers allows an easy separation catalyst-treated water, turning the biocatalyst in highly recyclable and reusable material.

Fique fiber was initially characterized by determining the percentage of ashes, total lignin, soluble and insoluble lignin, cellulose, hemicellulose, total solids and humidity, obtaining 0.94, 22.58, 0.5, 22.8, 36.3, 27.5 0.701 and 2.1%, respectively. Additionally, the isoelectric point (pzc), acidic and basic sites were measured.

Then, the fiber was subjected to a pretreatment in order to expose more active sites which on time serve as nucleation points for the deposition of nanoparticles. This was done by means of an ultrasound-assisted procedure, varying the sonication time (30, 60 and 90 minutes). The effect of the sonication was followed by measuring the contents of lignin, cellulose and hemicellulose within the fiber after each time. 60 minutes of sonication was found to be most beneficial time since the amount of lignin and hemicellulose present in the fiber was reduced, while conversely the amount of crystalline cellulose exposed increases.

In order to improve the chemical ability of the nucleation sites to interact with iron and then to form stable and size controllable nanoparticles, a cationization procedure was performed on the fiber; this was achieved by the exposure of the fiber to acidic and basic conditions promoting cationization. Three cationization times (1, 2 and 3 hours) were studied and their effect on the fiber chemical surface was evaluated using FTIR-ATR and XRD analysis. The results showed 3 hours as the optimal cationizing time.

Subsequent to the cationization of the fiber surface, a procedure of ionic/cationic exchange was implemented by exposing the fiber to a solution of iron salts in order to replace the cations in the fiber surface and anchoring the iron nanoparticles. Different conditions were studied in order to synthetize zero valent iron nanoparticles (nZVI) and others with different oxidation states. Finally, to track the impregnation process on the fiber, samples from the precursor solution were taken at different times and its iron concentration was determined by atomic absorption spectrophotometry (AAS). Complementary characterization performed by X-ray diffraction, BET surface area and electronic microscopy, among others.

Key words: Natural support (fique), iron nanoparticles, bionanocatalyst.

443764 Optimization of Heat Exchanger Network in Steam Assisted Gravity Drainage Oil Sands Surface Facilities

Abstract:

Steam assisted gravity drainage (SAGD) is the most viable and environmentally safe recovery technology for extracting heavy oil and bitumen. It is a thermal in-situ production process where parallel wells are drilled horizontally into an underground bitumen reservoir. Steam is produced at the central processing facility (CPF) and injected into the reservoir through the wells. The steam heats the bitumen to a point where gravity allows it to flow down to the lower well where the mixture of bitumen and water is then pumped to the surface. The water and bitumen are separated at the CPF. Typically, more than 90% of the energy consumed in producing a barrel of bitumen using SAGD is used to generate the steam that is injected into the bitumen reservoir. Fossil fuels is employed in the process, which causes GHG (greenhouse gas) emissions. The usage of hot and cold utility is reduced by increased heat exchange and optimization of the heat exchanger network, which results in reduction of GHG emissions. In this work, pinch analysis is performed to minimize energy consumption.

443774 Thermodynamic Anaysis of Dehydration of n-Propanol Using Heterogeneous Azeotropic Distillation

Abstract:

Heterogeneous azeotropic distillation is a widely popular process in chemical process industries for the separation of binary azeotropic mixtures into their constituent pure components. It involves the addition of an entrainer which is partially miscible and forms a heterogeneous azeotrope preferably with one of the original components. This is a powerful technique which combines several physical phenomena such as presence of azeotropes and liquid-liquid immiscibilities, and is used widely in industries to enhance difficult or intractable distillation separations. For the design of heterogeneous azeotropic distillation columns, a detailed study of the heterogeneous region is crucial to plan the distillation sequence, which requires experimental vapor-liquid-liquid equilibrium (VLLE) data.

The present study focuses on the investigation of the heterogeneous region of the well known industrially available azeotropic mixture n-Propanol-Water system using Propyl acetate as entrainer. The VLE data of the system was generated by means of modified Othmer still and the still is modified using a probe type ultrasonic homogenizer to account for the heterogeneity. Isobaric vapor- liquid- equilibria and vapor - liquid - liquid - equilibria data at atmospheric pressure have been obtained for the ternary system n-Propanol-water-propyl acetate and the constituent binary systems. The equilibrium data has been correlated using NRTL, UNIQUAC and UNIFAC models and compared with the experimental data.

443776 Study of Ionic Liquids As Entrainers for the Extractive Distillation of *N*-Propanol and *N*-Propyl Acetate Binary Azeotropic Mixture and the Isobaric VLE Studies of the System

Abstract:

Distillation is a widely accepted technique for the separation of binary liquid mixtures. Separation by simple distillation is not possible for binary mixtures that form azeotropes. One of the commonly used methods to separate azeotropes is extractive distillation. It is the process of adding an entrainer that can alter the relative volatility of the mixture and thereby breaking the azeotropic point. Ionic liquids on account of their physiochemical properties as well as the recyclability have been used as a preferred replacement for the conventional entrainers. Extractive distillation with ionic liquids has the advantages of both liquid solvent (easy operation) and solid salt(high separation ability). The corrosion and dissolution problems by using inorganic salts can be rectified by these non-volatile, chemically stable, high electrical conductive molten salts.

The azeotropic system under consideration in the present study is *n*-propanol - *n*-propyl acetate mixture. These compounds which constitute the products/reactants of esterification reaction of n-propanol form a minimum boiling azeotrope at 94.7°C. *n*-propyl acetate has been included in a list of volatile acetates as an ecologically benign CFC-free solvent for aerosol sprays.

Vapour liquid equilibrium(VLE) measurements of *n*-propanol and *n*-propyl acetate are performed at atmospheric pressure in a modified Othmer still which is a continuous distillation still in which vapour phase is recirculated. The most suitable ionic liquids for shifting of azeotropic composition of the system are selected using COSMO-RS model with suitable combinations of anions and cations. The isobaric VLE of the system with [EMIM][HSO_4] is obtained and compared with the existing data. The influence of alkyl chain length on the performance of ionic liquids is also analysed with regard to the previous studies. The effect of polarity is the most important factor on the behaviour of ionic liquid solvation in alcohols. Experimental VLE data were correlated using modified UNIFAC model.

443784 Measurement and Correlation of the Mutual Solubility of Diisopropylamine + Water and Triethylamine + Water Systems at High Pressure

Abstract:

The mutual solubility data of diisopropylamine + water and triethylamine + water systems were measured at the pressure up to 100 MPa and the temperature range was 280.2~316.2 K. Experiments were performed in a high-pressure autoclave with a moveable piston. The experimental data were correlated satisfactorily with NRTL activity coefficient model and the binary parameters were obtained.

443819 Improving High Shear Rate Friction Reduction Behavior of Polyacrylamide-Based Fracturing Fluids Using Xanthan Gum

Abstract:

The main component of slickwater in fracturing low permeability formations is a high molar mass polymer such as polyacrylamide (PAM), which acts as a drag reduction agent in turbulent flow. Polyacrylamide does not degrade in the environment and also suffers from mechanical degradation at high turbulent flow rates. Five different concentrations of PAM as well as mixtures of polyacrylamide/xanthan were prepared to investigate the possibility of improving shear stability of PAM. Drag reduction measurements were performed in a flow loop and friction factors at different flow rates were calculated using rheological parameters. Based on the observations, Xanthan (XG) is not as good a friction reducer as PAM, even at three-fold higher concentration. XG slightly improved the shear stability in mixed polymer solutions at low concentrations. But at high concentrations, XG significantly improves the friction reduction behavior of PAM at nonionic and ionic environments.

443836 Oxidative Dehydrogenation of n-Butane on Carbon Catalysts

Abstract:

Butenes and butadiene, which are one of the important industrial chemical products to produce synthetic rubber , are sythesized traditionally by oxidative dehydrogenation (ODH) of n-butane over various mixed metal oxides. Metal-free carbon nanomaterials has recently received much attention by researchers as a catalyst for the ODH reaction due to its environmental acceptability, corrosion resistance, and unique surface properties. For the reason, the reaction system was modelled in this study for the sythesis of butene and butadiene. For the modelling of the reaction system, the catalytic performance of several carbon catalysts, such as graphite and activated carbon, were initially tested. Then, the reaction mechanisms and kinetics were developed based on experimental results. Lastly, the mathematical modeling of the reaction system was optimized to maximize an overall yield of the main products. This approach allowed us to develop the optimized reaction system more efficiently.

This research was respectfully supported by Engineering Development Research Center (EDRC) funded by the Ministry of Trade, Industry & Energy (MOTIE). (No.N0000990)

443855 Modeling the Thermal Behaviors of an Ultracapacitor Module

Abstract:

Ultracapacitors, also known as supercapacitors, have the potential to meet the increasing power requirements of the sustainable energy systems for the grid-scale energy storage systems and electrified vehicles. As compared to batteries, ultracapacitors offer a higher power density, higher efficiency, and longer shelf and cycle life. Because the performance of an ultracapacitor depends on temperature, it is important to calculate accurately the thermal behavior of a single ultracapacitor cell and the module composed of multiple ultracapacitor cells for the efficient and reliable systems integration of an ultracapacitor in grid and transport applications.

In this work, modeling is performed to study the effect of the layout configugation of the ultracapacitor cells on the thermal behaviors of an ultracapacitor module. The ultracapacitor module is subject to the charge/discharge cycling with constant-current between the nominal voltage of the ultracapacitor module and the half of this voltage. The charge/discharge current values examined are 50, 100, 150, and 200 A. The validation of the modeling approach is provided through the comparison of the modeling results with the experimental measurements.

443856 Modeling the Electrolyte Flow Behaviors of a Zinc/Bromine Flow Battery Module

Abstract:

The zinc/bromine (Zn/Br_2) flow battery is a promising rechargeable power source for energy storage systems because of its inherent chemical simplicity, high degree of reversibility at the electrodes, good energy density, and abundant low-cost materials. The Zn/Br_2 flow battery system, intended for the grid-scale energy storage system, is to be made in modular form. A Zn/Br_2 flow battery module is composed of the multiple stacks consisting of bipolar carbon electrodes. Since the stacks require electrolyte circulation on both the negative and positive sides, two plumbing loops are provided. The flow loops contain an electrolyte storage tank and a circulation pump. A heat exchanger may be included for temperature control in the flow loop. The design of the electrolyte circulation plumbing determines the distributions of electrolyte flow and pressure within the stacks and the efficiency of the stack is dependent on the flow distribution throughout the stack. The modeling of the distributions of electrolyte flow and pressure within the module is needed to optimize the plumbing for electrolyte circulation and thereby to improve the performance of the module. In this work, modeling is performed to study the effect of the layout configuration of the plumbing for electrolyte circulation on the distributions of electrolyte flow and pressure within a 50 kWh Zn/Br_2 flow battery module composed of four stacks. It is seen that the plumbing layout for electrolyte circulation can have a significant effect on the flow rate distribution of electrolyte through the stacks.

443837 Blowdown Dynamic Modeling and Analysis of Pressure and Temperature

Abstract:

Vessel blowdown is one of the safety operation procedures that discharges flammable gas or liquid to flare system during emergency situation such as fire. When blowdown occurs, the liquefaction and freezing effect cause a pressurized vessel to be brittle, exposing it to possible hazard situation. Also the blowdown systems are recommended to be accordance with API(American Petroleum Institute) 521 code. API 521 requires pressure in vessel should be half of the initial pressure or be lower than 100 psig in 15 minutes after commencing blowdown in fire case. Therefore, accurate estimation over the change of temperature and pressure of the vessel during blowdown period is crucial for the design of blowdown system. For the reason, researchers have developed a few dynamic blowdown models so far. However, the model was found to be rather inaccurate under certain conditions. Therefore a new dynamic blowdown model was developed in this work. For the accurate analysis, the model algorithm considers both vapour and liquid phase in non-equilibrium condition that the temperatures of vapour and liquid phases are different. Futhermore, heat transfer from the vessel wall to discharge gas at either laminal or turbulent flow in the vessel was considered in the model to increase the accuracy of blowdown estimation. For thermodynamics, both Peng-Robinson and Soave-Redlich-Kwong equations of state were applied to the model, and the results were compared each other. Finally, the simulation results were compared with Haque et al's experimental and steady-state simulation results.

443974 Observer-Based Model Predictive Control for Time-Delay Systems

Abstract:

Model predictive control is one of the most widely used advanced process control technologies in industry.[1] It has many capabilities such as dealing with input/state constrains and uncertainties. Observer-based model predictive control was developed to estimate states information when the states are not measurable directly. Many chemical processes have time delays in either states or measurements, such as polymer production and some bioprocesses. [2] In this presentation, we are proposing an unknown input observer based model predictive control algorithm for time-delay systems. Unknown input observer is an observer design to eliminate certain types of unknown inputs in the system. Stability of unknown input observer-based model predictive control without time delay was provided by our previous study. Terminal cost and terminal region are designed for a class of non-linear time delay systems. The close-loop stability of the controller is provided. [3][4]

An example of bioreactor is provided which has uncertain time delays in measurements but with a known upper bound. Two measurements are provided which are online biomass and offline ethanol measurements. Model-plant mismatch is considered as unknown inputs in the system. State estimation and control are calculated using the observer and an optimization algorithm.

Reference:

[1] Marcus Reble et al., Model predictive control of constrained non-linear time-delay systems. IMA Journal of Mathematical Control and Information, 2010, 1-19

[2] Seung Cheol Jeong and PooGyeon Park, Constrained MPC algorithm for uncertainty time-varying systems with state delay, IEEE TRANSACTIONS ON AUTOMATIC CONTROL, 50, 2005, 257-262

[3] Mayuresh Kothare et al., Robust constrained model predictive control using linear matrix inequalities. Automatica, 32, 1996, 1361-1379

[4] Wook Hyun Kwon et al., A simple receding horizon control for state delayed systems and its stability criterion, Journal of Process Control, 13, 2003, 539-551

444020 Thermodynamic Modeling of Aqueous Lithium Sulfate and Its Solubility in the Presence of Na+, K+ and Mg2+ Sulfates

Abstract:

This work is an extension of the previous works done by our group towards development of a rigorous and comprehensive thermodynamic model for high salinity produced water that is essential for mass and heat balance calculations and simulation and design of novel treatment processes. Specifically this work presents a thermodynamic model that accurately represents all thermodynamic properties of aqueous lithium sulfate solution including the solubility of Li_2SO_4 in water and also in the presence of Na^+–K^+–Mg^{2+} sulfates. By explicit consideration of solution chemistry of each subsystem, symmetric electrolyte NRTL model is used to correlate available experimental data of all pertinent thermodynamic properties including activity coefficient, osmotic coefficient, excess enthalpy, and liquid heat capacity over the entire concentration range. This facilitates quantification of eNRTL binary molecule–molecule, molecule–electrolyte, and electrolyte–electrolyte interaction parameters. Also, solubility data are used to regress the thermodynamic constants required for modeling the liquid–solid equilibrium of lithium sulfate in water and in the presence of Na^+–K^+–Mg^{2+} sulfates. Displaying a significant prediction ability, the model accurately represents the thermodynamic behavior of the system from infinite dilution up to saturation and temperatures from 298.15 K to 573.15 K.

444044 Organic Solar Cells

Abstract:

This paper describes highlights the need of next-generation "organic solar cells". There is a continuing need for organic solar cells and arrays with very high specific power densities (1000-5000 kW/kg) for generating power in a new generation of ultra light space payloads example in NASA. An emerging technology with promise to meet these ambitious goals are organic solar cells based on very thin films of discrete layers. Theoretical efficiencies for multi junction solar cells are more ~70% as compared to ~30% for single junction devices. There is continuous development going on high mobility, band gap engineered organic donors and acceptors, as well as development of three junction devices and scaled up devices on lightweight flexible polymer substrates. The main interest in this paper is organic semi-conducting crystal: Pentacene, which has photo-conducting states. Organic solar cells manufactured today have low efficiency other than their stability issues. The efficiency of organic solar cells can be improved by either designing polymer materials that absorbs more of solar spectrum or by minimizing the charge collection efficiency by controlling the morphology of polymer films. Once fully developed, organic solar cells would revolutionary power the industry due to their flexibility and minimal weight

444055 A Refined Non-Random Two-Liquid Segment Activity Coefficient Model for Solubility Modeling

Abstract:

Accurate correlation and robust prediction for complex molecule solubility or solid-liquid equilibrium (SLE) in solvents and solvent mixtures have drawn much interest recently [Tanveer et al., *Chem. Eng. Prog.* **2014**, 110 (9), 37-47]. Solubility models such as the Non-Random Two Liquid Segment Activity Coefficient (NRTL-SAC) model [Chen and Song, *Ind. Eng. Chem. Res.* **2004**, 43, 8354-8362] are now routinely used in the industries. However, an important unfilled need is to model the solubility in oligomers/polymers/surfactants such as asphaltene nanoaggregates in petroleum fluids and drug molecules in lipid-based or solid dispersion formulation [Tung, *Org. Proc. Res. Dev.* **2013**, 17, 445-454]. This paper presents a refinement to the very successful NRTL-SAC model to better account for the contribution due to size differences of molecules. Specifically, the Flory-Huggins combinatorial entropy expression for conceptual segments in the model is substituted with the Staverman-Guggenheim expression for real molecules. We present the rationale for the refined model, the updated conceptual segment numbers for common solvents, and the newly identified conceptual segment numbers for common drug excipients. The model performance is illustrated by prediction of drug molecule solubility in various lipid solvents and solvent mixtures, compared to experimental data and previous work by other models. The refined NRTL-SAC model shows enhanced capability to correlate and predict drug molecule solubility in common solvents and excipients used in drug formulation.

444057 Explaining the Paradox of Asphaltene Precipitation with n-Alkanes

Abstract:

Explaining the paradox of asphaltene precipitation with n-alkanes
Md Rashedul Islam and Chau-Chyun Chen
Department of Chemical Engineering, Texas Tech University
Abstract
Paraffins, specifically alkanes, promotes asphaltene precipitation when they are blended with asphaltene-containing petroleum fluids. The alkane/oil volume ratio required for onset of asphaltene precipitation increases with the carbon number and yields a maxima. Experimental study indicates that the maxima lays at C7[1] or between C8 -- C10 [2]. However, precipitation yield decreases asymptotically with increase of the carbon number even if excessively high alkane/oil volume ratio is used [3]. For example, n-pentane be a better solvent for asphaltene than n-hexane in regard of both onset of precipitation and precipitation yield. However, high carbon number alkane beyond the maxima, n-hexadecane for instance, can be a better solvent at onset of precipitation point and a poorer solvent in regard of precipitation yield than n-pentadecane. A thermodynamic explanation for this paradox is necessarily important to find the compatibility of bitumen or crude oil with paraffinic hydrocarbons. In this work, this paradox is explained by the aggregation thermodynamics of asphaltene precipitation [4]. Both the enthalpy and entropy of mixing are investigated to explain the effect of paraffin carbon number on asphaltene precipitation. Athabasca and cold lake bitumen samples are studied.
References
[1] G. Hotier and M. Robin, "Action de divers diluants sur les produits pÃ©troliers lourds: mesure, interprÃ©tation et prÃ©vision de la floculation des asphaltènes," Oil & Gas Science and Technology, vol. 38, pp. 101-120, 1983.
[2] I. A. Wiehe, H. W. Yarranton, K. Akbarzadeh, P. M. Rahimi, and A. Teclemariam, "The paradox of asphaltene precipitation with normal paraffins," Energy & fuels, vol. 19, pp. 1261-1267, 2005.
[3] D. L. Mitchell and J. G. Speight, "The solubility of asphaltenes in hydrocarbon solvents," Fuel, vol. 52, pp. 149-152, 1973.
[4] M. Wang, Y. Hao, M. R. Islam, and C. C. Chen, "Aggregation Thermodynamics for Asphaltene Precipitation," AIChE Journal, Submitted.

444086 Optimization of Layered Graphene Oxide Nanostructures through Controlling the Membrane Deposition Rate Towards Size-Sieving Separation

Abstract:

Single-layer graphene oxide (GO) flakes were coated on anodized aluminum oxide (AAO) substrate through a simple vacuum filtration process, the effect of the deposition rate on the nanostructures of obtained membrane was investigated though pure water flux and organic vapor diffusion, which indicated GO membrane assembled with slow deposition rate might have more ordered lamellar structure and, correspondingly, have smaller effective pore size. In contrast, under fast deposition rate, GO flakes assembled in a more random way. Therefore, through controlling the membrane deposition rate, the nanostructure of layered GO membrane could be optimized for size dependent separation. We found after thermal reduction, 100 nm GO membrane (prepared @ 0.02 nm/s) has effective pore size smaller than 6.1 Å, which exhibit great potential in n-hexane/2,2-dimethybutane (DMB) separation.

444142 Robust Design of a Radiant Wall Burner for a High Temperature Environment Utilizing Computational Fluid Dynamics Simulations

Abstract:

Radiant Wall Burners are often utilized in Ethylene Cracking Furnaces. The conditions and fuel gases for a specific unit can vary considerably within the unit and across similar units. With burner air preheated to high temperatures, premixing the fuel can result in dangerous and destructive flashback, while complete nozzle-mixing of the fuel can result in burner instability at startup or turn-down conditions. These challenges can be solved through the generation of a robust radiant wall burner design, which is facilitated by executing Computational Fluid Dynamics simulations of many different geometric configurations.

444161 Novel Microwave Synthesis Carbon Dots from PINE Bark and Its Application for Heavy Metal Sensing

Monday, April 11, 2016
Exhibit Hall E (George R. Brown)

Saheed Sanni, Chemistry, Vaal University of Technology, Vanderbijlpark, South Africa, Puthalapattu ReddyPrasad, Chemistry, Vaal University of Technology, Vanderbiljpark, South Africa and Augustine E. Ofomaja, Chemistry, Vaal university of Technology, Vanderbiljpark, WV, South Africa

NOVEL MICROWAVE SYNTHESIS CARBON DOTS FROM PINE BARK AND ITS APPLICATION FOR HEAVY METAL SENSING

Sanni S.O., ReddyPrasad P., Ofomaja A.E.

Biosorption and Waste water Treatment Research Laboratory, *Department of Chemistry, Faculty of Applied and Computer Sciences, Vaal University of Technology, P. Bag X021, Vanderbijlpark-1900, South Africa.*

INTRODUCTION

Microwave pyrolysis for the production of carbon has the advantages of being cheap, produces high yield of carbon and is absolutely a green synthetic method [1]. In recent times, the use of microwave pyrolysis have been extended to the preparation of carbon dots. Carbon nanodots (C-dots) falls under the class of carbon nanoparticles with size below 10 nm[2]. C-dots are carbon materials which processes high water solubility and strong fluorescence depending on their size, edge, shape, surface ligands, and defects. Since C-dots preparation can be achieved from difference carbonaceous sources and by varying a range of parameters during their synthesis, the properties of the final product may vary depending on these factors. C-dots have sp^2 characteristics and possess several functional groups on their surfaces that makes them highly hydrophilic and easily functionalized with various organic, polymeric, inorganic, or biological spices[3]. Several synthesis methods that have been explored includes electrochemical synthesis, combustion/thermal/hydrothermal/acidic oxidation, supported synthesis, microwave/ultrasonic, arc discharge, laser ablation/passivation, and plasma treatment[4]. C-dots processes excellent such as water dispersibility, chemical stability and photostability, ease of modification, excitation-dependent multicolor emission, low toxicity, and good cell permeability, great efforts have been paid to their potential applications in biosensing, bioimaging, drug delivery, and other biological related aspects.

METHODS

The carbon dots (C-dots) with green luminescence was synthesized from microwave assisted pyrolysis of pine bark waste our carbon source at 800, 1000 1200 W for 1 Hr. The dots produced were characterized using FTIR, TGA, XRD, UV-Vis and Raman spectroscopy. The dots were applied for heavy metal sensing.

RESULTS AND DISCUSSION

The functional groups present on the C-dots (**Fig.1**) includes -OH stretching band mode at 3413 cm^{-1}, C-H stretching at 2934 and 2845 cm^{-1}, stretching peaks of C=O, C=C at 1731, 1604 and 1437 cm^{-1}. The XRD pattern (**Fig 2a &b**) of produced C-dots depicts sharp peak at 20.9^{o} and 26.8^{o}, which is highly crystalline cores and possess amorphous nature of C-dots. This peculiar peak is ascribed to 002 of the graphitic carbon, which in line with inter-planar spacing of 3.77 Å with higher spacing than bulk graphite at 3.34 Å. The thermogravimetric and differential thermal analysis display (**Fig 3**) that the C-dots possess thermal stability more than 100 ^{o}C and mass losses occurs afterwards. The thermal analysis depicts 3 endothermic peaks at 331, 542 and 735 ^{o}C, which are ascribed to oxygenated groups such as carboxyl, hydroxyl, carbonyl and phenols. The optical properties of the synthesize C-dots were explored using UV-vis absorption and photoluminescence (PL) spectroscopy. The UV-vis spectra display (**Fig 4**)intense absorption peak in the range of 260 to 290 nm, which is ascribed to π-π^* transition of the aromatic sp2 domain and also presence of multiple polyaromatic chromophores. A shoulder peak at 330 nm which is ascribed to n-π^* transitions of C=O and structural defects of the C-dots. PL spectrum of the C-dots (**Fig 5a**)exhibits an intense maximum emission peak at 432 nm upon excitation at 330 nm, which also shows characteristic yellowish green color under sunlight and intense blue coloration UV lamp (365 nm). The PL properties of as prepared C-dots was varied at (**Fig5b**) different excitation wavelength from 300 to 405 nm. There was red-shifting in emission peak as the excitation peak was increased from 300 to 405 nm and also simultaneously decrease in fluorescent intensity, which is ascribed to quantum effect and surface defects. This behavior of the prepared C-dots is dependent on excitation-fluorescence, which is accordance with other report [13]. The quantum yield of synthesized C-dots was measured to be 17% at maximum excitation at 330 nm

Fig.6 shows the Raman spectrum of the PB-GQDs, which has been usually used to confirm the quality of the prepared PB-GQDs. There is two major Raman peaks at 1370 cm^{-1} (D band), 1580 cm-1 (G band). The peak of G band corresponds to the Sp^2 bonded carbon atoms in a two-dimensional (2D) hexagonal lattice and the peak of G band is associated with the vibrations of carbon atoms with dangling bonds in the termination plane of tangled graphite or glassy carbon. The D/G ratio band ratio is ID/IG: 0.68 is confirm the formation of graphene like structure from synthesized pine bark.

Fluorescence detection of Cu^{2+} ion

The detection of Cu^{2+} was performed at room temperature in phosphate buffer (PBS) solution pH 4.3. 1.0 mL of PB-CQDs solution was added into 0.5 mL of PBS buffer, followed by the addition of 1.0 mL different concentrations of Cu^{2+}. The fluorescence emission spectra were recorded after reaction for 5 min at room temperature (~28 °C).

Selectivity and sensitivity studies

Moreover sensitivity and selectivity is additional significant parameter to assess the performance of the sensing system. Consequently, we examined the fluorescence intensity changes in the presence of representative metal ions such as including Cu^{2+}, Hg^{2+}, Ca^{2+}, Al^{3+}, Ba^{2+}, Mn^{2+}, Ni^{2+}, Pb^{2+}, Co^{2+}, Zn^{2+}, Fe^{3+}, $Cs^{2+}$and Mg^{2+}, of each at a concentration of 15.0 µg mL^{-1} were reacted with a PB-CQDs solution. Fig. 7 A displays the relative change in PL intensity of PB-CQDs in the presence of various metal ions. Fe^{2+}, Mn^{2+} and Hg^{2+} ions cause the slight PL changes (defined as the relative change of PL intensity in 60–80% as compared to blank) which can be attributed to the nonspecific interactions between the carboxylic and/or amine groups and the metal ions. Among these metal ions, Cu^{2+} displays the strongest PL quenching effect on PB-CQDs, assigning to the special coordination between the Cu^{2+} ions and the phenolic hydroxyl and/or amine groups of PB-CQDs which has been widely used for the detection of Cu^{2+} ions or colored reactions in traditional organic chemistry. Such a specific fluorescence quenching effect may originate from the strong interactions between Cu^{2+} ions and the surface groups of PB-CQDs which transfer the photoelectrons from PB-CQDs to Cu^{2+} ions. These results clearly demonstrate that the PB-CQDs-based Cu^{2+} sensor is highly selective to Cu^{2+} over the other metal ions. Fig.8. Fluorescence responses of CCDs to the addition of different metal ions under acetate buffer (pH 4.3). The concentration of each metal ion is 15 µg $mL^{-1.}$ F_0 and F correspond to the fluorescence intensities of PB-CQDs at emission wavelength of 440 nm excited at 320 nm in the absence and presence of metal ions, respectively.

As shown in Fig. 3, the fluorescence intensity of PB-CQDs gradually decreased with the Cu^{2+} concentration. The fluorescence intensity displayed liner responses with the Cu^{2+} concentrations, with two linear ranges of 0.0□14.0 µg mL^{-1} and 15.0□25.0 µg mL^{-1} (Fig.3B). The linear equations are $F/F0$ =0.9827□0.8485C_{Cu2+} (R = 0.993) and $F/F0$ = 0.7565–0.0033 C_{u2+} (R = 0.996), respectively.

REFERENCES

[1] W. B. Lu, X. Y. Qin, S. Liu, G. H. Chang, Y. W. Zhang, Y. L. Luo, A. M. Asiri, A. O. Al-Youbi, X. P. Sun, Economical, green synthesis of fluorescent carbon nanoparticles and their use as probes for sensitive and selective detection of mercury(II) ions, Anal. Chem. **2012**, 84, 5351-5357.

[2] J. Jiang, Y. He, S. Li, H. Cui, Amino acids as the source for producing carbon nanodots: microwave assisted one-step synthesis, intrinsic photoluminescence property and intense chemiluminescence enhancement. Chem Commun **2012**, 48, 9634–6.

[3] Y. Wang, A. Hu, Carbon quantum dots: synthesis, properties and applications, J. Mater. Chem. C **2014,** 2, 6921–6939.

[4] C. Liu, P. Zhang, X. Zhai, F. Tian, W. Li, J. Yang, et al., Nano-carrier for gene delivery and bioimaging based on carbon dots with PEI-passivation enhanced fluorescence, Biomater. **2012,** 33, 3604-13.

Figures:

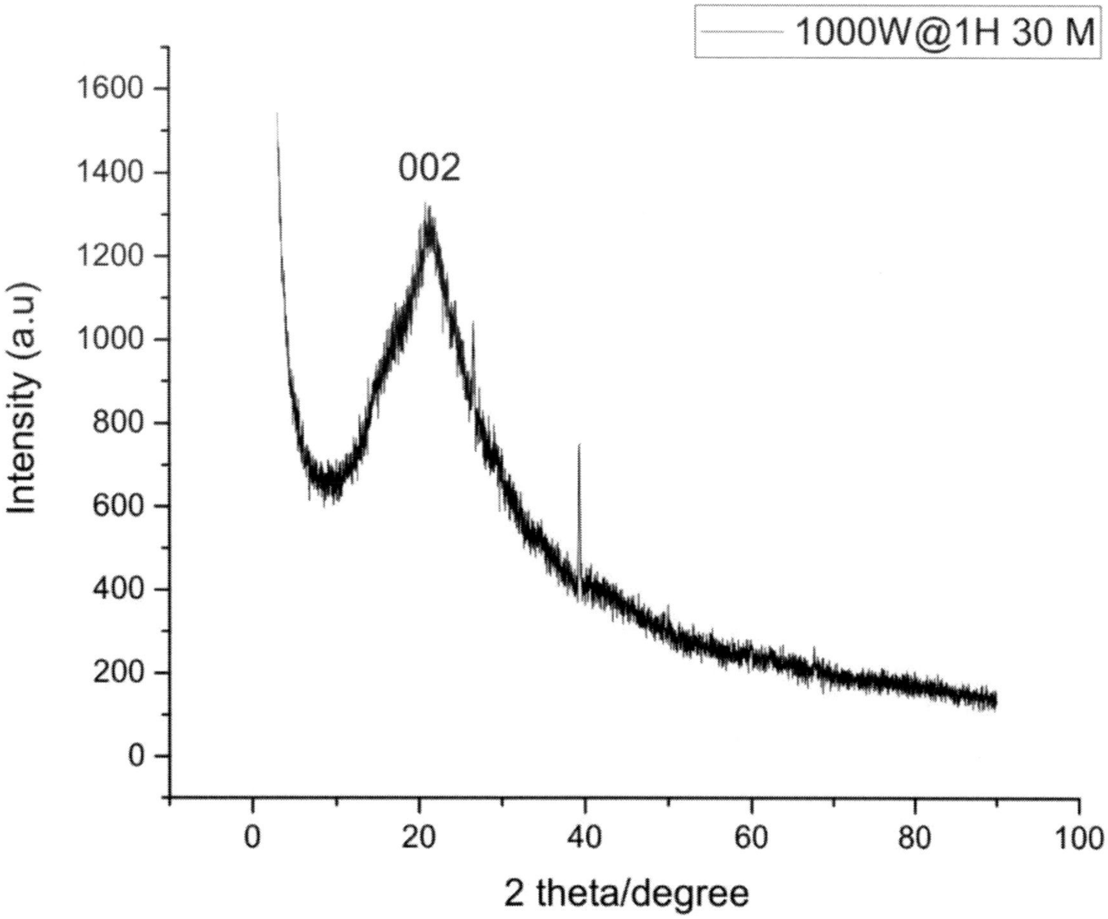
1000W@1H 30 M
002
Intensity (a.u)
1600
1400
1200
1000
800
600
400
200
0
0
20
40
60
80
100
2 theta/degree

Fig 1. FTIR spectra of as-prepared C-dots from Pine Bark

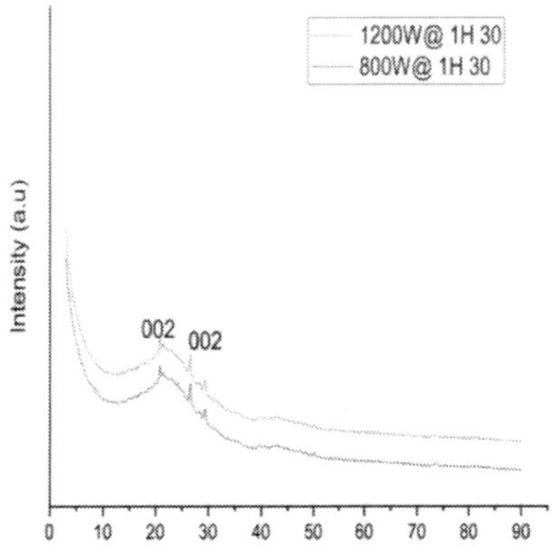

(a)

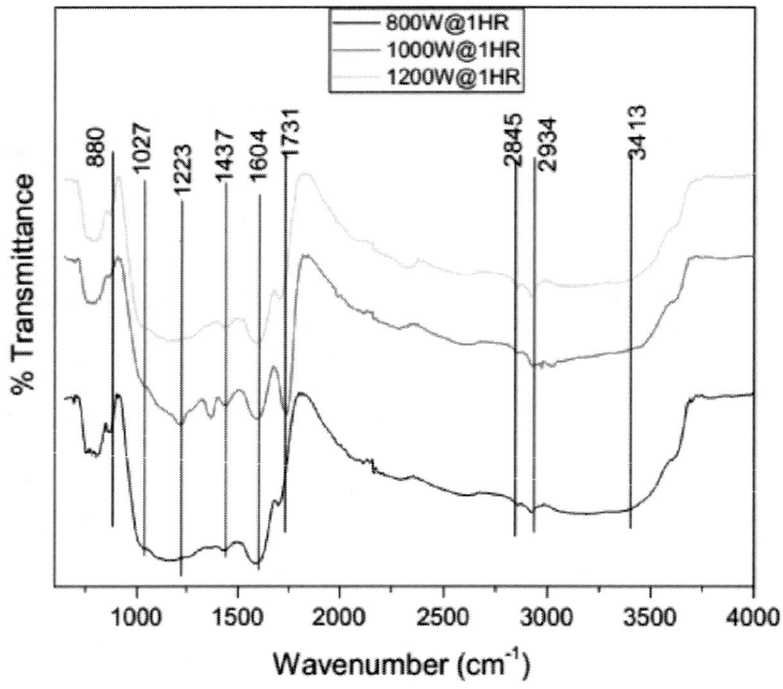

(b)

Fig. 2a & b. XRD pattern of C-dots.

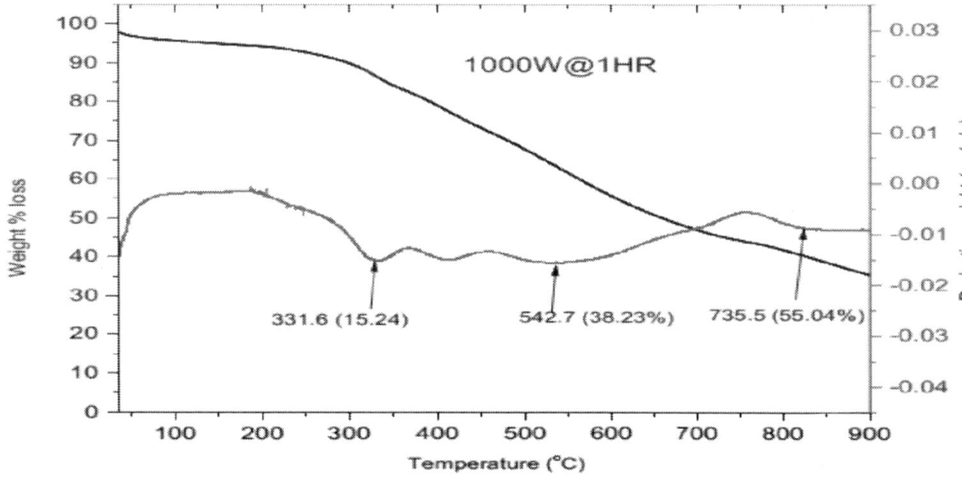

Fig 3. Thermal curves for C-dots under N_2 atmosphere.

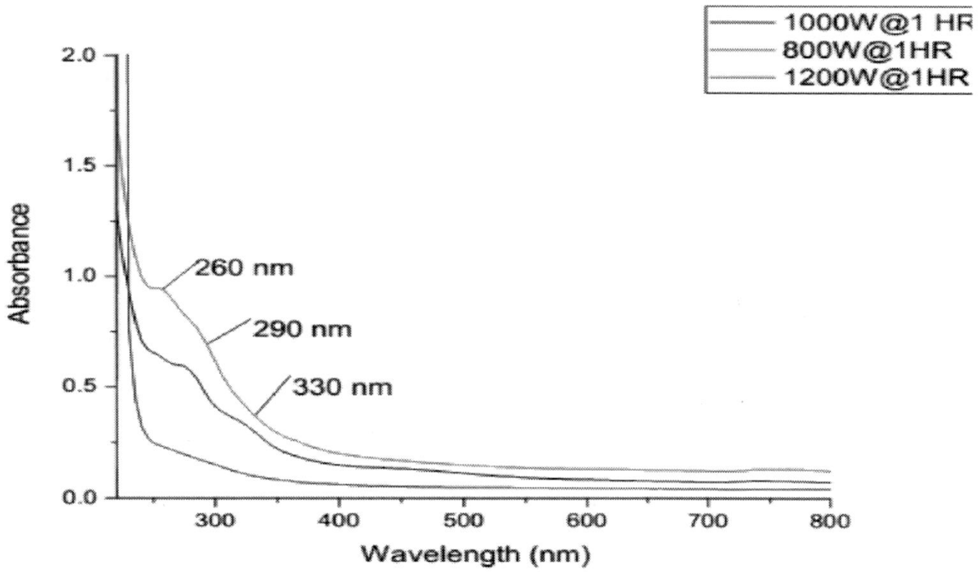

Fig 4. UV-vis absorption of C-dots.

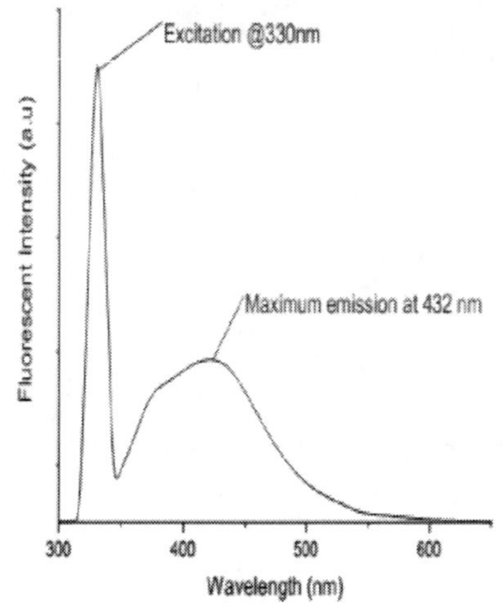

(a)

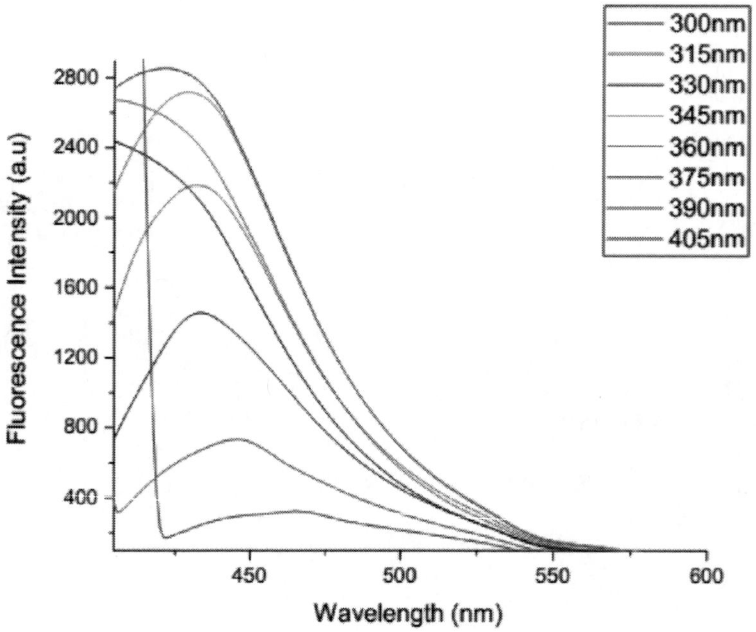

(b)

Fig 5a. PL excitation and emission wavelength (b) PL emission spectra at different excitation wavelength from 300 to 405 nm in 15 nm increments.

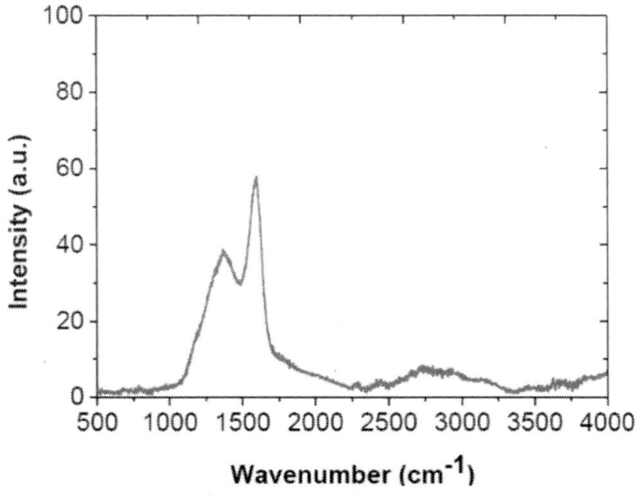

Fig. 6. Raman Spectra of PB-CQDs

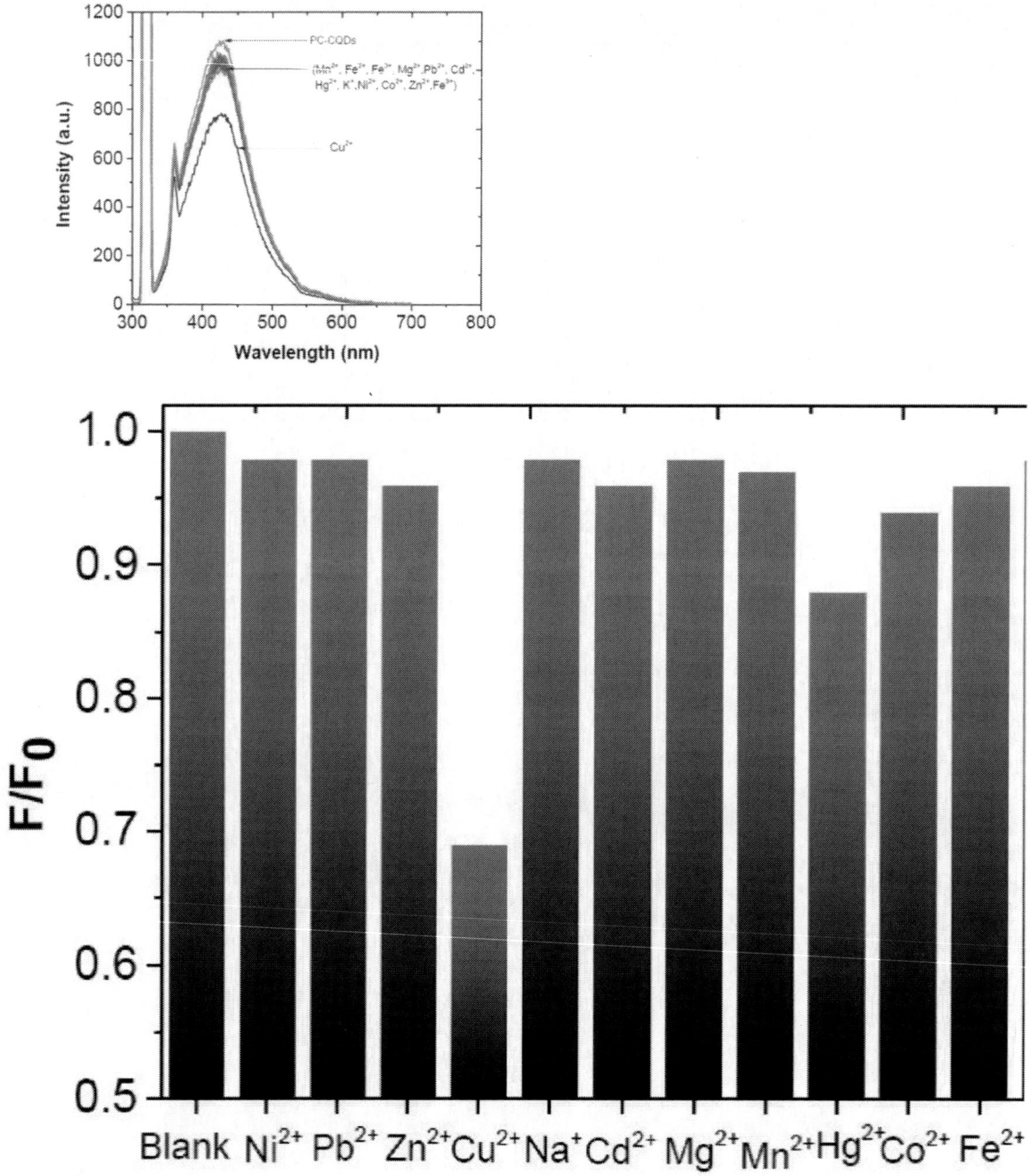

Fig.7. Fluorescence responses of CCDs to the addition of different metal ions under acetate buffer (pH 4.3).

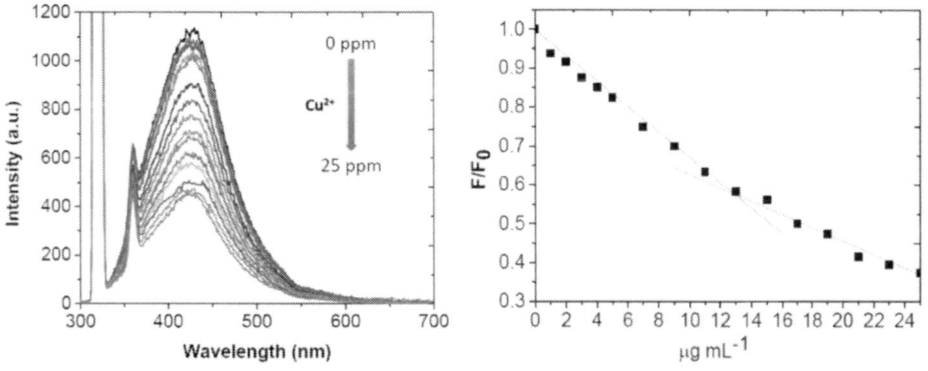

Figure 8. Fluorescence titrations of the y-CDs with Cu^{2+} from 0 to 25 μg mL^{-1}. (Influence of Cu^{2+} concentration on the fluorescence intensity of PB-CQDs (top to bottom: 0, 1, 2, 3, 4, 5, 7, 9, 11, 13, 15, 17, 19, 21, 23 and 25.0 μg mL^{-1})

444269 Functionalization of OH-Terminated Nanoparticles By Cyclic Azasilanes

Abstract:

Colloidal suspensions of functionalized nanoparticles have been suggested for enhanced oil recovery, formation damage remediation, flow assurance, and water treatment operations. Surface modified nanoparticles can also be used as tracers in reservoir studies, active media in thermal management fluids, and fillers in high performance elastomers. Conventional functionalization of nanoparticles with alkoxysilanes and chlorosilanes is limited by needs for catalytic amount of water for the reaction to proceed, requirement of multiple hours to fully saturate the surface, and release of by-products upon reaction completion. Monolayer deposition with a high density of functional groups represents a particularly difficult challenge.

Our ongoing studies indicate that application of cyclic azasilanes that have a strained ring structure with adjacent Si and N centers provides highly efficient functionalization of hydroxylated silica, alumina, magnetite, and diamond nanoparticles which is catalyst-free and by-product-free. The data on synthesis and characterization of functionalized nanoparticles as well as possible reaction mechanism of functionalization will be discussed.

444093 New Development on Hoist Scheduling with the Relaxation of Cyclic Precondition

Abstract:

Multi-Stage Material Handling (MSMH) processes are designed to manufacture materials requiring a series of treatments and in industry, they are broadly used due to their capability to handle large work orders and compatibility with materials with different recipes. A typical example is electroplating process. A hoist handles the transportation of each job among units (stages) according to its processing recipe. The hoist movement plan is predetermined through optimization and this plan is repeated to perform continuity. Obviously, the ultimate goal is to maximize the productivity within each movement plan, which is equivalent to minimize the cycle time to finish one work order. The optimization process is generally called Cyclic Hoist Scheduling (CHS), since every movement plan is a cycle due to its repeatability. This precondition in term of hoist operation and unit usage is general for most cases. However, it could block the really optimal results.

In this study, this precondition has been relaxed for multi-capacity units, thus the unit usage can be more flexible. Thus, the real optimality can be found through CHS. To deal with the cycle connection issue from the relaxation of the precondition, a new module Cycle Connection Identification (CCI) is proposed. Obtained hoist schedule will be checked. Once a schedule with cycle connection problems is identified, a new cycle will be scheduled forcing the smooth connection with the first cycle. The objective of scheduling a new cycle is to match the beginning of the first cycle and the ending of the second cycle. If the two-cycle group is still with cycle connection issue, more cycles will be scheduled till the cycle group is free from cycle connection problems. Each cycle is called a sub-cycle and the constructed cycle group without cycle connection problems is called a full cycle. Note that in each sub-cycle, the work order and all transportation operations are completed.

In this paper, hoist scheduling models with/without such a precondition have been developed and proposed as mixed integer linear programming. The performance of the two models is tested through case studies and the difference is compared and demonstrated in detail.

444359 Development of Novel Emulsified Acid Using Waste Oil

Abstract:

In carbonate reservoirs, hydrochloric acid (HCl) is commonly used for stimulation purposes. The main disadvantage for using HCl is its high and rapid reactivity allowing for less control on acid reaction with formation. In addition, HCl may react with the tubing causing severe corrosion and damages. Several slow-reacting acid systems, though HCl-based, have been developed to retard the HCl reaction and control its reactivity to allow deeper penetration in carbonate reservoirs before it breaks out and reacts with formation rock. However, the selection of the hydrocarbon phase of emulsified acid is critical to the creation and stability of the emulsified acid system. Diesel oil, crude oil and solvents such as xylene have been reported to be successful in the field.

In this work, alternative and less expensive source of oil is developed to replace the current practices. Emulsified acid is a water-in-oil emulsion where oil is the continuous phase and acid is the dispersed one. Industrial and filtered waste oil was used to prepare the emulsified acid of this work. Extensive work has been carried out to study the chemical composition of the waste oil, namely SARA and to select the proper emulsifier. Emulsions were prepared using 15% HCl with emulsifier concentration ranging from 0.5 to 2.0 vol% and 70:30 of acid-to-oil ratio. Droplet size distributions were measured and used to benchmark with other emulsions. Emphasis was given to the stability of the emulsion especially at high temperatures. In order to apprehend the effect of corrosion inhibitor on the emulsion stability, its concentration was varied (up to 0.3 vol%). Later, rheological analyses of the waste oil emulsion were compared to the commonly used diesel-based emulsified acid.

Results showed the potential use of waste oil to prepare high temperature emulsified acid (up to 100°C). The addition of corrosion inhibitor to the emulsified acid caused a negative effect on stability resulting in a less stable emulsion. On the other hand, addition of cationic emulsifier helped to formulate a more stable emulsion. Apart from that, the optimum concentration of the emulsifier was at 2.0 vol%. The rheological data confirms the stability of the new emulsified formulation at higher temperatures. This paper summarizes the findings of using waste oil emulsion and recommends it for field applications. The newly proposed emulsified acid introduces waste oil as a replacement of diesel.

The novelty of using waste oil will help to significantly reduce the expenses of formulating emulsified acid. Since waste oil is basically waste it has a good impact on the environment, that being said, instead of throwing it, the industry will be able to recycle it.

444364 Molecular Simulation of Controlling Ions and Their Role in Low Salinity IOR for Carbonate Reservoirs

Abstract:

Carbonate geology represents around 50% of hydrocarbon-bearing reservoirs in the world and more than 70% in some areas such as the Middle East. Water flooding has not been very efficient in carbonate reservoirs due mainly to the low water wetness. Modifying the ionic composition and salinity of the injected seawater demonstrated that the oil recovery from these carbonate reservoirs could be improved. Although the underlying mechanisms are still a subject of debate, the improved oil recovery has been mostly attributed to the increased water-wetness of the rock through the interactions of specific ions like Ca^{2+}, Mg^{2+}, SO_4^{2-}, and Cl^- , which are present in seawater with the rock surface under favorable conditions

The study of hydrated gas-phase metal ions provides a link between intrinsic chemistry of the isolated ion and its chemistry in solution. The structures and binding enthalpies of a variety of gas-phase and solution mono- and divalent ion hydrates containing up to 12 water molecules have been studied computationally. A DFT and molecular dynamics simulations were employed to study hydrated structures of the form $[\mathbf{M}(H_2O)_n]^{2+}.mH_2O$ for calcium and magnesium and $[\mathbf{M}(H_2O)_n]^{+}.mH_2O$ for sodium, in which there are n water molecules in the first coordination shell and m water molecules in the second coordination shell. The hydration, dehydration, and interactions of metal ions with Cl^{-1} and $SO^{-2}{}_4$ have been studied to better understand the mechanism of low salinity IOR. Density functional (DFT) theory using the B3LYP functional, and 6-311++g** basis set has been used. The charge of the metal ion decreases monotonically as the number of water molecules increases in the complex. Extensive calculations of the energetic contribution of individual structural characteristics as well as the energetic trends in binding energy are developed, which will lead to better understanding of how low salinity water IOR works.

Keywords: Hydration, carbonate reservoir, DFT, binding energy, Low Salinity

444278 Insight into the Mechanism of Toxic Chromium (VI) from Wastewater Using Magnetite Coated PINE Powder

Abstract:

INSIGHT INTO THE MECHANISM OF TOXIC CHROMIUM (VI) FROM WASTEWATER USING MAGNETITE COATED PINE POWDER

Pholosi A., Ofomaja A., Naidoo E.B.

Biosorption and Wastewater Treatment Research Laboratory, Department of Chemistry,

Faculty of Applied and Computer Sciences, Vaal University of Technology, P. Bag X021, Vanderbiljpark, 1900 South Africa.

INTRODUCTION

A knowledge of the interaction between pollutant molecules or ions and active sites on an adsorbent surface is an important phenomenon in many industrial processes such as catalysis, adsorption and other separation techniques. The use of lignocellulosic materials for Cr(VI) adsorption has gained interest over the years **(Miretzky and Cirelli, 2010)**. The adsorption properties of lignocellulosic materials has been attributed to acidic functional groups on their surfaces which can take part in ion-exchange, complexation, chelation and hydrogen bonding with pollutant species in solution **(Vithanage et al., 2015)**. Two mechanisms for lignocellulosic adsorption of Cr(VI) have been proposed; (i) an electrostatic mechanism in which at low pH, the protonated acidic functional groups on the biomaterial carrying a positive charge attracts the negatively charged Cr(VI) species **(Demirbas, 2005)** and (ii) the adsorption-coupled reduction mechanism in which Cr(VI) is reduced to Cr(III) either in solution (Direct reduction) or after being adsorbed (Indirect reduction) by electron-donor groups of the biomaterial that have lower reduction potential values than that of Cr(VI) **(Gao et al., 2008)**. The Cr(III) ions formed in solution may then form complexes with Cr(III)-binding groups on the biomaterial surface or released into bulk solution by repulsion between Cr(III) and other positively charged groups on the biomaterial **(Miretzky and Cirelli, 2010)**.

METHODS

Sodium hydroxide treated pine coated with magnetite (NTP-NC) was prepared by co-precipitation method. The effect of solution pH was performed in a batch system in which 0.5 g of the adsorbent was contacted in six 250 cm^3 beakers containing 75 cm^3 of 75 mg/dm^3 Cr(VI) solutions set at pH 1, 2, 4, 6, 8 and 10 at 26 °C and agitated at 200 rpm for 2 h and the clear solution analysed for Cr(V/I) left using UV-Vis spectrophotometry at 460 nm

RESULTS

The results of the effect of initial solution pH on the adsorption of Cr(VI) onto NTP-NC is shown in Fig. 1. To confirm the operating mechanism, the change in $[H^+]$ concentration at each initial pH, the oxidation/reduction potential before and after adsorption and the amounts of Total Cr, Cr(VI)and Cr(III) present in solution at each pH was monitored for NTP-NC and the results displayed in Fig1 . The change in $[H^+]$ concentration ($\Delta[H^+] = [H^+]_{final} - [H^+]_{initial}$) at the end of the adsorption process for each initial solution pH was recorded and plotted against the initial solution pH. The results showed that the change in $[H^+]$ concentrations at the end of the adsorption process where higher for lower initial solution pH's (where lower amounts of Cr was found in solution) than for higher initial solution pH's (where higher amounts of Cr was found in solution) and the change in $[H^+]$ concentration became small and almost constant after initial solution pH 4. This is consistent with the assumption that the removal of Cr from solution is associated with $[H^+]$ ion consumption and that as the concentration of $[H^+]$ ions reduces in solution, the Cr removal reduces. Cr(VI) ion in acidic solution has a positive redox potential value (above 1.3 V at standard conditions) and in the presence of an electron donor becomes unstable and can be reduced to Cr(III) **(Park et al., 2008)**. The results show that the initial ORP values (before adsorption) for the Cr solutions at pH's of 1, 2, 4, 6, 8, and 10 were 339, 280, 165, 55, -53 and -172 mV for NTP-NC respectively. The positive values of ORP at lower solution pH's indicates higher possibility for reducing Cr(VI), while the lower and negative ORP values indicates lower possibility for reducing Cr(VI). On addition of the adsorbents into the solutions, a change in ORP were observed . The values of ORP at the end of the adsorption process became 330, 268, 30, 9 -29 and -61 for NTP-NC respectively. A reduction in ORP at the end of the adsorption stage signifies that reduction has occurred in the system. **Yu et al. (2014)** showed the amount of Cr(VI) removed from aqueous solution in contact with zero-valent iron (nZVI) could be related to the change in ORP and pH of the solution. The authors observed a decrease in ORP from 400 to 100 mV and a corresponding 10 unit pH increase as Cr(VI) was removed from solution.

Finally, analysis of the prepared Cr solutions at the various solution pH's showed that the concentration of Total Cr obtained from ICP and that of Cr(VI) using UV-vis measurements were very similar, suggesting that Total Cr was almost equal to Cr(VI) in solution. After contacting the various solutions with the adsorbents, the concentration of Total Cr in the pH 1 solution was observed to have reduced to 21.56 mg/dm^3 for NTP-NC, while the Cr(VI) concentrations was 10.66 mg/dm^3 for NTP-NC. With solution pH 2, the concentration Cr(III) left in solution was lower than that of solution pH 1 for both samples. The reason is that at pH 1 there is an abundance of H^+ ions in solution which accumulates on the functional groups on the adsorbent making the surfaces of the adsorbent positively charged and repelling away the positively charged Cr(III) ions from the adsorbent surfaces **(Suksabye et al., 2009)**. At solution pH 2, it is believed that there is less accumulation of H^+ ions at the adsorbent surfaces which led to reduced electrostatic repulsion and increased adsorption of Cr(III) from solution. This increase in Cr(III) adsorption is also confirmed by the reduction in Total Cr in solution between solution pH 1 and 2. The concentrations of Cr(VI) left in solution at pH 2 was higher than at pH 1 for NTP while for NTP-NC the Cr(VI) concentrations were lower at pH 2. For solution pH's higher than 2 and up to 10, the residual Cr(III) in solution becomes very low and almost constant. This is consistent with the small change in $[H^+]$ in solution which is almost constant at this pH range.

FTIR Evidence for the adsorption of Cr(III)

When Cr(III) is formed in solution via adsorption-coupled reduction mechanism it may undergo hydrolysis and/or complexation depending on the solution pH and the total Cr(III) concentration. Hydrolysis may lead to the formation of various Cr(III) species with an increase in solution pH which gives rise to a reduction in its solubility and increasing the chances of precipitation of the Cr(III) species. At solution pH 1 – 4, Cr^{3+} and $CrOH^{2+}$ are the major species which can be adsorbed by negatively charged ligands on the biomaterial surface. On the other hand, at solution pH range of 4 – 10, Cr(III) species (hydroxo complex), such as $CrOH^{2+}$, $Cr(OH)_2^+$ and $Cr(OH)_3$, are the dominant forms in the aqueous system and at high pH is readily transformed into soluble complex, $Cr(OH)_4^-$. When NTP-NC was contacted with Cr(VI) solution for 2 hr, the final solution pH was increased to pH 2.53. The FTIR spectra for NTP-NC (Fig. 2) shows the presence of characteristic Fe-O peaks of bulk Fe_3O_4 at 553.26 cm^{-1}, 564.36 cm^{-1}, 570.21 cm^{-1} and 576 21 cm^{-1}. Functional groups associated with the pine includes the broad band at 3250.99 cm^{-1} representing –OH of cellulose, 2905.17 cm^{-1} representing C-H stretching, while the peaks at 1627.69 and 1602.87 cm^{-1} shows the COO^- and C=O bonds respectively. After 2 hr contact with Cr(VI) solution, the peak at 564.34 cm^{-1} was observed to have split into several new peaks (Fig. 3b (ii)). These emerging peaks are characteristic of Fe-O bonds of maghemite (γ-Fe_3O_4), for example the characteristics peaks of maghemite was observed at 595.21, 685.65 cm^{-1} and 582.70, 634.91 cm^{-1} suggesting a change in the iron oxide phase (Fe_3O_4 to γ-Fe_2O_3) due to the conversion of Cr(VI) to Cr(III) by magnetite nanoparticles. According to **Peterson et al. (1997)**, Cr(III)(hydr)oxide produced is precipitated on the magnetite surface. In this study, α-Cr_2O_3 peak were observed at 604.63 cm^{-1} indicating that Cr(III) was formed during the adsorption process. Finally it was also observed that the carboxylate group at 1635.55 cm^{-1} was shifted to 1637.73 cm^{-1} while a new peak was observed at 1737.72 cm^{-1} suggesting that complexation between Cr(III) and Carboxylate ions also occurred.

REFERENCE

Demirbas, A. Energy Source 27 (2005) 1449–1455.
Miretzky, P., Cerelli, A.F. A review. J. Hard. Mater. 180 (2010) 1-9.
Vithanage, M., Rajapaksha, A.U., Ahmad, M., Uchimiya, M., Dou, X., Alessi, D.S., Ok, Y.S. J. Environ. Manage. 151 (2015) 443-449.
D. Park, C.K. Ahn, Y.M. Kim, Y.-S. Yun, J.M. Park, J. Hazard. Mater. 160 (2008) 422–427.
P. Suksabye, A. Nakajima, P. Thiravetyan, Y. Baba, W. Nakbanpote.. J. Hazar. Mater. 161 (2009) 1103–1108.
R.F., Yu, F.H., Chi, W.P., Cheng, J.C. Chem. Eng. J. 255 (2014) 568–576

M.L. Peterson, A.F. White, G.E. Brown, G.A. Parks Environ. Sci. Technol. 31 (1997) 1573–1576.

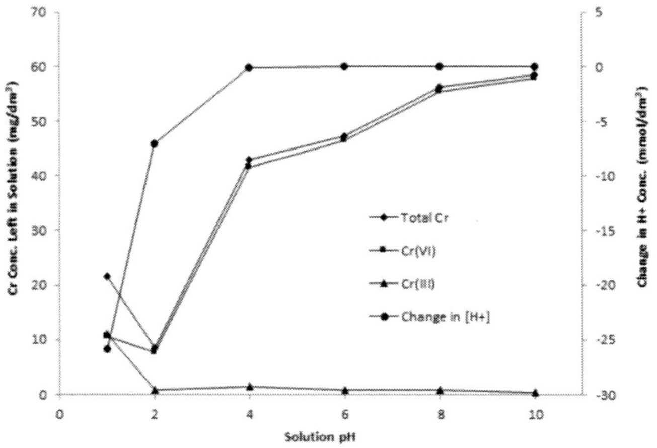

Fig. 1 Effect of solution pH on the adsorption of Cr(VI)

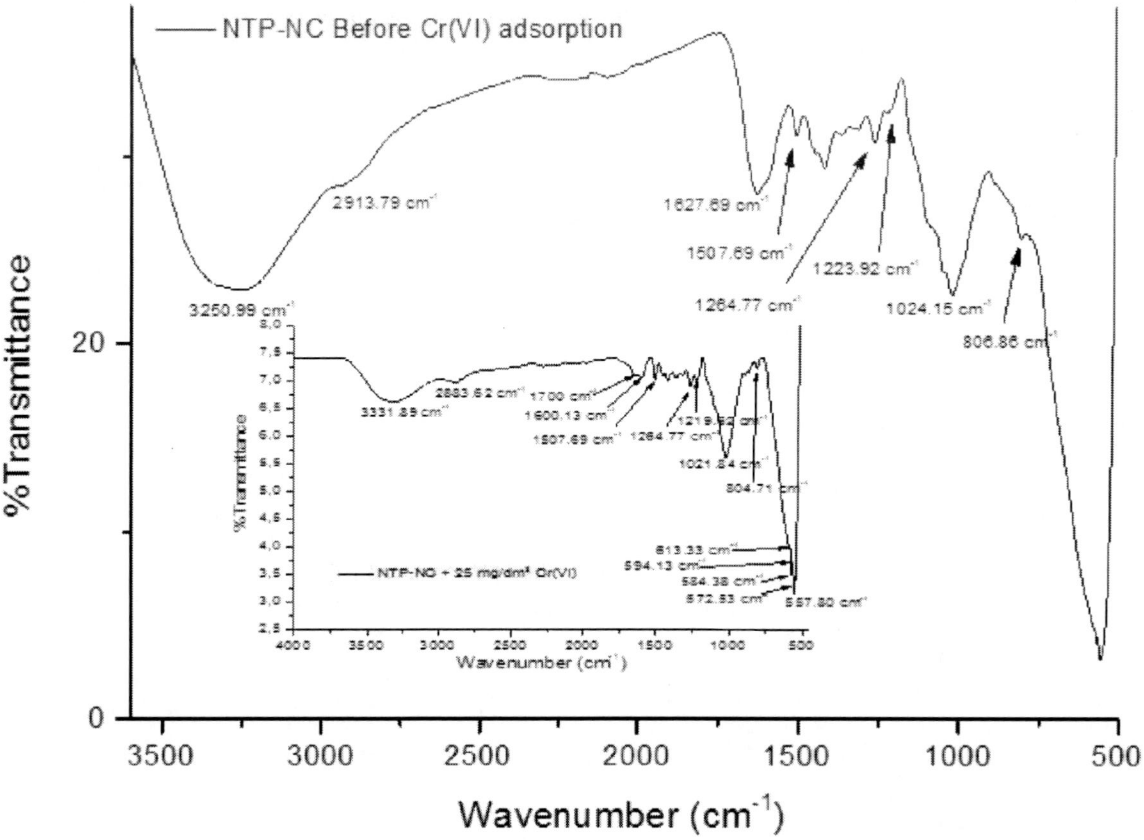

Fig, 2: FTIR of NTP-NC

444280 Copper-Phyllosilicate Core-Shell Nanocatalysts with Balanced Active Sites for Carbon-Oxygen Hydrogenolysis Reactions

Abstract:

Copper-Phyllosilicate Core-shell Nanocatalysts with Balanced Active Sites for Carbon-Oxygen Hydrogenolysis Reactions

Hairong Yue, Weimao Fang, Changmin Liu, Jiankang Ying

College of Chemical Engineering, Sichuan University, Chengdu, China

Email: hryue@scu.edu.cn

Abstract

Hydrogenolysis of carbon-oxygen (C¨CO) bonds (e.g., esters, ethers, furfural, and CO_2) has emerged as a versatile synthetic tool in organic methodology, as it could produce a variety of products (e.g., chemicals, fuels, and polymers). Copper-based catalysts have been intensively explored for hydrogenation reactions as the copper sites account for the selective hydrogenation of carbon-oxygen bonds and relatively inactive for the hydrogenolysis of carbon-carbon bonds. Work on understanding the active sites of copper catalysts for hydrogenation reactions indicted that both Cu^0 and Cu^+ species were crucial to the activity of Cu-based catalysts. However, researchers were not able to establish a relationship between activity and Cu^0/Cu^+ active species since the conventional Cu-based catalysts is the deactivation by metal particle growth and unstable surface Cu^0 and Cu^+ active species in the strong reductive H_2 and oxidizing carbon-oxygen atmosphere.

We renctly present on efficient approaches for fabrication of a series of copper-phyllosilicate core-shell nanocatalysts and nanoreactors with balanced and stable Cu^0 and Cu^+ active species. We chose the hydrogenation of dimethyl oxalate, ethylene cyclicarbonate and as probe reactions for a better understanding of properties of active sites and the structure-activity relationship. The results indicated a synergy between copper and the oxide components and the balanced surface Cu^0 and Cu^+ species can greatly improve the catalytic C-O hydrogenolysis performance of the catalysts. These nanocatalysts with balanced and stable Cu^0 and Cu^+ active species, confinement effect, intrinsic high surface of Cu^0 and Cu^+ and unique tunable tubular morphology, has potential applications in the high-temperature hydrogenation reactions.

Keywords: core-shell nanocatalysts, Hydrogenation reaction; copper catalyst

References

1. Yue, H.; Zhao, Y.; Ma, X.; Gong, J.: Ethylene glycol: properties, synthesis, and applications. *Chem Soc Rev* 2012, 41, 4218-4244.

2. Gong, J.; Yue, H.; Zhao, Y.; Zhao, S.; Zhao, L.; Lv, J.; Wang, S.; Ma, X.: Synthesis of Ethanol via Syngas on Cu/SiO2 Catalysts with Balanced Cu^0-Cu^+ Sites. *J Am Chem Soc* 2012, 134, 13922-13925.

3. Zhao, S.; Yue, H.; Zhao, Y.; Wang, B.; Geng, Y.; Lv, J.; Wang, S.; Gong, J.; Ma, X.: Chemoselective synthesis of ethanol via hydrogenation of dimethyl oxalate on Cu/SiO_2: Enhanced stability with boron dopant. *J Catal* 2013, 297, 142-150.

4. Yue, H.; Ma, X.; Gong, J.: An Alternative Synthetic Approach for Efficient Catalytic Conversion of Syngas to Ethanol. *Acc Chem Res* 2014, 47(5):1483-1492

5. Yue, H.; Zhao, Y.; Zhao, L.; Lv, J.; Wang, S.; Gong, J.; Ma, X.: Hydrogenation of dimethyl oxalate to ethylene glycol on a Cu/SiO2/cordierite monolithic catalyst: Enhanced internal mass transfer and stability. *AIChE J* 2012, 58, 2798-2809.

6. Yue, H.; Zhao, Y.; Zhao, S.; Wang, B.; Ma, X.; Gong, J.: A copper-phyllosilicate core-sheath nanoreactor for carbon¨Coxygen hydrogenolysis reactions. *Nat Commun* 2013, 4, 2339.

Stimuli-responsive Rubber in Oil and Gas Applications

Rostyslav Dolog, Sankaran Murugesan, and Valery Khabashesku

Baker Hughes, 14990 Yorktown Plaza Dr, Houston, Texas, 77040, USA

ABSTRACT

A problem that exists today in the oil & gas industry is the loss of mechanical properties by reactive rubber seals due to swelling, and the limited space in downhole well applications to accommodate larger swellable packing elements.

This paper presents cost-effective, reinforcing, stimuli-responsive fillers, such as certain classes of fly ash, for use in rubber composite materials in various oil and gas applications. Apart from cost effectiveness, these fillers can also enable stimuli-responsive effects, as well as improved mechanical properties (e.g. modulus) for rubber composites when swelled in water. Fly ash (Class C) is a common component of Portland cement which was used in the literature to achieve a stimuli-responsive effect when added to Hydrogenated Nitrile Butadiene Rubber (HNBR) as filler. Exposure to water brines can cause a filler to promote an increase in pH which enables the hydrolysis of HNBR nitrile groups to carboxyl groups, thus creating an ionomer. This process is followed by neutralization of the new ionomer with metal ions, and formation of ionic crosslinks, further improving mechanical properties of the rubber.

This paper demonstrates that incorporation of fly ash as a reinforcing filler allows development of swellable seals that can retain, or even increase, their modulus and stiffness upon swelling. These characteristics allow avoiding the loss of mechanical properties by seals due to swelling and also eliminating the need for large swellable packing elements that occupy additional space downhole.

KEYWORDS: Elastomers, stimuli-responsive, fly ash, composites

INTRODUCTION

There is a growing demand for high performance elastomers in the oil & gas industry. Rubber compounds are used for various sealing applications, packer elements, stators for pumps, etc. In general, rubber formulations are comprised of elastomer matrix, curing system, antioxidants, plasticizers, along with reinforcing filler, which provides the compounds with their strength. The reinforcing filler is usually represented by carbon black and, to a lesser extent, silica. Along with other factors, filler mechanical properties and concentration generally define hardness, modulus and tensile strength of the elastomer compound. Throughout the years, many different materials have been tried as alternative fillers for rubber composites, including nanomaterials (CNTs, graphene, nanofibers), recycled rubber particles, and ash/fly ash. In the case of ordinary carbon black, the filler serves not only for reinforcement, but also as a cost reducer, because raw elastomer is often a more expensive component. Mechanical properties of the composites are affected by the aspect ratio of the filler, quality of the dispersion, and interface between blend components. The latter determines the effectiveness of load transfer from polymer matrix to the reinforcing filler. High performance compounds can be designed by incorporating well dispersed fine particles with a high aspect ratio and good compatibility with the polymer matrix.

In this context, the fly ash is a cheaper, low reinforcing filler alternative to carbon black, which will help reduce cost of the compounds for "non-demanding" sealing applications requiring large amounts of raw materials. Fly ash is produced in large quantities as a byproduct of coal burned at power plants, and can be considered as a recycled material. Fly ash costs several cents per pound, while carbon black, depending on the grade, is usually one order of magnitude more expensive. Hemmings et al[1] demonstrated that fly ash can be used as filler for polymeric composites with enhanced mechanical properties. Adoption of this technology in the oil and gas industry could change the design approach for reactive rubber compounds. Further functionalization of fly ash through a cost-efficient process could ensure compatibility with the rubber matrix, while promoting filler dispersion and ensuring an efficient load transfer. Such functionalization's include, but are not limited to, surface modification with fatty acids, dicarboxylic acids or phosphoric acid. The reasoning for surface modification is to attach functional groups with aliphatic chains to fly ash particles to make them compatible with aliphatic rubber chains, or to attach acidic groups on their surface that can interact with nitrile groups of HNBR. Initial attempts towards chemical functionalization of fly ash have been made by Seal et al for treatment of offshore oil spills.[2]

Apart from cost effectiveness, composites with fly ash can show stimuli-responsive properties. It has been shown that adding Portland cement to the HNBR matrix improves its mechanical properties in the swollen state by water uptake.[3] Exposure to water causes Portland cement to grow in volume and mass, and promotes an increase in pH which enables the hydrolysis of HNBR nitrile groups to carboxyl groups, thus creating an ionomer. This process is followed by neutralization of the new ionomer with metal ions and formation of ionic crosslinks, further improving mechanical properties of the rubber. Fly ash (Class C) is a common component of Portland cement, which contains a large variety of metal oxides, and is capable of increasing the

---

[1] Filler Comprising Fly Ash For Use In Polymer Composites. Hemmings et.al., US 6,916,863 B2, July 2005.
[2] Hydrophobic High Surface Area Zeolites Derived from Fly Ash for Oil Spill Remediation, Seal et.al., Environ. Sci. Technol., 2013, 47 (11), pp 5843–5850
[3] Stimuli-Responsive Cement-Reinforced Rubber, Musso et. al., ACS Appl Mater Interfaces. 2014 May 14;6(9):6962-8.

pH of the water solution by hydrolysis, which has a potential to achieve a stimuli-responsive effect when added to HNBR as filler.

Degradation of tensile properties is a common problem for swellable packers and other sealing elements. By selectively using certain classes of fly ash as effective reinforcing filler, one can develop swellable seals that can retain or even increase their modulus and stiffness upon swelling. Therefore, there is a potential to minimize or even avoid the loss of mechanical properties by water swellable reactive packing elements due to swelling. This provides an opportunity to decrease the size of these elements, which are often represented by rather long sections, and hence occupy a significant amount of space downhole. In this paper we have studied the reinforcing effects of fly ash fillers with HNBR compounds as well as stimuli responsive properties of these compounds.

EXPERIMENTAL SECTION

Rubber compounds T1, T2, T3, and T4 are elastomers based on a custom-built peroxide curing system. Formulation F2 was prepared by adding 15 wt. % of fly ash class C to compound F1, which is a commercial swellable NBR sulfur-cured compound.

T series compounds were prepared using a C.W. Brabender® internal mixer operated at 60 RPM for 20 minutes at ~160°F and then molded into sheets and cured at 340°F for 30 minutes in a hot press. Compound F2 was prepared by mixing fly ash with compound F1 in the Brabender mixer for 6 minutes at 60 RPM at ~100°F. Compound F1 is a commercial water swellable formulation that was used as received. F series compounds were cured at 300°F for 60 minutes in a hot press.

Tensile properties of rubber samples were tested using dog bone specimens, according to the ISO37 standard, at room temperature and at 350°F using an MTS® testing system.

Swelling tests in 3% KCl brine and LVT 200 oil were performed to assess the effect of fillers on the chemical resistance of elastomers. Standard one-inch rubber buttons were tested at 200°F at ambient pressure for 6 days. Thermal properties were studied using TGA Q50 and DSC Q2000 instruments.

To analyze the stimuli-responsive effect introduced by fly ash fillers, rubber buttons were aged for 10 days in 3% KCl in the oven at 200°F and then removed from the solution, dried for 30 minutes in the oven and then for 3 hours at room temperature. Samples were later tested in the MTS testing system in compression mode.

Viscoelastic properties of stimuli-responsive samples, based on water swellable compounds, were accessed using the RSA G2 dynamic mechanical analysis (DMA) instrument. This experiment was conducted in a temperature ramp mode of 2°C/minute at 1Hz frequency and 0.2% amplitude strain. A Nicolet 6700 Fourier Transform Infrared (FTIR) spectrometer was used to confirm neutralization of carboxylic groups by metal cations.

RESULTS AND DISCUSSION

The reinforcing properties of fly ash were evaluated at room temperature (Table 1) and at 350°F (Table 2) and compared with N550 carbon black. Sample T3, reinforced with fly ash, demonstrated inferior tensile properties and hardness compared to a compound filled with carbon black. However, at room temperature, fly ash was capable of delivering a reinforcing effect to the rubber compound without sacrificing its elasticity, e.g. the ultimate elongation of elastomer compound remained virtually unchanged after incorporation of 50 parts per hundred rubber (PHR) of fly ash, while its modulus and tensile strength have increased. Therefore, potentially, one can substitute carbon black with high loadings of fly ash in certain applications, thus further decreasing the cost of the rubber product. The summary of mechanical properties is provided in Table 1, E_x denotes modulus of a compound at certain strain (x), i.e. stress at strain.

Sample	N550 CB (PHR)	Fly ash Class F (PHR)	Tensile strength (psi)	E_{25} (psi)	E_{50} (psi)	E_{100} (psi)	Elongation (%)	Hardness Shore A
T1	0	0	829	71	97	121	472	48
T2	50	0	3402	165	273	744	290	68
T3	0	50	1176	94	127	166	467	54

Table 1. Room temperature tensile properties and hardness of elastomer samples

The relatively low increase in modulus and hardness of T3 compared to T1 at room temperature can be explained by a low aspect ratio of fly ash and weak interactions between blend components. The latter problem can be solved by functionalization of fly ash particles.

The general data trends from the high temperature tensile tests closely resemble the results of the tensile test conducted at room temperature. However, at high temperature, rubber specimens break at much lower elongation, which leads to a great decrease in the measured tensile strength. This behavior is typical for elastomers and limits their application to a certain maximum service temperature and is also due to the fact that high temperature tests are conducted at conditions that push the limits of the material. The modulus values at 15% (E_{15}) and 20% (E_{20}) strain are not provided herein, because half of the tested specimens were broken before they reached such strains.

The superiority of carbon black as an efficient filler became even more pronounced at high temperature. However, incorporation of fly ash doubled the tensile strength and increased the elongation at break by one-third when compared to the unfilled elastomer.

Sample	E_5 (psi)	E_{10} (psi)	E_{15} (psi)	E_{20} (psi)	Tensile strength (psi)	Elongation (%)
T1	31	46			91	23
T2	61	130	194	261	832	39
T3	42	73	98	119	173	30

Table 2. Summary of high temperature (350˚F) tensile data for elastomer samples

Thermal properties of samples T1-T3 were assessed with Differential Scanning Calorimetry (DSC) and Thermogravimetric analysis (TGA). TGA allows evaluating thermal stability of samples and to verify the actual composition of a sample from the ash content. The results of the DSC and TGA analysis are summarized in Table 3. During TGA, samples were heated to 600°C at 10°C/minute rate in a nitrogen (N_2) environment, and subsequently heated at 10°C/min to 600 °C in air to burn off carbon-based fillers. This allows decomposing of organic matter first by applying heat, and then burn the remaining carbon-based material in air atmosphere.

The initial degradation of the polymer started at 200°C and progressed to complete decomposition of aliphatic chains in the temperature range of 420-500°C with a highest rate of decomposition at ~460°C. Sample T2 incurred the second rapid mass loss after 600°C due to burning of the carbon black, which comprised most of the ash content at 600°C in a N_2 environment. The DSC experiment demonstrated the glass transition temperature of the elastomer remained largely independent of the type of filler - the respective DSC heating scan was conducted with a 10°C/minute temperature ramp after precooling the sample to -50°C.

Sample	W_{600} (%)	W_{800} (%)	$T_{deg}1$, °C	$T_{deg}2$, °C	T_g, °C
T1	2.4	2.4	452.8	-	-29
T2	29.8	2.7	458.7	616.5	-29
T3	32.8	32.7	447.3	-	-28

Table 3. TGA/ DSC results for elastomer samples

Table 3 contains ash contents (W) at 600°C and 800°C, degradation temperatures (T_{deg}) which represent the temperature of the highest rate of decomposition during TGA experiment, and glass transition temperature (T_g) of the samples, as determined by DSC scan.

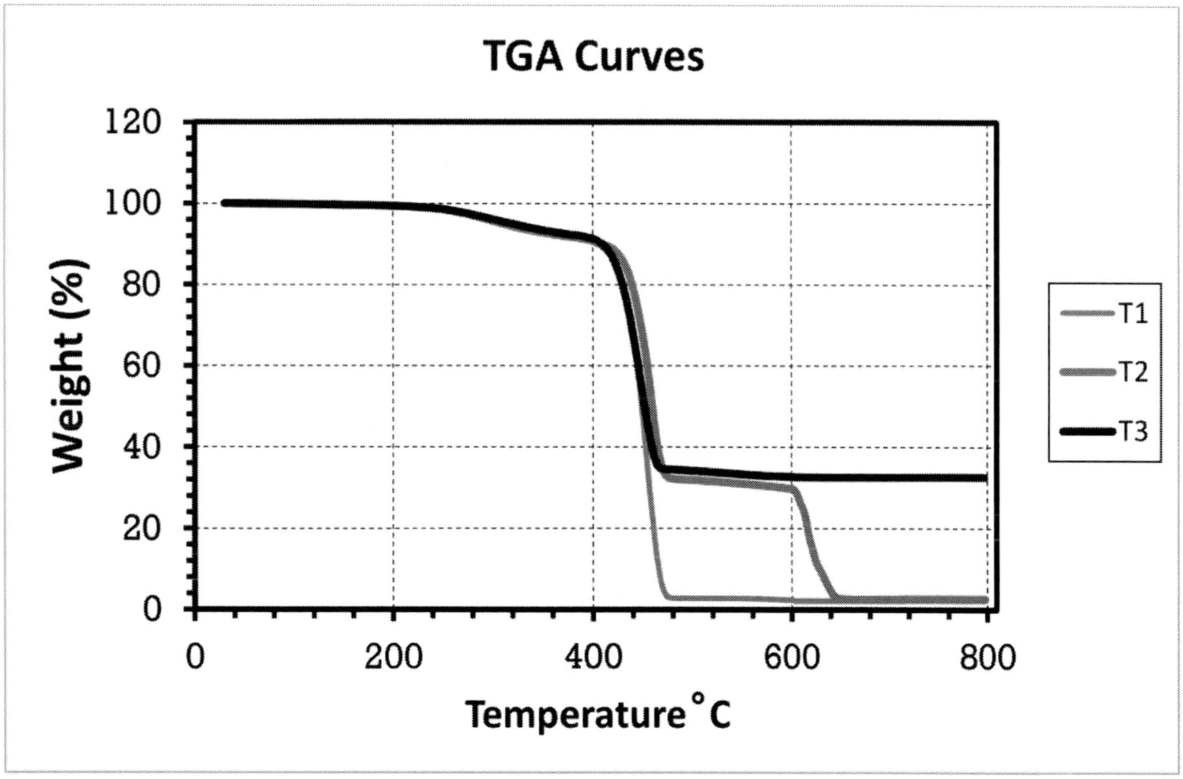

Figure 1. TGA Curves for samples T1, T2, and T3

Sample \ Testing fluid	3% KCl brine	LVT 200	Specific gravity
T1	0.6	13.0	0.96
T2	0.3	11.6	1.13
T3	0.8	9.1	1.18

Table 4. Volume increase (%) of elastomer samples due to swelling for 6 days at 200°F

Musso et. al. developed a stimuli-responsive HNBR composite reinforced with Portland cement that was increasing its stiffness and modulus when swelled in water[3]. Fly ash (Class C) is a common component of Portland cement, which contains the components necessary to achieve a similar effect. Therefore, one can avoid the loss of mechanical properties by water swellable packing element seals upon activation. To test this hypothesis, we studied the effect of aging in water brines on the properties of elastomers with fly ash fillers.

Rubber samples were molded and cured into standard size rubber buttons (cylinders of ~0.25" height and ~1" diameter). These buttons were aged for 10 days in 3% KCl brine in the oven at 200°F, removed from the solution, dried for 30 minutes in the oven and for 3 hours at room temperature. The swelling of both compounds was close to ~1% by volume (Table 4). Samples were later tested in an MTS machine in compression mode. Each button was compressed to 70% strain, buttons remained intact during the experiment, and visually all appeared to have a certain level of compression set at the end of the experiment. The results of the experiments are summarized in Table 5 and Figure 2. Only one button was tested for each sample. The provided compression modulus values (C_n) represent values of stress at a percentage strain (n), e.g. C_5 is stress at 5% compression strain.

Compound T4 was reinforced by fly ash Class C, which contains larger weight fraction of CaO than fly ash Class F (~25% vs ~10-15% respectively) and is expected to produce a higher pH and closely resemble cement that was used to prepare stimuli-responsive rubber in the original paper by Musso et.al.[3] The results demonstrate that exposure to water (3% KCl) at high temperatures can improve the modulus for the compounds filled with fly ash Class C (T4), while the control sample (T1) showed the opposite behavior. After aging, the compressive modulus at 50% strain (e.g. stress at 50% strain) of control sample T1 decreased by 19%, as compared to the unaged sample, while compressive strength of sample T4, reinforced by fly ash, increased by 15%. The difference in modulus values is even more drastic at 70% compression strain (-35% and 31%, respectively). This is in agreement with the hypothesis stated earlier.

Sample	Filler	Compressive modulus (psi)				
		C_5	C_{10}	C_{25}	C_{50}	C_{70}
T1	None	19	42	127	468	1810
T1 swollen	None	16	36	109	380	1185
T4	Fly Ash class C (50 PHR)	21	46	134	434	1628
T4 swollen	Fly Ash class C (50 PHR)	29	57	153	501	2125

Table 5. Summary of compressive modulus properties after 3% KCl soaking and 200°F drying

To further investigate the stimuli-responsive property of fly ash filler, the latter was incorporated into a water swellable compound, which provided for significantly more interaction of filler with water to occur. Sample F2 comprises unmodified commercial water swellable NBR formulation F1 mixed with 15 wt.% of fly ash Cass C. Sample F2 was processed into a 0.04" slab, then a rectangular sample was cut from the slab and swollen in 3% KCl solution until 100% volume increase. The F2 sample was then dried overnight in the oven, which lead to de-swelling of the sample due to solvent evaporation. Mechanical properties of the F2 sample were assessed with DMA prior to the swelling test, in a swollen state, and after drying - the data summary is shown on Figure 3.

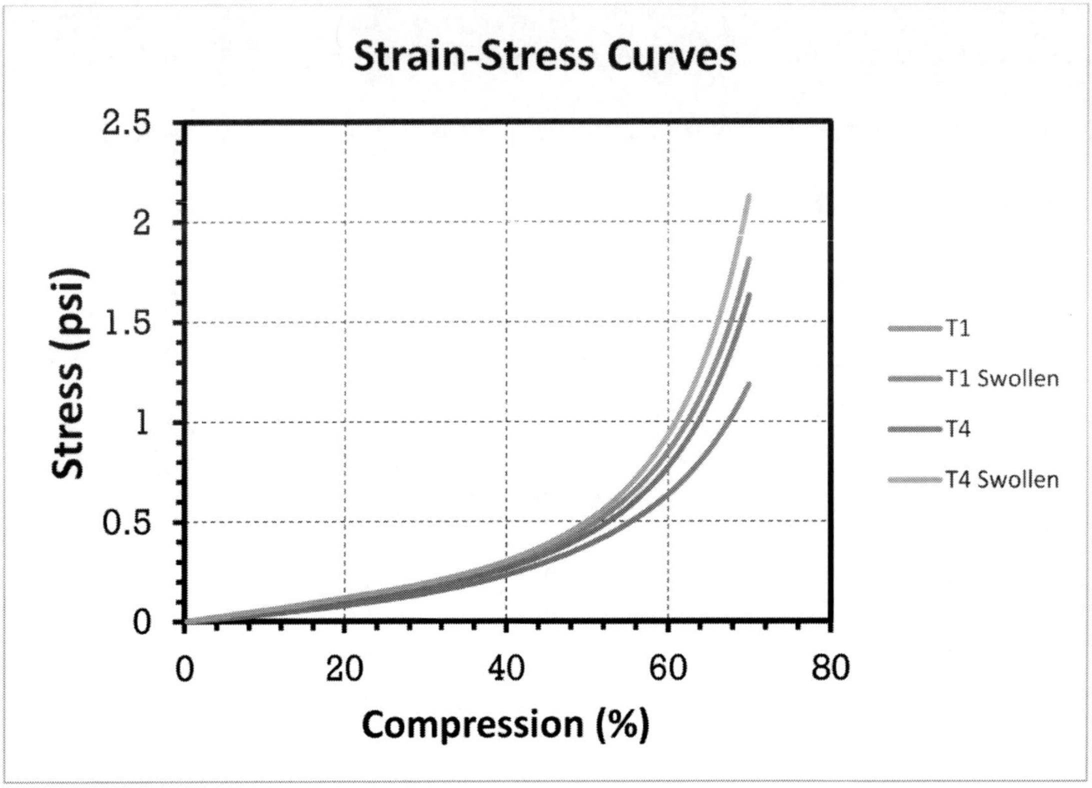

Figure 2. Stress-strain curves

The storage modulus of sample F2 after swelling and subsequent drying was about an order of magnitude higher than that of original sample F2 measured prior to swelling experiment. The storage modulus of the dried F2 sample was also less sensitive to the increasing temperature, contrary to the weakening original sample F2. The increase in the storage modulus of swollen F2 with temperature is simply due to the drying of the sample. The obtained result demonstrated that fly ash can essentially act as a stimuli-responsive material.

The significant improvement in the modulus is believed to be due to the neutralization of the swellable component of the commercial F1, and potential hydrolysis with subsequent neutralization of NBR elastomer. This is confirmed by the shift in T_g to a lower temperature range after drying as illustrated in Figure 3.

Carboxylic groups were detected in the original sample F2 prior to swelling, hence, they cannot be attributed solely to hydrolyzed HNBR, but rather to the super-absorbent polymer present in water swellable rubber compound (see Figure 4). FTIR confirms neutralization of carboxylic groups because a significant increase of the peak near ~1540 cm^{-1} and a lowering of peak intensity at ~1720 cm^{-1} have been observed. This neutralization is believed to promote the formation of ionic clusters in the compound with fly ash considered to be the source of cations. The neutralization of carboxylic groups could as well be attributed to K^+ cations present in the KCl brine solution, however, in that case, the respective peak in FTIR spectra would appear at higher wave numbers, approaching ~1600 cm^{-1}.

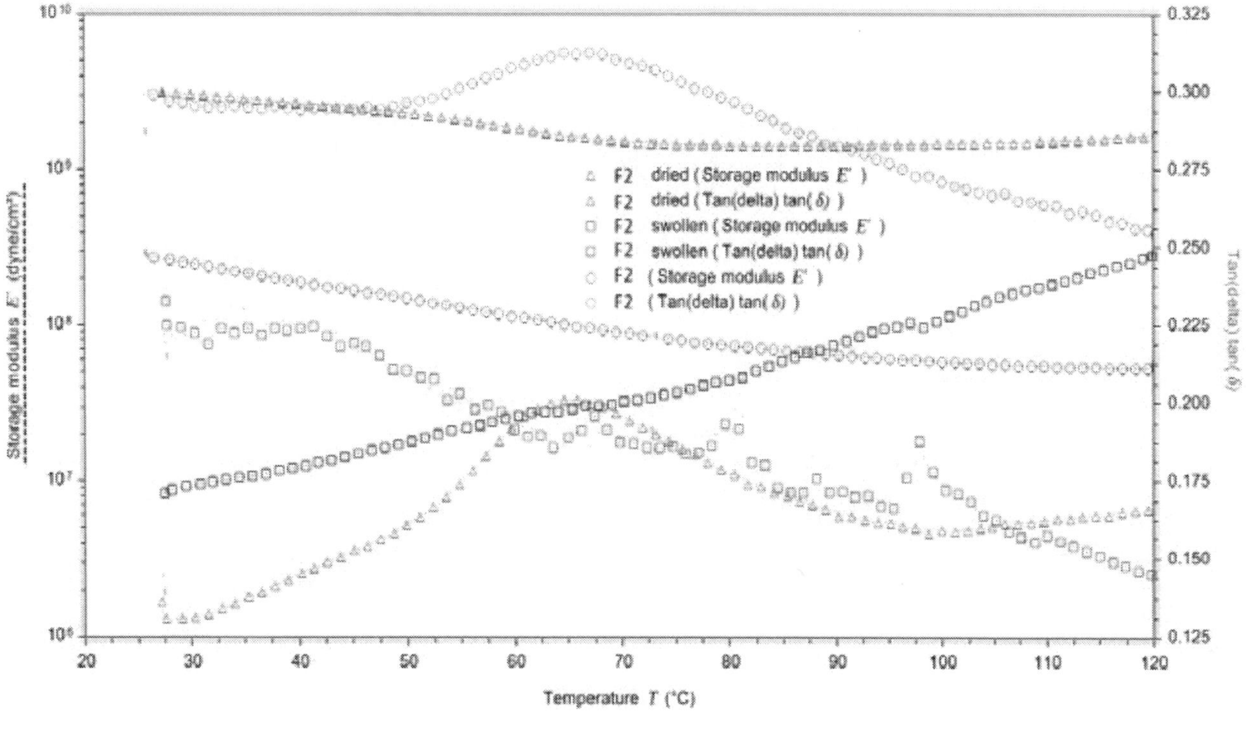

Figure 3. DMA data on F2 sample

It is important to mention that sample F2 exhibited a certain degree of brittleness after drying, which is undesired in rubbers, but this can be managed by incorporation of common ionomer plasticizers.

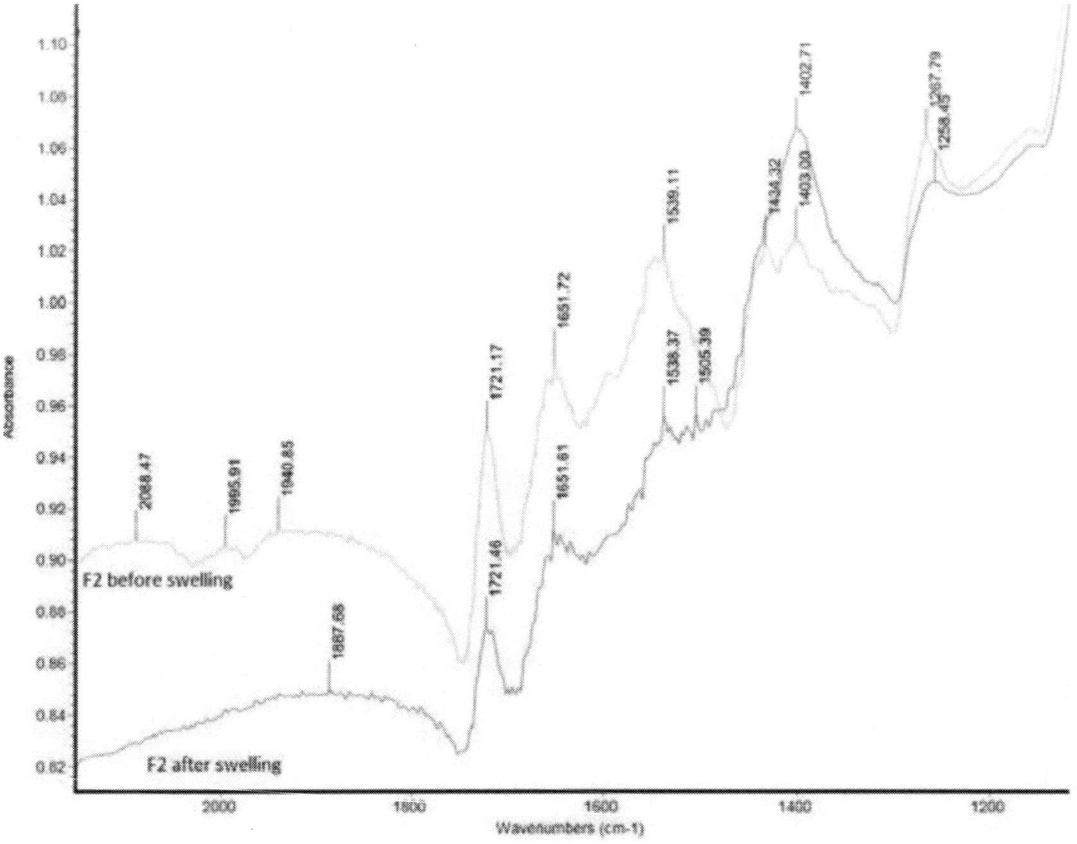

Figure 4. FTIR spectra of F2 sample taken before and after swelling

Conclusions

Fly ash demonstrated a great potential for consideration as cost effective, low reinforcing filler for downhole oil and gas applications. Although the reinforcing properties of fly ash are inferior to carbon black, incorporation of fly ash can improve the mechanical properties of HNBR elastomer compared to the unfilled raw crosslinked elastomer at both room temperature and at 325°F.

Furthermore, composites of HNBR with fly ash Class C demonstrated a stimuli-responsive effect by significantly increasing their modulus upon swelling in water brine. This effect was even more pronounced in composites of fly ash with swellable water compounds, which provided an easy access of filler to water upon swelling. A ten-fold increase in modulus was achieved in such fly ash reinforced sample, after swelling in 3% KCl water brine and drying in the oven overnight. The FTIR confirmed neutralization of carboxylic groups present in the investigated system, which contributed to overall improvement of mechanical properties of the compound upon swelling.

444310 Formal Modeling of Accidents and Automatic Generation of Tutoring Scenarios for Intelligent Safety Training of Process Operators

Abstract:

With the emphasis of accident-free operations and prevention of catastrophic events, safety-training demands for the chemical plants have been on the increase globally. However, the research results on modeling of accidents and involving human operators and the advanced theory and technology for (safety) operator training have been shared limitedly among domain experts and commercial developers. Even with the recent trend of simultaneously delivering the operator training system with the construction of a new process plant, in commercial operator training systems (OTS), safety training used to implement quite limited and obvious scenarios (and the burden was almost up to the trainer), partially due to the difficulty in modeling propagations of probable accidents. By the regulatory requirements or proactive investment by companies, effective scenarios for training of safety and abnormal situation management are necessary to be improved in the aspects of the size of volume as well as the training depth. Thus, considering the emergence of enhanced virtual reality technologies and inexpensive gadgets like HMD, supporting the immersive training environment, the future efficiency of the training heavily relies on the quantity and quality of safety scenarios, which would be essential in guiding the entire training process.

In this research we first overview a government-sponsored research project developing an enhanced safety training system for chemical process operators, using haptic devices in addition to enhanced audio and visual equipment. The initial version of the system runs with a fixed set of training scenarios developed for predetermined accident scenarios. However, with a limited set of training scenarios, the trainees simply memorize and rote learn the steps, required to be mastered to proactively solve the problems with an advent of possible accidents. In addition to that, without formal modeling, the system cannot give timely feedbacks and the performance of the trainee cannot be measured in real-time. Then to enhance the contents and usefulness of the training, the knowledge of process operations is collected and systematically stored as a form of knowledge base, in addition to the accident scenarios obtained by executing Fault Tree Analysis and HAZOP study.

By using a formal modeling language like UML and Petri nets, the accident scenarios and operators' problem-solving tasks are modeled hierarchically and classified into distinguishable subtasks. Then a new training scenario is generated on-line through automatic synthesis based on planning using the knowledge base. Safety training scenarios are generated intelligently for the tutoring of a single operator's job function as well as effective collaboration among a group of operators for real-scale malfunction diagnosis and intervention at the initial propagation stage of an accident. As case study, safety trainings for city gas governing stations are executed based on systematic scenario generation and analyzed for improved effectiveness and field applicability of the system. Reduction of time and cost in the training scenario production and real-time customized training based on new operator's level of knowledge and skill are also discussed in the framework of intelligent tutoring systems.

444376 Ozone Source Apportionments of Local Anthropogenic Emissions

Abstract:

Ozone is one of six principal air pollutants designed by US EPA. NOX and VOC emissions from anthropogenic sources are the most important factors to control ozone pollutant.

In this report, photochemical dispersion models are used to estimate the effect of ozone from local anthropogenic emission sources. EKMA-type diagrams are generated for NOX- and VOC- effects. Furthermore, APCA (Anthropogenic Precursor Culpability Assessment) technology is used to estimate ozone source apportionments.

EKMA-type diagrams show that 10-30% emission reductions could reduce ozone concentrations to meet standard. Further APCA results show the average ozone contributions from regions and emission categories. APCA results explain different effects for different monitoring stations.

444397 Flare Minimization and Emission Characterization during Start-Ups of Chilling Train in Ethylene Plants

Abstract:

Flaring is a crucial safety operation to protect chemical plant personnel and equipment during plant start-up. However, it generates huge amounts of off-spec products, and causes negative environmental and social impacts, as well as results in tremendous raw material and energy losses which could be saved to generate much more needed products. Thus, cost-effective start-up flare minimization strategies which is a double-win practice, simultaneously benefiting both the industrial and environmental sustainability, are becoming more and more important and attractive to the industries.

The chilling train is a very important section for the ethylene plants and accounts for more than 50% of the total start-up time and flare emissions of the whole ethylene plants. Within this section, the charge gas from the upper stream process is refrigerated to a very low temperature (lower than -160 °C) to separate hydrogen and methane. Meanwhile, the left charge gas is liquidized for cryogenic separation. The chilling train system consumes over 40% of the refrigerating capacity of an ethylene plant. During the chilling train start-up, different operational conditions with different compositions and flowrate of precooling process will affect both product separation efficiencies, start-up stability time and flare emission.

Traditional start-up scenarios are developed based on experience and improved by trial and error, which is inefficient and dangerous. In this study, a general methodology on flare minimization for chilling train start-up operations via rigorous dynamic simulation is presented. The methodology starts with setup and validation of chilling train steady-state and dynamic simulation models. The initial state of start-up in the real plant is achieved through systematically transforming from the validated dynamic model, and several operation conditions with different compositions and flowrate of precooling stream are examined to identify an optimal initial state of start-up based on the obtained initial state. Thereafter different flare minimization strategies for the chilling train start-up are conducted on the optimal initial state model. Also, emission source characterization under different flare minimization strategies for chilling train start-up is investigated. Deep insights of the start-up strategies, emission source distribution and dynamic emission profiles are provided. This study provides detailed technical support for both the industry and environmental agencies on evaluating and developing cost-effective flare minimization strategies in the future.

Experimental Study on Najas Minor of Producing Bio-Oil

Abstract: Due to the diminishing of petroleum reserves and the environmental degradation caused by consumption of fossil energy , many attentions have been paid to the application of biomass energy which is acceptable as an abundant renewable energy resource. Najas minor grows in dense clusters with highly branched stems of which a small fragment could propagate into a thrive one.In this study, we researched the properties of Najas minor to make biofuels .Firstly,the najas minor after heavy metal ions degradation were fed into pyrolyser to prepare biofuels. Significant differences of the product was observed compared with only Najas minor pyrolysed.
KEY WORDS: Najas Minor; catalytic pyrolysis; bio-oil

1.Introduction

Energy demands and environmental problems caused by fossil energy have been a matter of growing concern.Bio-oil has been considered to be a promising way for producing renewable bio-fuels or chemicals[1-3].In recent years,many countries pay attention to the importance of biomass that is an abundant renewable energy resource.Biomass can be defined as any organic material mainly consisting of carbon, hydrogen,oxygen, nitrogen, and some other components in small proportions[4].Although the biomass has been studied for nearly a century,most studies mainly concentrated in Lemna minor, Lemna gibba, Spirodela polyrhiza and Landoltia punctata.We still know little about the biological characteristics and potential application value of other varieties of biomass[5].Najas minor,a submersed herb,grows in dense clusters with highly branched stems of which a small fragment could propagate into a thrive one.Najas Minor is native to Europe, Asia and North Africa from the Netherlands to Morocco east to Japan and the Philippines,including China,Siberia, Central Asia, Iran, Turkey, Ukraine, Germany, France Italy and a host of other Countries.

Catalysts used for catalytic cracking are basically divided into the following categories: molecular sieve, alkali metal salt, metal oxide, etc.Different catalysts have different catalytic effects.The use of metal nanoparticles has shown to be very important in recent industrial applications[6].The application of precious metal nanometer materials in catalysis,which has caused the extensive concern of researchers in recent 30 years,has become a hot research topic in the field of catalysis.Gold nanoparticles have received a great deal of attentions due to their special optical and electrical performances, catalytic properties and remarkable quantum size effect, surface effect etc[7]. The synthesis of metal nanoparticles can be classified broadly into wet and dry methods[8]. The cost of production is relatively high both materially and environmentally. There is an unequivocal need to develop more cost-effective and environmental benign methods to alternate these existing methods[9].Researchers in this field turn to use organisms to produce inorganic nanostructures extracellularly[10].

Although there has been a large volume of work on the synthesis of gold nanoparticles[11-13], and obtainning bi-oil by abundant biomass, there are relatively fewer attempts to combine these two methods.In this work, gold nanoparticles were obtained by aqueous chloroauric acid solution with the of Najas Minor at room temperature. The direct pyrolysis and catalytic pyrolysis experiments of Najas Minor were carried out to observe the effects of catalysts on pyrolysis process of Najas Minor.

2. Materials and Methods

2.1. Experimental samples

The Najas Minor collected from Tianjin city and the solutions of Au(III) were used in this study.The stock solutions of Au(III) were prepared by dissolving analytical grade chloroauric acid

($AuCl_3 \cdot HCl \cdot 3H_2O$, Kewei Reagent, China) in deionized water.

2.2.Experimental apparatus and procedure

Pyrolyzer with GCMS,shown in Fig. 1,is mainly used for analyzing the composition of polymer.Pyrolyzer with GCMS includes three parts :quartz pyrolysis tube,pyrolyzer and GCMS.

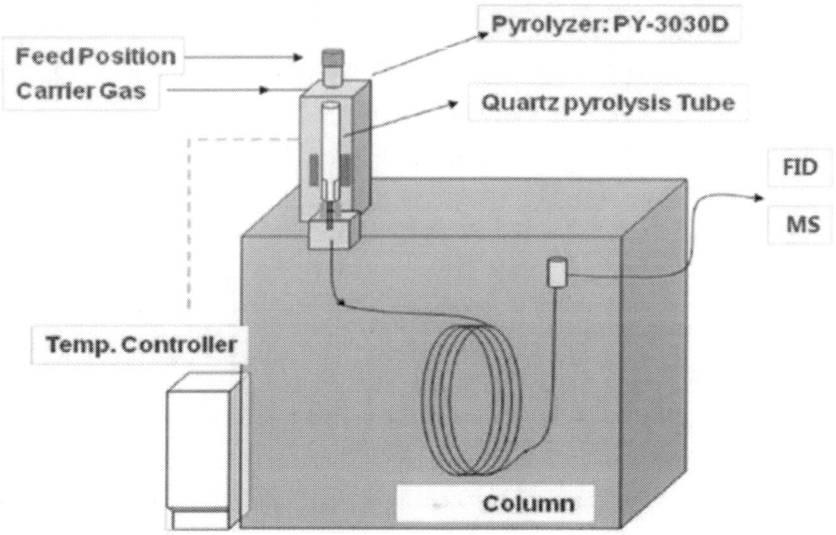

Fig. 1 Diagram of Pyrolyzer with GCMS

The quartz pyrolysis tube (shown in Fig.2) length is 12cm.The sample end diameter is 5 mm,and the pyrolysis products end diameter is 3mm.This special structure is beneficial to the fixation of the reactant in the tube.The top and bottom of the reactants are both supported by quartz wool to prevent reactants blowed by the carrier gas ,or fine particles into the chromatographic column.All pyrolysis experiments were carried out on the EGA/PY-3030D pyrolyzer(Frontier Lab,Japan) .The samples were filled in the special quartz pyrolysis tube to reduce the amount of valuable catalyst and reduce the cost of research work.

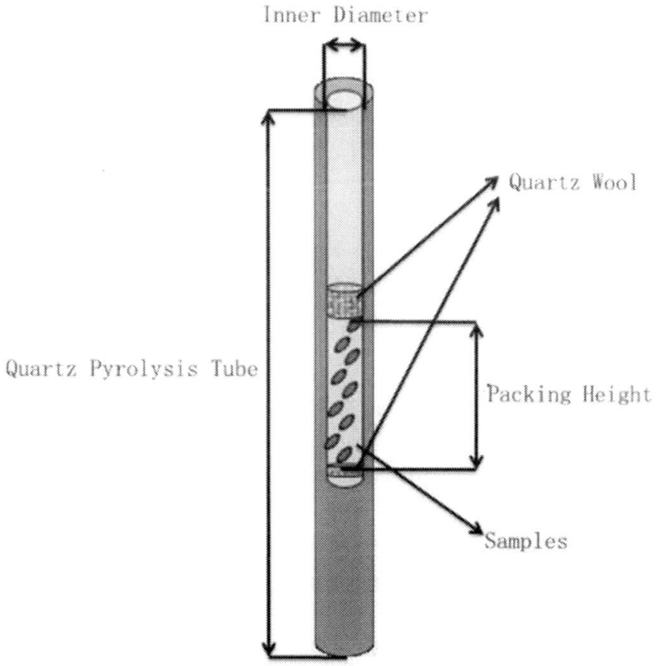

Fig. 2 Diagram of Quartz Pyrolysis Tube

Approximately 0.5g Najas Minors were used in each experiment.The samples were impregnated in 50ml deionized water and 50ml solution of chloroauric acid(0.0793mmol/L) for a night,followed by drying at 80℃ until the weight didn't change.Each experiment used 2.5mg Najas Minors as reactants.Each sample was heated, under the control of heating controller, from ambient temperature to 600 ℃ repeatly and kept at 600 ℃ for 2 h.

3. Results and discussion

3.1.MS Analysis of the Products during the Pyrolysis

Under the same pyrolysis conditions, bio-oil obtained by direct pyrolysis and catalytic pyrolysis were analyzed by GCMS,and the main chemical compounds were shown in Table1 and Table2.

Table 1 Indentification and quantification of main chemical compounds by direct pyrolysis

NO.	peak area%	Formula	Compound
1	12.67	$C_3H_6O_2$	2-Propanone, 1-hydroxy-
2	3.67	C_7H_8	Toluene
3	3.28	C_7H_8O	p-Cresol
4	3.81	$C_5H_{10}O$	Pentanal
5	3.53	C_8H_8O	Benzofuran, 2,3-dihydro-
6	2.17	C_8H_8N	Indole
7	5.93	$C_{20}H_{40}O$	3,7,11,15-Tetramethyl-2-hexadecen-1-ol
8	2.52	$C_{22}H_{42}O_2$	Phytol, acetate
9	5.25	$C_{16}H_{32}O_2$	n-Hexadecanoic acid
10	8.99	$C_{18}H_{30}O_2$	9,12,15-Octadecatrienoic acid, (Z,Z,Z)-

Table 2Indentification and quantification of main chemical compounds by catalytic pyrolysis

NO.	peak area%	Formula	Compound
1	6.01	C_7H_8O	p-Cresol
2	1.75	$C_7H_7NO_2$	3-Hydroxypyridine monoacetate
3	4.58	C_8H_7N	Indole
4	3.05	$C_{10}H9N$	1-Phenyl-1-cyclopropanecarbonitrile
5	3.90	$C_{12}H_{22}O_{11}$	.alpha.-D-Glucopyranose, 4-O-.beta.-D-galactopyranosyl-
6	8.04	$C_{20}H_{40}0$	3,7,11,15-Tetramethyl-2-hexadecen-1-ol
7	2.82	$C_{20}H_{40}$	2-Hexadecene, 3,7,11,15-tetramethyl-, [R-[R*,R*-(E)]]-
8	3.99	$C_{15}H_{28}$	1-Pentadecyne
9	2.12	$C_{15}H_{28}$	3,7,11-Trimethyl-2,4-dodecadiene
10	8.75	$C_{16}H_{32}\,0_2$	n-Hexadecanoic acid

3.2 discussing

According to the chemical category and common characteristics, the products were divided into hydrocarbons, alcohols, acids, ketones, nitrogen compounds and other compounds.Nitrogen compounds detected by GC/MS analysis ,mainly come from the pyrolysis of amino acids in Najas Minor.Table 3 showed the distribution of pyrolysis products.As shown in the table,the hydrocarbons ,alcohols and nitrogen compounds in biological oil of catalytic pyrolysis were significantly increased compared with direct pyrolysis.The content of hydrocarbons increased from 12.4% to 34.6%.The content of alcohols increased from 9.1% to 16.6%.The gold nanoparticles loaded in surface of Najas Minor, was conducive to the generation of hydrocarbon and alcohol in pyrolysis.By analyzing the GC/MS results, the main hydrocarbons in bio-oil by catalytic pyrolysis were long chain alkanes,and olefins.The main components in gasoline and diesel is hydrocarbons. As a result, whether the biomass can be effective as future power fuel oil, depends largely on the hydrocarbon content of bio-oil after further treatment.

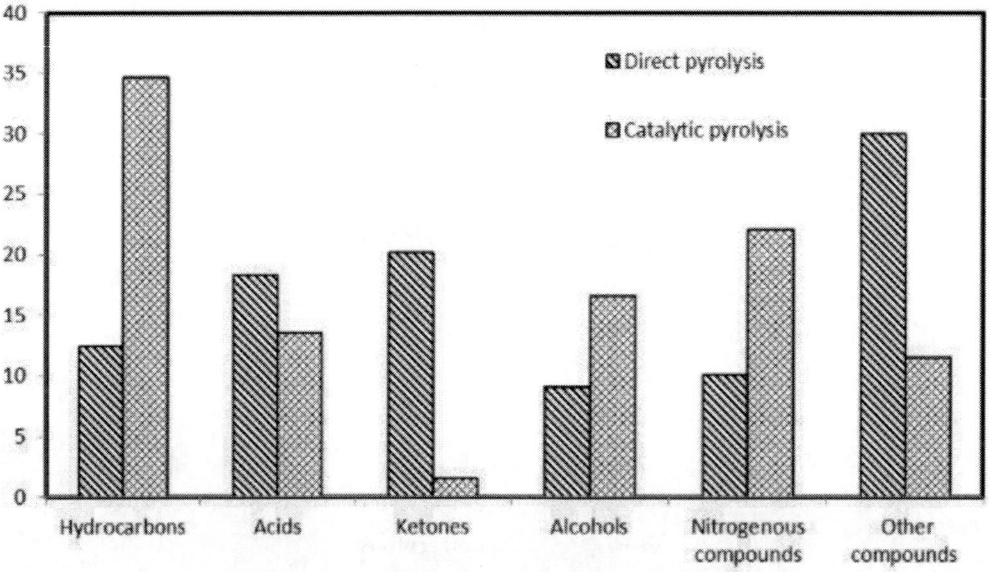

Fig.3 Distribution of pyrolysis products

4.Conclusion

Biomass was explored as a potential biomass energy resource. The experimental study on the direct and catalytic pyrolysis of Najas Minor had been presented, using Pyrolyzer with GCMS,to observe the catalytic effects of gold nanoparticles.In this study,remarkable catalytic effects for pyrolysis were observed.The pyrolysis of Najas Minor with gold nanoparticles,provided more hydrocarbons, alcohols and nitrogen compounds.

参考文献:

[1]Gong F, Yang Z, Hong C, et al. Selective conversion of bio-oil to light olefins: Controlling catalytic cracking for maximum olefins[J]. Bioresource Technology, 2011, 102(19):9247-54.

[2]Zhang Z, Bi P, Jiang P, et al. Production of gasoline fraction from bio-oil under atmospheric conditions by an integrated catalytic transformation process[J]. Energy, 2015, 90:1922-1930.

[3]Hui S, Yuqi D, Jinzhao D, et al. Transesterification of sunflower oil to biodiesel on ZrO2 supported La2O3 catalyst.[J]. Bioresource Technology, 2010, 101(3):953–958.

[4]Yaman S. Pyrolysis of biomass to produce fuels and chemical feedstocks[J]. Cheminform, 2004, 35(31):651-671.

[5]Hejný S. Ellias Landolt The family of Lemnaceae—A monographic study—Vol.1 biosystematic investigations in the family of Duckweeds (Lemnaceae) (vol. 2)[J]. Folia Geobotanica Et Phytotaxonomica, 1993, 28(1):50-50.

[6]Gardea-Torresdey J L, Tiemann K J, Gamez G, et al. Gold Nanoparticles Obtained by Bio-precipitation from Gold(III) Solutions[J]. Journal of Nanoparticle Research, 1999, 1(3):397-404.

[7]Marie-Christine D, Didier A. Gold nanoparticles: assembly, supramolecular chemistry, quantum-size-related properties, and applications toward biology, catalysis, and nanotechnology.[J]. Chemical Reviews, 2004, 35(16):293-346.

[8]Liu B ,, Xie J ,, Lee J Y, et al. Optimization of high-yield biological synthesis of single-crystalline gold nanoplates.[J]. Journal of Physical Chemistry B, 2005, 109(45):15256-63.

[9]Raveendran P, Fu J, Wallen S L. Completely "Green" Synthesis and Stabilization of Metal Nanoparticles[J]. Journal of the American Chemical Society Jacs, 2003, 125(46):págs. 13940-13941.

[10]Absar Ahmad , Satyajyoti Senapati , MIK, et al. Extracellular Biosynthesis of Monodisperse Gold Nanoparticles by a Novel Extremophilic Actinomycete, Thermomonospora sp.[J]. Langmuir, 2003, 19(8):3550-3553.

[11]Selvakannan P, Mandal S, Phadtare S, et al. Water-dispersible tryptophan-protected gold nanoparticles prepared by the spontaneous reduction of aqueous chloroaurate ions by the amino acid[J]. Journal of Colloid & Interface Science, 2004, 269(1):97-102.

[12]Gittins D I, Caruso F. Biological and Physical Applications of Water-Based Metal Nanoparticles Synthesised in Organic Solution[J]. Chemphyschem, 2002, 3(1):110 - 113.

[13]And A G, Murphy C J. Seed-Mediated Synthesis of Gold Nanorods: Role of the Size and Nature of the Seed[J]. Chemistry of Materials, 2004, 16(19):3633-3640.

444353 A Combined Rheology and Stability Study on Emulsified Acid for High Temperature Applications

Abstract:

HCl emulsion has been a popular formulation to stimulate carbonate reservoirs due to the various advantages that emulsified HCl has over ordinary HCl; including but not limited to deeper penetration and less corrosion damages. Emulsified HCl has even more potential advantages if utilized properly. The reaction kinetics, stability and rheology of emulsified HCl plays a significant role on achieving successful acidizing treatment.

The objective of this paper is to experimentally study the factors affecting the stability of emulsified HCl. The emulsion was prepared using 15wt% HCl and diesel with cationic emulsifier. The acid solution-to-diesel ratio was 70:30. The effect of stirring intensity, concentration of the emulsifier, addition of corrosion inhibitor, temperature, addition rate of acid phase to oil phase and mixing time of the emulsion have been studied. Stirring intensity was varied from 600 to 1500 rpm. Emulsifier dosage ranged between 5 to 20 gpt. Temperature used for heating, to mimic reservoir condition, was as high as 120°C.

The results showed that there is an optimum condition for emulsifier dosage and for mixing time of the emulsion. The optimum emulsifier dosage was found to be 10 gpt with 5 minutes of mixing time. As for the addition rate of acid phase to oil phase, 1.4 ml/min of drop-wise addition was found to give a very stable emulsion. The effect of stirring intensity on the emulsion stability was proportional, the higher the rpm speed the better the stability. Corrosion inhibitor, on the other hand, played a role as a destabilizer. With regards to heating temperature, it was found out that the higher the temperature, the faster the emulsion breaks.

The novelty of this study is the comprehensive approach in combining stablity and rheology of emulsified acid for high temperature applications. The particle size analysis and solution morphology help in better understanding HCl emulsion and how it works.

COMPARATIVE ANALYSIS OF MOLECULAR STRUCTURE IDENTIFIABILTY BASED ON SIGNATURES AND DESCRIPTORS

Zelimir F. Kurtanjek
University of Zagreb, Croatia

Abstract

This work is focused on molecule computer representation aimed for molecule predictive model (QSAR) and inverse modelling. Two different approaches are compared, i.e. the molecule structure coding based on graph theory (Faulon et al.[1] extended valence) and the chemical based descriptors[2]. Applied are 1-3rd level molecule graph fragments and 630 2D molecular descriptors. The molecule test sets are alkanes, alkenes, acetones, aromatics, organic acids and halogenated hydrocarbons in the range of C_1-C_{12}, selection of biomolecules (including aminoacids) and a set of binary ionic liquids (cations: imidazole, pyridinium, quinolinium, ammonium, phosphonium). The comparative comparison of molecule description based prediction potential are statistically evaluated by octanol-water (PWA) partition coefficients (for hydrocarbons), viscosity of ionic liquids and biological impact on acetycholinesterase inhibition. The evaluation is based on two distinct levels. On the first level the comparative analysis based on internal and external correlation between "scores" in multidimensional spaces of the separate first, second and third order valance spaces, and the molecular signature space. Firstly, the potential of information content applicable for inverse modeling is tested on ability for molecule classification by use of linear discrimination (LDA) based on the singular value decomposition (SVD) of corresponding molecular fragments and descriptors. For chemical-biological property inverse modelling proposed is application of constrained genetic algorithm (GA). Tested are linear structure multivariate (chemometric) models such as principal component regression and partial least squares and nonlinear decision trees (random forest). Applied are singular value decompositions for reduction of the molecule signature and the descriptor spaces up to the significance level of p=0.05. The selected level of significance is presumed to be in coherence with experimental data of molecule physical and chemical properties. The main focus is on extraction of the main molecular fragments (partial graphs extracted from molecular signatures) and specific molecule descriptors. Importance of application of open source software modules provided by Bioclipse[3] and R software packeges[4] for computational-statistical numerical algorithms is emphasized.

Key words: Molecule signatures, descriptors, inverse modelling, decision tree

Introduction

Advancement of high-throughput experiments have lead to „big data" databases applied in pharmaceuticals, medicine, new material science, catalysis, and new industrial processes. The role of chemoinformatics as a science for computer encoding of molecule properties and algorithms for data mining and inference are the essential steps in the development. Computer encoding of molecular properties has been a part of chemical engineering from the beginning of process design. It is mostly based on molecule group contributions, but has been extended by application of molecular descriptors (1-2-3 D), molecule signatures, and application of molecular quantum mechanics. Modeling is faced with the problem of „high dimension curse". For example, about 5 000 various molecular descriptors can be determined for each molecule, although most of them are highly correlated. It results in an ill-conditioned problem of modelling with large uncertainties in parameter and poor model prediction accuracy outside modelling training sets. Great improvement has been achieved by application by use of model ensembles, like random forest of many individual decision trees. These models are highly flexible, can be trained on discontinuous and nonlinear functional relations, can encompass various data formats like numerical and categorical variables, and due to large number of randomly selected individual predictors in principle do not have tendency to overfitting. However, when the object of modelling is extraction of important molecular features, due to autocorrelation between descriptors random forest, produced lists of importance are large sets of poorly distinguishable variables. The focus of this work is to apply feature regularization for optimization and reduction of list of variable importance. Compared are two methods for structure encoding, molecular descriptors and molecular signatures of first and second order.

Modeling and Factor Regularization

In this work are selected two distinct data sets of organic molecules from open source data bases (J. Sangster, 1989) and ionic liquids ILs (The UFT/ Merck and NIST, 2015), and lists of the selected molecules are given in Table 1. As target properties for organic molecules are selected the corresponding molecular classes and partition octanol-water (POW) coefficients, and dynamic viscosity in the range of 30-80 ^{0}C of ILs is selected The molecules are coded into SMILES and used as input lists into software for evaluation of molecular signatures. Most popular open source software tools are DRAGON (R. Todeschini, et al., 2009) and PADEL (Chun Wei Yap, 2010). Here are used 1-2 D descriptors generated by PADEL. For the same list of molecules are determined first and second order molecular signatures or extended valances (J.P. Faulon, 2003). The signatures are two dimensional graph based structural molecular fragments. The vertices are atoms and the edges are corresponding bonds (lengths or heights), and are here only used of the first and second level. A molecule is characterized by a complete set of signatures as input data for modelling. Since the models are based on large number of variables (molecular features) possibly lager than number of samples (molecules), usually denoted as p>>n models. They are intrinsically „ill conditioned" leading to good predictions on train data but very poor predictions on test data. Also, produced list of variable importance are very dense "clusters are not distinguished" with large uncertainties (broad confidence intervals). This makes the problem of inverse modelling aimed to predict a set of candidate molecules for given set of target properties, such as in a case of high dimension combinatorial task for selection of ILs cations and anions, very unreliable. In order to reduce model dimensionality and improve robustness and parsimonious of important molecular features here is applied modelling with feature regularization. The model is based on ensemble of random trees (CART models, L. Breiman et al. 1998) with included feature regularization.

Data sets of organic molecules	**Range of number of atoms**				
Molecular classes	C	H	O	Cl	F – I - Br
Alkanes	5-12	12-30			
Alkenes	4-9	6-18			
Acetones	6-24	6-26			
Halogenated hydrocarbons	3-10	6-22	1		
Organic acids	1-16	2-32	2		
Esters	4-11	6-22	2		
Aromatics	1-12	1-21		3	2 – 2 - 1
Data sets of ionic liquids (ILs)					
Cations	**Anions**				
1,3-Dimethylimidazolium (+)	PF_6 (-) Cl (-) CH_3COO (-)				
1-Methylpyridinium (+)	BF_4 (-) $MeSO_4$ (-) $EtSO_4$ (-)				
1,1-Dimethylpyrrolidinium (+)	Tf_2N (-) CF_3SO_3 (-)				

Table 1. The molecular groups of the organic and ionic liquid molecules selected for random forest models and extraction of the most important molecular features. The properties for the organics are obtained from J. Sangster (1989) and for the ionic liquids from Merck and NIST data bases (2015). Molecular descriptors and signatures of ILs ions are recorded separately and adjoined as a single record.

The modelling is formulated as an optimization (functional and multivariable minimization) problem with two parts of the objective function (T. Hastie et al., 2009).

$$\min F_{object.}(X,T,\omega) = SSE(y,\hat{y}) + \phi(T,\omega) \tag{1}$$

The first part is the sum of square errors (SSE) based on difference between observed y and prediction $\hat{y}$ based on input data by an ensemble T of interrelated CART models (L. Breiman et al. 1998), The second part is a selected metrics ϕ for the model complexity depending the sum of leaf probabilities for population of individual CART models and their total number N_T .

$$\phi(T,\omega) = \gamma \times N_T + \frac{1}{2} \times \lambda \times \sum_i^{N_T} \omega_i^2 \tag{2}$$

The ensemble output prediction is a sum of predictions of individual T_i tree CART models:

$$\hat{y} = \sum_i^{N_T} T_i(X) \tag{3}$$

A very efficient open source code Extreme Gradient Boosting XGB (T. Chen, 2016) is available in R language (R Developm-ent Core Team, v. 3.2.3, 2015). The optimization algorithm is consecutive, starting with first CART model which is appended with a next tree boosted by errors in predictions of previous tree. At each step optimization is solved by linearization of the objective function to quadratic terms. Number and structure of trees as well as the parameters for regularization need to be found by trial and error and minimal number of trees by folding and cross validation error on the training set. The critical step is validation on untrained set to have acceptable model error and avoid model overfitting with minimal model complexity.

Results and Discussion

Tested are models for prediction of the seven molecular classes (Table 1.) based on molecular descriptors and molecular signatures (first and second level). Results for molecular descriptor feature importance are shown on Fig. 1. The effect of regularization is very pronounced. Out of hundreds of descriptors for the molecular classification needed are only few, the most important being BCUTw.1h. On contrary, when an ensemble of independent CART trees is applied, i.e. random forest model, level of accuracy of prediction is the same, but the list of variable importance is densely populated with numerous descriptors.

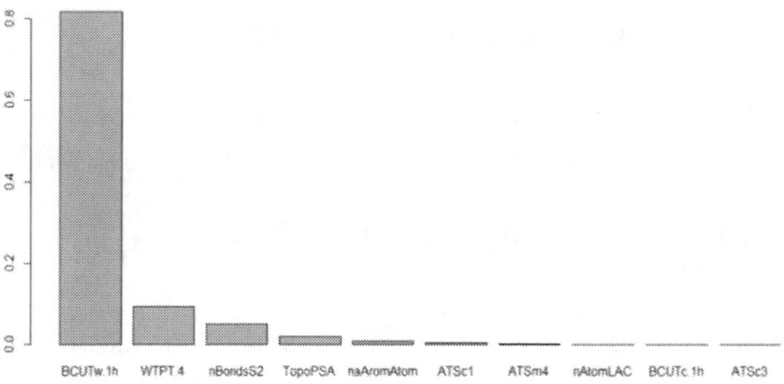

Fig 1. Variable importance molecular descriptor factors for the inverse modeling of the selected molecular classes. The first five descriptors are: BCUTw.1h (based on a weighted version of the Burden matrix which takes into account both the connectivity as well as atomic properties of a molecule), WTPT.4 (Randić's weighted path), nBondS2 (bond count), and naAromAtom (number of aromatic atoms).

The same classification problem was tested by the first and second order molecular signatures. The important signatures are depicted on Fig. 2-3. Accuracy of prediction on trained sets of molecules compared to descriptors is about the same, though the importance factors are not as distinctive as in case of the descriptors.

significantly better when molecular descriptors are applied. The

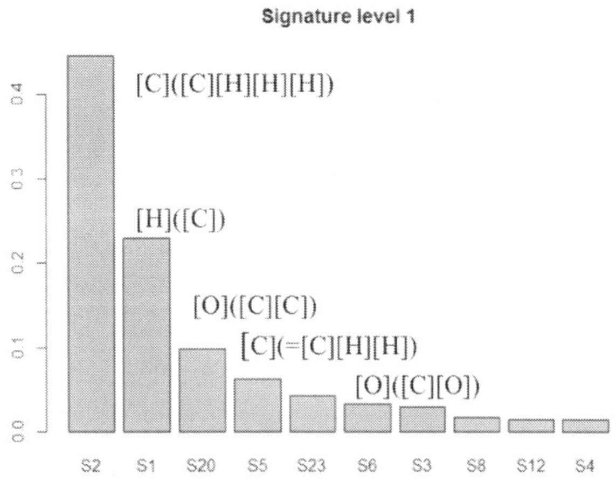

Fig 2. Relative importance of the individual first level signatures for inverse prediction of the selected molecular classes. The first five most important signatures are coded according to SMILES (Simplified Molecular Input Line Entry System).

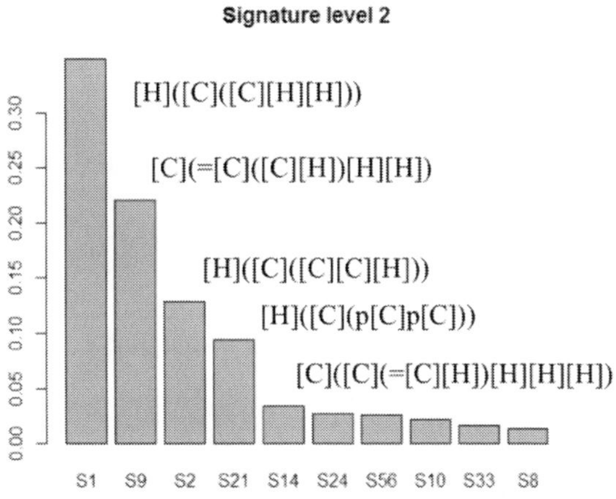

Fig 3. Relative importance of the individual second level signatures for inverse prediction for the selected molecular classes. The first five most important signatures are coded according to SMILES (Simplified Molecular Input Line Entry System).

However, model accuracy on untrained set of molecules, depicted in Fig. 4, is significantly lower for the model with molecular descriptors compared to signatures. The average error with descriptors is about 5% while for signatures about 11%. The "box and whiskers" plots are obtained by error evaluation of ten randomly generated untrained sets.

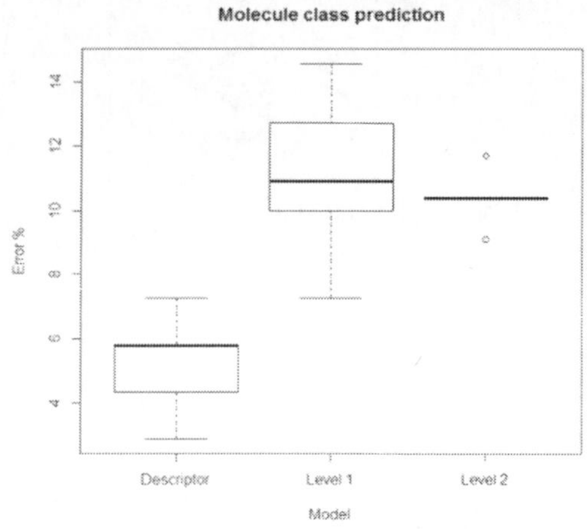

Fig 4. "Box and whiskers" plot for the model validation (relative errors) for inverse prediction of the molecular classes based on the set of untrained molecular descriptors and the first and second level of molecular signatures.

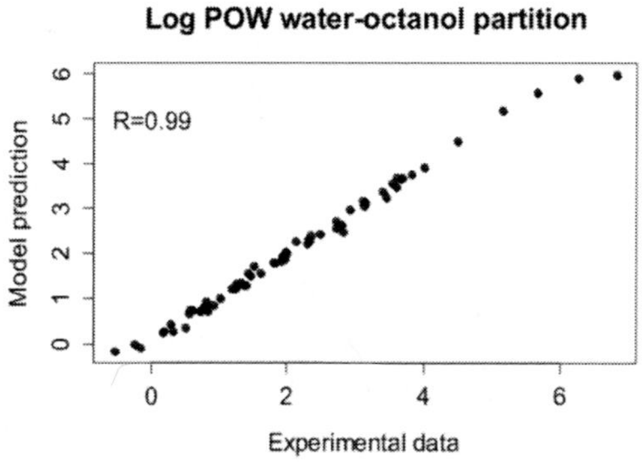

Fig 5. Predictions of the partition octanol-water coefficients POW, for the specified molecular classes of organic molecules, are based on the random forest model with the molecular descriptors.

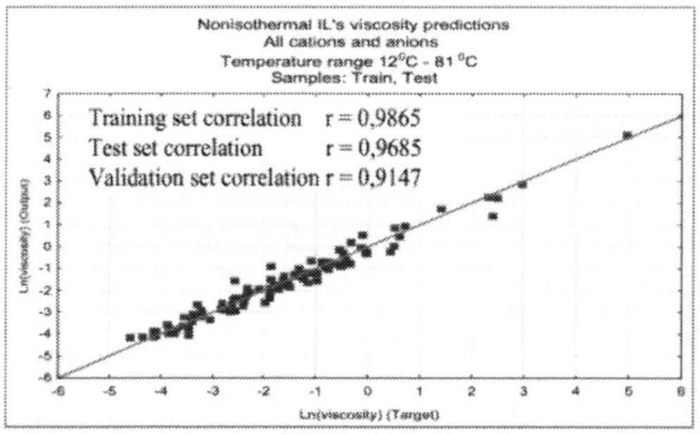

Fig 6. Predictions of dynamic viscosities for the specified groups of ionic liquids (ILs) based on the random forest model with the molecular descriptors. The viscosities cover the temperature range from 12 – 81 ^{0}C and five orders of magnitude.

In Fig. 4 can be observed that dispersion of errors (quantiles) for the second level molecular signatures is negligible, though two outliers are present.

For prediction of molecular properties applied are random forest models based on molecular descriptors. For all of the organic molecules, included in the seven classes, evaluated are octanol-water partition coefficients POW, shown in Fig. 5. The range of POW values is very high, on log scale fro 0 to 6, but the obtained accuracy of the model predictions is high (Pearson correlation R=0.99). The same approach is applied for prediction of ionic liquid viscosity data in temperature range from 12 to 81 ^{0}C. Since temperature effects on viscosity are intrinsically nonlinear, they represent a modelling challenge. However, the decision trees are very efficient for accounting nonlinearities, though nonlinearity is reflected in non-orthogonal decision planes, rendering trees with significantly increased number of branches. For the range of 6 order of magnitude the accuracy of prediction on trained set is R = 0.98 and on untrained set R = 0.91.

Conclusions

Compared are ensemble decision trees models for molecular feature class and property prediction based on extensive list of molecular descriptors and first and second level of extended valances or signatures.

The aim was to test effects of regularization on ensemble of CART decision tree models. Proven is effectiveness of the extreme gradient boosting algorithm in cases with extensive list of (molecular) features.

Obtained is increased accuracy on untrained sets of data, and more importantly, obtained are parsimonious lists for ranking of variable importance lists. This enables further significant reduction in development of robust and simple decision tree models.

The reduction in number of molecular features applied for decision tree models may greatly simplify the challenging problem of inverse modeling based on molecular descriptors and/or signatures.

References

1. J. Sangster, J. Phys. Chem. Ref. Data, **18**(3)(1989) 1111-1227
2. The UFT/ Merck Ionic Liquids Biological Effects Database, www.il-eco.uft.uni-bremen.de
3. NIST, Ionic Liquids Database Standard Reference Database #147, <ilthermo.bouder.-nist.gov /ILThermo/mainmenu.uix>
4. R. Todeschini and V.Consonni: "Molecular Descriptors for Chemoinformatics", (2 volumes), WILEY-VCH,Weinheim (Germany) 2009.
5. Chun Wei Yap, PaDEL-Descriptor: An Open Source Software to Calculate Molecular Descriptors and Fingerprints, J. Comp. Chem., (2010):**32**: 1466-1474.

6. J.-L. Faulon, D.P. Visco Jr., R.S. Pophale, The signature molecular descriptor. 1. Extended valence sequences vs. topological indices in QSAR and QSPR studies, J. Chem. Inf. Comput. Sci. **43** (2003) 707–720.
7. J.-L. Faulon, C.J. Churchwell, J.D.P. Visco, The signature molecular descriptor. 2. Enumerating molecules from their extended valence sequence, J. Chem. Inf. Comput. Sci. **43** (2003) 721–734.
8. L. Breiman, J. H. Friedman, R. A. Olshen, C.J. Stone, „Classification and Regression Trees", CRC, Boca Raton, 1998.
9. T. Hastie, R. Tibshirani, J. Friedman, The Elements of Statistical Learning, Data Mining, Inference and Prediction, Springer Verlag, Second Edition, New York, 2009.
10. T. Chen, T. He and M. Benesty (2016). xgboost: Extreme Gradient Boosting. R package version 0.4-3. https://CRAN.R-project.org/package=xgboost
11. R Development Core Team, , R: A language and environment for statistical computing. R, Vienna, Austria,www.R-project.org 2015.

444791 Dissolution and Precipitation of Minerals Under Different Alkali Concentrations in a Reactive CO_2 Batch System

Abstract:

The primary focus of this research is to study the dissolution and precipitation of minerals and salts from Berea sandstone and brine samples taken from the Great Salt Lake in a reactive CO_2 batch system. Various experiments were constructed to measure dissolution and precipitation reaction rates under different alkali concentrations. Changing the alkali concentrations of the Great Salt Lake brine samples produced changes in dissolution and precipitation of minerals from the Berea sandstone and the Great Salt Lake brine samples. Data from ICP-MS was used to quantify mineralogical changes in the samples and Micro-CT was used to measure the porosity change in the Berea sandstone samples. The results from these analytical instruments showed patterns of mineralogical dissolution and precipitation in the post-experimental fluid that correlate to the change in alkali concentrations.

444825 Significance of Energy Intensity and GHG Emission Reduction Towards Sustainability of a Petrochemical Plant

Abstract:

Sustainability is an important pillar in today's petrochemical industry for improving process efficacy by reducing specific energy consumption, greenhouse gas (GHG) emissions, and wastewater. SABIC (Saudi Basic Industries Corporation) like other giant petrochemical companies has set stringent sustainability targets and investing huge resources to transform its plants from the present position to the first quartile in each category.

SABIC is the world largest producer of ethylene oxide and ethylene glycol. For each ton of glycol produced, 0.2 to 0.5 tons of CO_2 is emitted as GHG. The significance of sustainability "in general" while the methodology adopted to reduce energy intensity, carbon footprint, and wastewater "in particular", will be discussed thoroughly in our talk.

444827 Syngas to Ethylene Glycol: A Technology Overview

Abstract:

SABIC is the world largest MEG producer petrochemical company and most of its MEG is being produced through the hydrolysis of EO. Ethylene oxide (EO) is a simplest form of epoxides and due to its molecular structure, acts as a versatile intermediate for a variety of chemicals such as; monoethylene glycol (MEG), di- and triethylene glycols, ethoxylates, ethanol amines, and glycol ethers. The growing demand of MEG owing to its downstream utilization in polyester industry has prompted the research of an efficient, cost effective, and environmentally benign EO production process.

The major feedstock for MEG are ethylene and syngas. The heavy dependence of MEG producers on ethylene and its scarcity in some parts of the world promoted the need of alternate feedstock for highly demanded MEG. The technological developments utilizing syngas as a feedstock will be discussed along with the commercial analysis and future growth in EO/EG business.

444834 Stability and Rheology of Emulsified Acid in Waste Oil: Effect of Emulsifier Concentration, Mixing Speed, Temperature and Droplet Size

Abstract:

Acid in oil emulsion is used in the industry for stimulation purposes. In particular, HCl as the acid phase and diesel as the oil. Emulsifying HCl has numerous benefits over regular HCl. Perhaps the main reason is due to deeper penetration. Apart from that, less corrosion damages are caused since the external phase is hydrocarbon. Several studies showed the success of replacing diesel as an external phase with other hydrocarbon oil, such as crude oil and xylene.

This work utilizes the extra hydrocarbon left unused –or sometimes dumped- from refineries. This unused hydrocarbon is referred to as waste oil. First the chemical composition of waste oil was studied using HPLC, SARA fractionation as well as FTIR. This helped in selecting the proper emulsifier agent for the waste oil. The HCl-waste-oil emulsion was prepared using 15wt% HCl with a ratio of 70:30 of acid-to-oil. Extensive work was done to ensure reproducibility of data. This study was done to experimentally investigate the effects of distinct important variables on thermal stability and apparent viscosity of the HCl-waste-oil emulsion. These variables are; emulsifier concentration, mixing speed and the addition of corrosion inhibitor. All experiments were conducted at a high temperature 120°C. To assess the quality of the HCl-waste-oil emulsion, droplet size was measured as a function of temperature (up to 100 °C).

The results showed that the HCl-waste-oil emulsion is a shear-thinning fluid. Power-law model was applicable seamlessly to all of the apparent viscosity data for all of measured temperatures. The optimum conditions were found to be; 0.5vol% and 600 rpm for emulsifier concentration and mixing speed, respectively. Apart from that, the emulsions prepared with corrosion inhibitor were less stable compared to the ones without it. However, the addition of corrosion inhibitor tended to generate monodispersed emulsion but bigger size droplets for temperatures 25°C and 50°C and polydispersed droplets for higher temperatures. Droplet size measurements showed an average size of 1.47 to 3.09 μm. Overall, the lab results showed a promising potential for the HCl-waste-oil emulsion to be used in the field.

This work takes into account; reducing the budget of acidizing job and at the same time environmental concerns. It makes use of the unwanted hydrocarbon from refiners. And, it saves the environment by using the waste oil instead of dumping it.

444994 Multiscale Simulation of Proteins Motion in the Microchannel

Abstract:

Protein adsorption is an important phenomenon, which plays a prominent role in chemical, industrial and medical areas. In this work, the dynamic behavior of the egg-white lysozyme has been investigated in a micro-channel with a multiscale modeling approach. The Molecular Dynamic simulation, free energy computation and Langevion Dynamics (LD) simulations have been combined to study lysozyme adsorption process on Au (111) surface. The gold surface has strong attraction to proteins. The protein-surface adsorption or desorption free energies is calculated. From the case studies of LD simulations, the surface has effects on the protein dynamic adsorption process. By the comparison of LD, MD and the experiments, the time step is concluded as an important factor to study protein adsorption behavior at the water-surface interface. Fine time step at the scale of 1 picosecond is necessary to provide enough accuracy to resolve proteins' dynamics trajectories.

445004 Grand Challenges in Future Energy and Chemicals: Climate Change and Sustainability As Drivers for New Technology and Innovation

Abstract:

Global demand for energy will double by 2050, driven by increasing population and affluence especially in emerging economies. While fossil fuels will continue to underpin the energy supply for the near future, concern over impact on climate change continues to grow, driving interest in expanded use of renewables and development of carbon capture and sequestration or re-use solutions. In addition to CO_2, water use and air emissions are critical issues to be addressed in future and current energy supply, as is energy storage for effective use of renewables. This presentation will provide an overview of what industry is doing today to address these issues, including the factors and metrics used in decision making.

445029 Electrospun Nanocomposite Fibers for the Removal of Heavy Metals in Water

Abstract:

Uniform electrospun nanofibers with small diameters were successfully prepared by systematically optimizing the electrospinning parameters including polymer concentration, salt loading, applied voltage, tip-to-collector distance, and flow rate. These parameters are shown to affect the fiber morphology and the addition of iron (III) nitrate salt was found to be able to reduce the fiber size. Due to the conductive nature of iron salt in polymer solution, even a relative lower voltage of 6 kV could generate a stable fluid jet and yield the nanofibers. The thermal properties of the composite fibers were investigated by thermal gravimetric analysis (TGA) and differential scanning calorimetry (DSC). Rheological studies were also conducted. The composite nanofibers were further carbonized to carbon nanofibers. Applications of the nanofibers in environment remediation have shown promise.

445030 Heavy Metal Detection in Treated Water Via a Fabric Nanocomposite Sensor

Abstract:

We present the conception of an electrochemical sensor for on-site/portable water analysis. This flexible sensor was developed to quantify the amount of heavy metals (in real-time), alleviating the wait time, large sample size, possible contamination, and expensive analytical equipment associated with current procedures for water analysis. This sensor is composed of a multiwalled carbon nanotube (MWNT) functionalized polymer mat, for a flexible yet conductive sensor. The MWNTs are then functionalized with a cyclo-oligomeric calixarene, which is shown to selectively form a supramolecular complex with metal ions. Upon complex formation, the charge carriers are drawn away from the carbon layer and the detection is prevalent at appropriate levels. The optimization of MWNT and calixarene functionalization, as well as the selectivity and sensitivity of the sensor will be discussed.

447119 Development R&D of Manoir Industries and Emphasis on Coating

Abstract:

The Senior VP Technical Support & Projects will present the development R&D of Manoir Industries and especially the coating (it is our lattest development) as well the services we can provide such as metallurgy expertise, advice, convector cleaning, site welding, welding supervision in order to increase the furnace productivity. So it is 60% R&D and 40% services. Manoir Industries manufactures cast components for the petrochemical market and develops high temperature alloys such as the Manaurites.

444478 Biogas Reforming with Addition of Carbon Dioxide to Syngas over Ni-Based Bimodal Pore Catalyst

Abstract:

Methane dry reforming of (MDR) has been widely studied due to its important applications in producing of syngas ($CO + H_2$), removal of two greenhouse gases, and upgrading of biogas (mainly composed of CH_2 and CO_2) into value added chemicals. Compared to the noble metal catalysts, nickel-based catalysts are suitable for industrial scaled-up MDR due to their high activity and low cost. However, nickel-based catalysts have been encountering a significant challenge that is the active metal sintering and coke deposition, leading to the catalyst deactivation and then the poor stability. Nevertheless, it has been reported that the deactivation of nickel catalysts can be suppressed by adding promoters like strong Lewis bases with the enhancement of chemisorb CO_2, such as MgO and CaO, and like lanthanide elements with the capacity of oxygen storage and release, such as CeO_2 and La_2O_3. Al_2O_3 is a commonly applied catalyst carrier own to its good pore size dispersion, high specific surface area and high mechanic strength.

Herein, we report that the bimodal NiCeMgAl catalysts with different NiO-loading were synthesized by the refluxed co-precipitation method and were evaluated for MDR performance by using real biogas from anaerobic digestion of food wastes. The sample containing 15wt% NiO-loading (Ni_{15}CeMgAl) was found to be active enough at 750 °C with a high CH_4 conversion of around 90%. Ni_{15}CeMgAl catalysts kept their bimodal porosity after reduction and MDR process. The evolution of the Ni_{15}CeMgAl catalyst before and after MDR process was investigated by SEM, TEM, XRD, and TGA techniques. The correlation between the structure evolution and catalytic performance change was also discussed.

444551 Continuously Adjustable, Molecular-Sieving "Gate" on 5A Zeolite for Distinguishing Small Organic Molecules By Size

Abstract:

Zeolites/molecular sieves with uniform, molecular-sized pores are important for many adsorption-based separation processes. Pore size gaps, however, exist in the current zeolite family. This leads to a great challenge of separating molecules with size differences at ~0.01 nm level. Here, we report a novel concept, pore misalignment, to form a continuously adjustable, molecular-sieving "gate" at the 5A zeolite pore entrance without sacrificing the internal capacity. Misalignment of the micropores of the alumina coating with the 5A zeolite pores was related with and facilely adjusted by the coating thickness. For the first time, organic molecules with sub-0.01 nm size differences were effectively distinguished via appropriate misalignment. This novel concept may have great potential to fill the pore size gaps of the zeolite family and realize size-selective adsorption separation.

16AIChE
Spring Meeting, Houston, TX

The Principle of Intrinsic Reality
Solving the World's Woes through Chemical Engineering and Economics

Jed Moyano Bellen, SCPM, PMP
Hyundai Engineering & Construction Co. Ltd.
9 Temasek Blvd., #27-02 Suntec Tower 2, Singapore 038989
jed.bellen@hdec.co.kr

Prepared for Presentation at
American Institute of Chemical Engineers
2016 Spring Meeting
11th Global Congress on Process Safety
Houston, Texas
April 10-14, 2016

UNPUBLISHED

The Principle of Intrinsic Reality
Solving the World's Woes through Chemical Engineering and Economics

Jed Moyano Bellen, SCPM, PMP
Hyundai Engineering & Construction Co. Ltd.
9 Temasek Blvd., #27-02 Suntec Tower 2, Singapore 038989
jed.bellen@hdec.co.kr

Keywords: Principle of Intrinsic Reality, Chemical Engineering, Economics, Sustainability

Abstract

Human beings in profound rationality always seek for explanations of the phenomena around them. To understand these phenomena, human beings have developed systematized body of knowledge as a practical way of dealing with life's challenges. Among the systematized body of knowledge developed are chemical engineering and economics.

An avant-garde way of looking at each of these disciplines leads to the Principle of Intrinsic Reality. This principle relates the fundamental laws of the two disciplines to glean insights on the constraints imposed by conservation laws on economic systems at both the macro and micro levels.

It can be argued that if mass and energy in the universe are constant, then logically diminishing returns and marginal utility are bound to be observed. This logic is quite evident in a microorganism culture which follows the Monod growth model.

It can also be thought that conservation laws imply scarcity of resources. As such, it enjoins everyone to find optimal and efficient use of resources in the context of trade-offs, production decisions and sustainability. It even calls for the development of new technologies to produce more outputs from less inputs (i.e. shifting of production possibility frontier to the right). In this way, the Principle of Intrinsic Reality leads to the imperative for sustainable optimization.

Furthermore, it is deemed that the conservation laws can explain the phenomenon called "polarity of wealth." As mass and energy when harnessed create wealth, then the constancy of mass and energy implies boundary for wealth creation. In effect, wealth can neither be created nor destroyed but can only shift from one entity to another thereby creating polarity in wealth distribution. This assertion can be the key towards more equitable restructuring of economic systems to close income gap and thereby result to stable and resilient populace.

In conclusion, this paper envisions for a paradigm shift from traditional money-centric view of chemical engineering and economics to a synergistic convergence of disciplines which can powerfully tackle the world's woes and thereby effect sustainability.

1. The World in which We Live Today

The past few years saw unprecedented global challenges which have put into question existing economic models and scientific paradigms. Among these challenges are the dystopian global economy, climate change, environmental degradation and terrorism. The magnitude of these challenges has grown to a profound level that newer ways of thinking and perspectives become necessary. With that end in mind, this paper attempts to look at the fundamental laws of chemical engineering and economics in new ways to recognize our intrinsic reality from which these pressing challenges can be tackled at large.

2. The Principle of Intrinsic Reality

It is an observable fact that advancement in science increasingly blurs boundaries of various disciplines in tackling proactively the pressing global issues or problems. Within this purview, two seemingly diverging disciplines namely chemical engineering and economics can be taken together to glean insights on our intrinsic reality and what it means for our economic and ecological systems on a larger scale.

In the author's contention, intrinsic reality is grounded on the fundamental laws of chemical engineering and economics. These laws are the Conservation Laws (Chemical Engineering) and Law of Diminishing Returns (Economics). It can be argued that the Law of Diminishing Returns is observable because of the constraints imposed by the Conservation Laws. As such, this leads to the Principle of Intrinsic Reality which can be stated thus;

> *In any system, both at the macro- and micro-levels, the limiting factor for growth is the constancy of potential inputs.*

Potential inputs mean not the actual inputs to processes but the constant material or energy available which can be harnessed to create wealth. The reasoning or proof of this principle is presented in subsequent discussions.

2.1 Proof Based on the Monod Growth Model

The best illustration to understand the Principle of Intrinsic Reality is the Monod Growth Model. Monod in his 1949 seminal paper presented an empirical relation between exponential growth rate and concentration of essential nutrient. The data was fitted into hyperbolic equation:

$$R = R_K\left(\frac{C}{C_1 + C}\right)$$

In this equation, R is the growth rate during the exponential phase; R_K is the rate limit for increasing concentrations of C; C is the concentration of the nutrient; and C_1 is the concentration of nutrient at which the rate is half the maximum.

Figure 1: Graphical Representation of Monod Growth Model

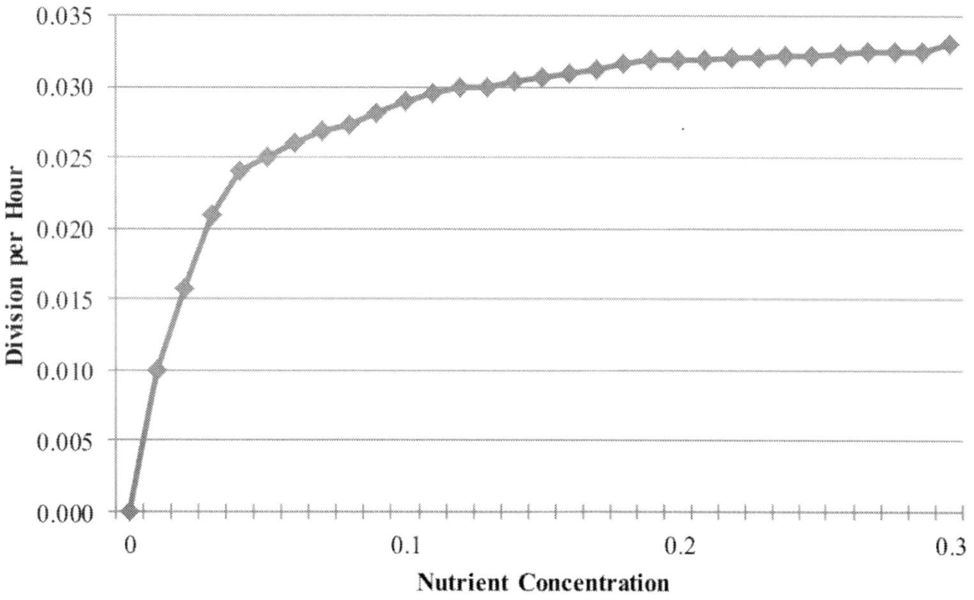

Monod (1949) further pointed out that the limiting factors commonly found are exhaustion of nutrients, accumulation of toxic metabolic products and changes in ion equilibrium.

The foregoing discussion has been observed at the micro-level with the bacterial culture. At the macro-level (i.e. the world at large), analogous phenomena are also observed to occur because of the same constraint imposed by the constant quantity of resources (i.e. material and energy). One example of these phenomena is that human beings cannot grow indefinitely. If they do, then how can the world sustain that growth with limited available resources (e.g. foods) to support it?

Another analogous phenomenon is in population growth. Thomas Malthus (1789) predicted that when population growth exceeds agricultural production, the population level is forced to return to subsistence-level conditions. This phenomenon as Malthus (1789) believed is due to the restriction of available resources supporting population growth, which others have claimed to be exploding as the earth becomes over populated.

In a surprising manner, a similar phenomenon can be observed in production processes in which the increase in a single factor of production assuming all other factors constant, results not to more marginal output but to a reduced one. This was attributed by classical economists, Thomas Malthus and David Ricardo, to the decreasing quality of inputs while neoclassical economists attributed this to the disruption in the entire production process as additional units of input are added to a fixed amount of capital. But in view of the Principle of Intrinsic Reality, the phenomenon is viewed intrinsic to the production system in which case infinite amounts of mass or energy cannot be processed with perfect efficiency. Hence, the law of conservation of mass and energy comes into the fore.

It can be deduced from the discussion that there is an underpinning intrinsic reality to the more obvious phenomena in the world from the micro-scale (bacterial growth) to the macro-scale (human growth and production processes).

2.2 Proof Based on Diminishing Marginal Utility

Another way to prove the Principle of Intrinsic Reality is through the diminishing marginal utility observable among humans and perhaps other organisms as well. Marginal utility, as defined in economics, is the gain from an increase or the loss from a decrease in the consumption of goods or services. The diminishing marginal utility is an observable phenomenon in which the utility or satisfaction an individual gains diminishes as he or she consumes more of the same unit of goods or services. Graphically, this can be represented as in Figure 2.

Figure 2: Graphical Representation of Diminishing Marginal Utility

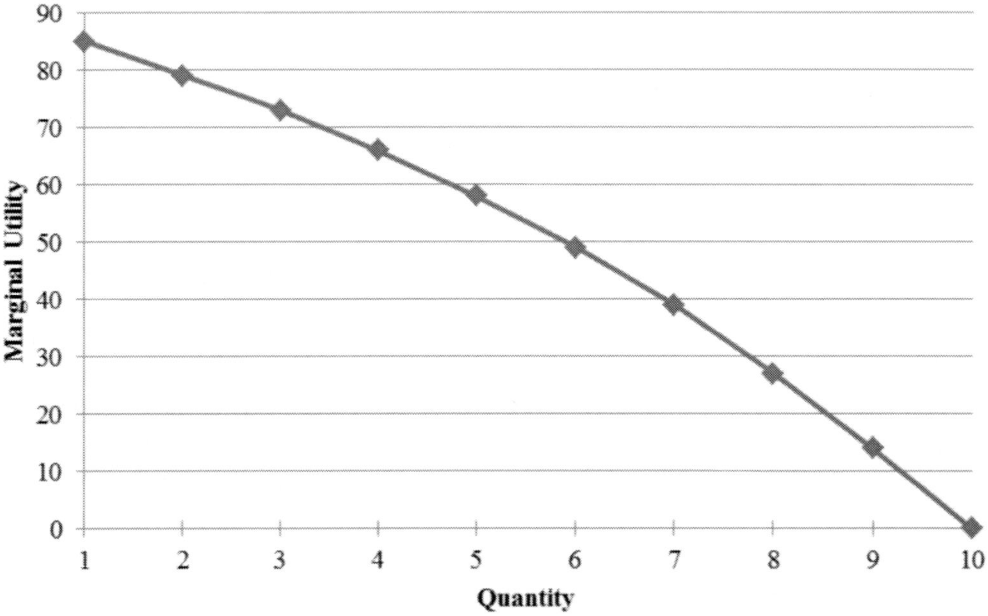

It can be deduced from this phenomenon and in support of the discussion in section 2.1 that human beings cannot grow indefinitely due to the fact that in our intrinsic reality, there is always a limit so as the constancy of available resources (i.e. mass and energy) can support our existence.

This means that in view of our intrinsic reality, we cannot have infinite greed otherwise an upsurge of unintended consequences are bound to happen (i.e. environmental degradation, terrorism and eventually demise of humanity).

Therefore, recognizing our intrinsic reality is the first step towards solving the world's woes and thereby resulting to sustainability and security.

2.3 Expanding the Production Possibility Curve through R&D

The caveat here is that the constant mass or energy (i.e. the limits) leads to trade-offs. Trade-offs is encountered when personal or business decisions are made. In making these decisions, one has to ensure that he or she becomes better off with his or her chosen option. We also realize in our life that we cannot get all that we want and we have to make trade-offs so long as it places us in a better or preferred position. This is more of a philosophical discourse, but it goes to show that the concepts of chemical engineering can be as useful in life and in business as it is in chemical processing.

To tackle these trade-offs, the Production Possibility Frontier in economics can be used. In view of the Principle of Intrinsic Reality where growth is limited by the fixed amount of material and energy, the Production Possibility Frontier needs to shift to the right as shown in Figure 3.

Figure 3. Graphical Representation of Expanding the Production Possibility Curve

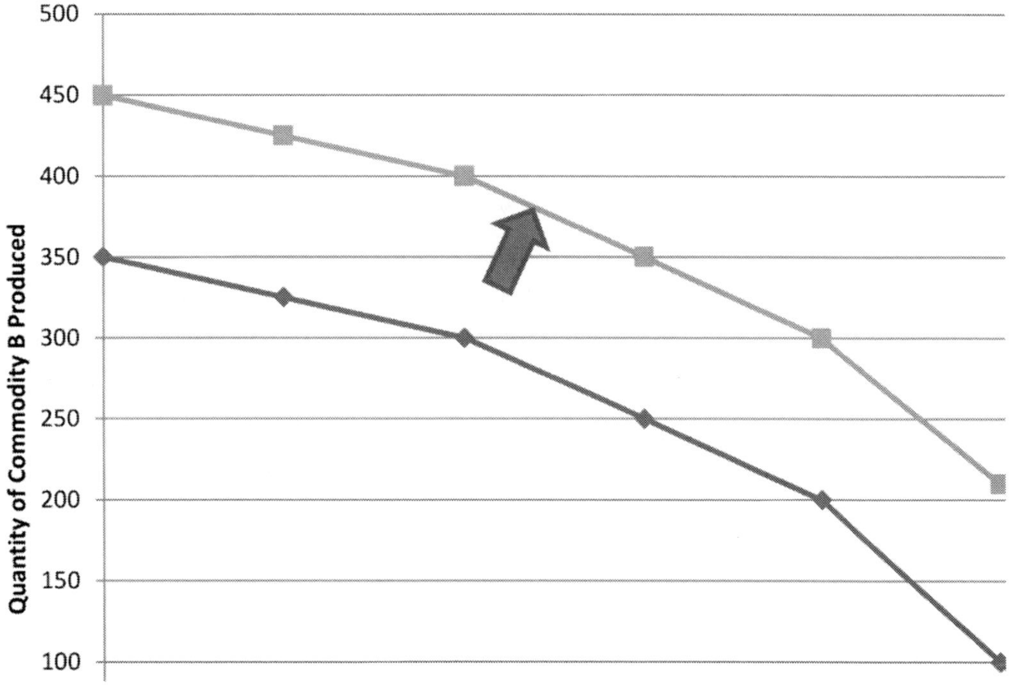

This shift has an implication on the practice of chemical engineering as Research and Development (R&D) is needed to achieve the said shift. R&D enables the economy to produce more of the two commodities (i.e. Commodity A and Commodity B) resulting in economic growth. The key challenge here to Chemical Engineering, particularly on Chemical Engineering R&D, is to look at the bigger picture where economic growth emanating from R&D output is pivoted on sustainability; considering its impact to the environment and the society at large.

3. Paradigm Shift in the Practice of Chemical Engineering

"Money" or "Profit Maximization" has been at the center of the practice of chemical engineering (Smith 2005). Various mathematical optimization methods are used to design systems, processes or components of processes with the end in mind of optimizing profits. This mental model however, will no longer work in our situation today as we face economic, ecological, social and spiritual crisis (Scharmer 2010).

To address this, Scharmer (2010) pointed out seven acupuncture points to shifting capitalism into a regenerative ecosystem economy. Within this purview, the practice of chemical engineering has a greater role to assume. For example, Scharmer (2010) suggested that the design of all production and consumption cycle be completely earth-to-earth, i.e. without necessitating for landfill and that it has to co-evolve with the natural ecosystem. In this regards, the way we do optimization should be revamped into sustainable optimization. By sustainable optimization, it means optimization of the entire system and adding in environmental as well as societal objectives to the equation.

In this definition, two things need to be considered. The first one is systems thinking which emphasizes the understanding of the entire system so the interaction among components will be known and unintended consequences mitigated (Senge 2006). This is a holistic approach to problem solving and finds enormous use in process design.

It should be noted however that the construction of processing plant, whether it is a greenfield or a brownfield project, is rooted on projected market demand. In some instances, economic volatility necessitates companies to experiment with multiple pathways to survive. In this circumstance and within the purview of systems thinking, design thinking comes into the fore. Design thinking is a collective principle emphasizing empathy for users, doing prototyping and tolerance for failure (Kolko 2015). This principle finds enormous use in chemical engineering designs such as in product development, process development, systems improvement and the like. This is also useful for simplifying or humanizing chemical engineering designs with the end in mind of contributing to sustainability. It also has to be noted that of paramount importance in using this tool is imagining the future with the view in mind of status quo and the likely outcome for sustainable development.

The second consideration is the inclusion of environmental and societal objective functions to the objective functions (economic criteria) of profit, cost, energy and yield (Edgar, et al. 2001) in optimization. The environmental objectives can be the impact or risk to environment and the societal objectives can be the impact or risk to society or common good.

These are just few examples of the needed paradigm shift in chemical engineering practice in consideration of the principle of intrinsic reality. But there can be more. A picture of the envisioned new paradigm for the practice of chemical engineering is shown in Figure 4.

Figure 4. New Paradigm in the Practice of Chemical Engineering

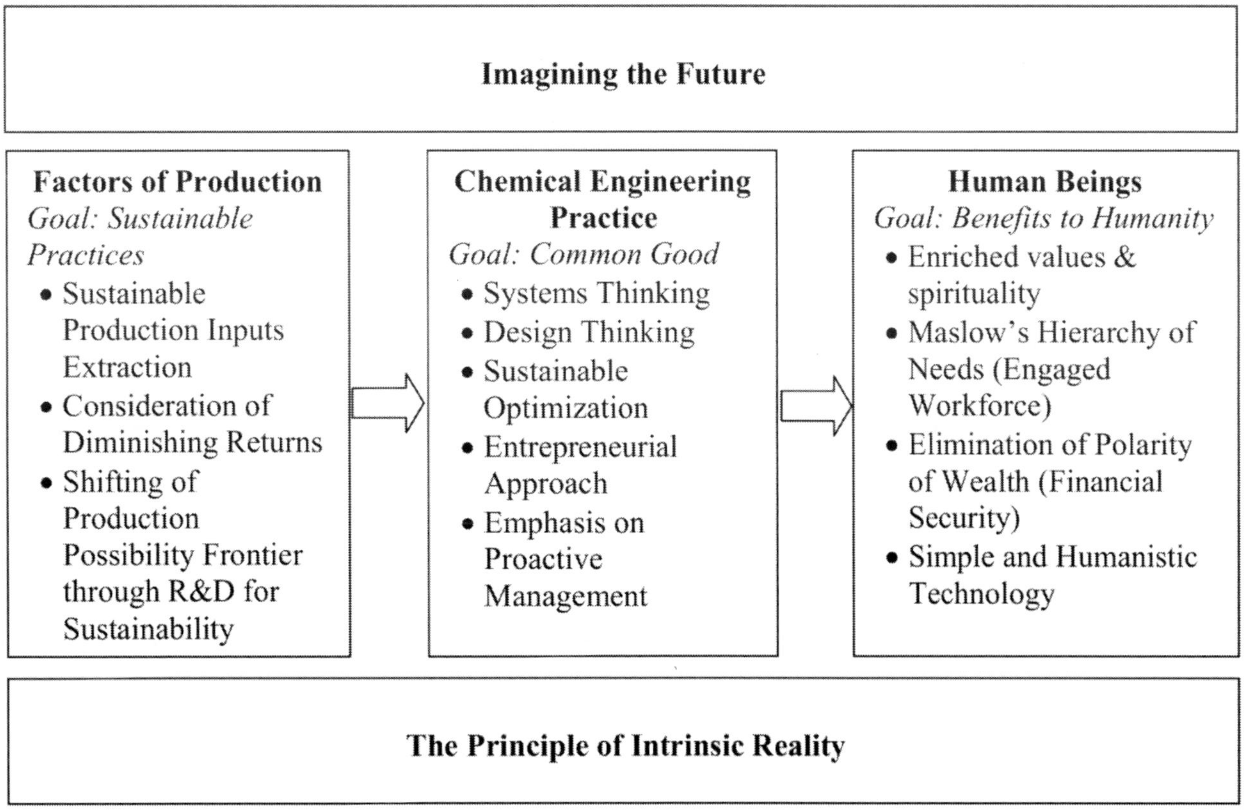

It should be realized that at the core of our economic systems is chemical engineering which can drive the economy into positive direction with the least unintended consequences. Related to that is capitalism in which upgrading of its operating systems can affect the practice of chemical engineering to a substantial degree. A discussion of this is provided in the following section.

4. Wealth Creation and Its Imperative for Sustainability

The development of this paper was inspired by the book "Freakonomics" in which the authors Steven D. Levitt and Steven J. Dubner (2005) asserted that economics can explain almost everything. It came to mind that if economics as a basic science can explain almost everything, then why not chemical engineering which was founded on chemistry, the central science?

Reflecting on it deeper, while explanations clear our doubts, we need more than that so we can tackle the world's woes. As such, this author thought that perhaps combining the two disciplines of chemical engineering and economics can be a potent force towards this end. How this can possibly be done is discussed in this section.

The starting point here is a review of the economic history of the world. In his book "A Farewell to Alms", Gregory Clark (2007) reviewed the economic history of the world in which he divided it into two periods; the period before 1800 and the period after. This is graphically represented in Figure 5.

Figure 5. Shares of Major Economies in World GDP through Time

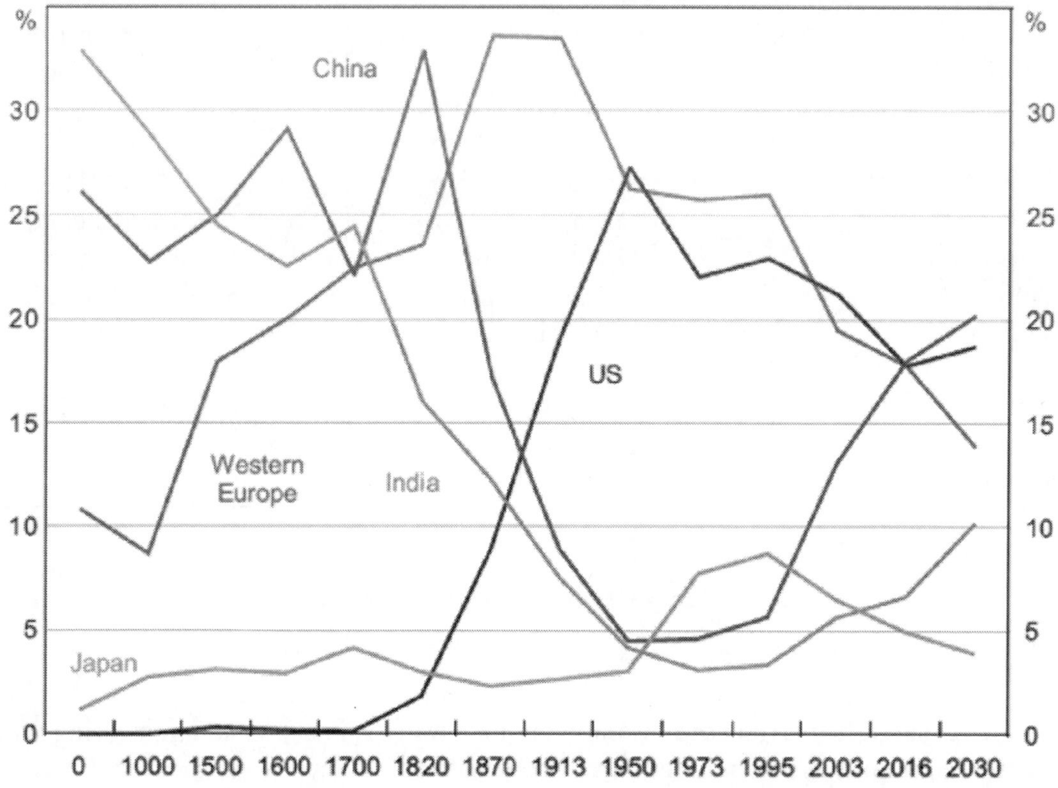

Note: Based on purchasing power parity (PPP) valuation of country GDP

Sources: International Monetary Fund (IMF), *World Economic Outlook* database, April 2011; Maddison (2006)

Thompson (2012) interpreted this graph by stating that "everything to the left of 1800 is an approximation of population distribution around the world and everything to the right of 1800 is a demonstration of productivity divergences around the world." In other words, everything to the left before Industrial Revolution depicts economic power by the sheer size of population and everything to the right after the Industrial Revolution shows the effect of mass production on country's economic output.

Clark (2007) on the other hand called the era before the industrial revolution, i.e. ca. 1800, as Malthusian era in which income per person varied across societies and epochs but growth is not observed. As Clark (2007) explained, this was due to a mechanism called Malthusian trap in which short-term increase in income through advancement in technology is dragged down by population growth. He further explicitly stated:

> "Economics as a discipline arose in the dying decades of the Malthusian era. Classical economics was a brilliantly successful description of this world. But the torrent of goods unleashed by the Industrial Revolution not only created extremes of wealth and poverty across nations, it also undermined the ability of economic theory to explain these differences."

In view of this statement, Figure 5 is reinterpreted in this paper. The contention of this author is that *polarity of wealth* can be gleaned from the graph. *Polarity of wealth* means the uneven concentration of wealth in the global economic system as some parts of the world are richer than others. The polarity can also be observed to shift as some richer parts of the world decline in economic power and some poorer parts of the world rise in economic wealth.

A historical example can be gleaned from Figure 5. India and China up until 1500 CE had 50 and 60 percent share of the world's economy, respectively. In 1 CE, China had a population of almost 60 million, while the area of the present day United States had a population of 680,000. In the late 18th century, the Industrial Revolution rendered countries with large population, just countries with the largest populations.

Looking at North America, it took them 1800 years to overtake China's economic output. However, by 1950, the US economic output was three times as great as China, even though its population is three times less. Comparing China with UK in 1913, the UK had a population of 45 million and China had 437 million but the economic output was almost the same. Analyzing what happened in Europe and later in the United States, Europe was on the lead even before 1800 (Nye 2012).

In 2010 according to data, China overtook Japan as the second biggest economy, though the numbers may not be strictly comparable. Japan's data are seasonally adjusted while those for China were not. Angus Maddison suggested that China and India were the biggest economies in the world for most of the past 2000 years. That is why economists find it a mystery why they fell so far behind and now they are flourishing ('China's economy' 2010). This may not be mysterious considering the *polarity of wealth* as was mentioned earlier and that this polarity can be argued to emanate from the Principle of Intrinsic Reality.

The reasoning here is that the world has constant mass and energy. Assuming that all these mass and energy can be harnessed to create wealth, it could mean that there is a limit for wealth creation. Theoretically, the limit can be expanded indefinitely through R&D but experience demonstrates that this was not achieved yet or getting anywhere near being achieved. As such, the constant mass and energy vis-à-vis the limit for wealth creation can be thought of as the reason for the polarity of wealth, i.e. wealth can only shift from one entity to another.

The implication of this assertion can be demonstrated by the fact that the 1% richest have accumulated more wealth than the rest of the world combined (Oxfam International 2016). The world has never seen 100% of its population to be wealthy. As such, time and again, equitable distribution of wealth is called for and mechanisms to achieve such_should be made a priority in government policies.

In a more concrete way, Scharmer (2010) suggested measures to address this growing inequality which is consistent with the mindset espoused by the Principle of Intrinsic Reality. These measures are as follows:

- Creation of a New Class of Common Property Rights

The creation of common property rights is intended to address the cost to society of private economic undertakings. For example, a private company is extracting natural resources from a certain place. While the private company can profit from that endeavor, the community in that place might suffer the undesirable consequences (e.g. flooding due to forest denudation). To ensure sustainability and institutionalize the voice of future generations in this case, a trust can be established which will own the rights and distribute funds (e.g. from cap-and-trade system for carbon emissions) to citizens (one share per person), communities and the government for investment in public infrastructure.

The trust is an effective means because it does not function like a company (driven by profits) or government (driven by special interest groups) but that it is accountable for long-term sustainability of the specific commons it manage for future generations.

- Creation of a Human Right to Basic Income

One way to ensure equitable distribution of wealth is to create a human right to basic income for every human being. This right when combined with free or affordable access to healthcare and education would create a playing field where everyone has a chance to pursue aspirations and dreams. In other words, it enables individuals to put their real creativity for the benefit of larger community. Another intention of this measure is to free up paid labor (a form of dependence) so that human beings can enhance their creativity and entrepreneurial skills.

- Inclusive Capitalism

To eradicate the *polarity of wealth*, capitalism should be made inclusive so that money and capital flows are directed to all sectors of the economic system. Currently, our money system leads to accumulation of oversupply of money and capital in high financial but low environmental and social gain areas. At the same time, there is an undersupply of money and capital in areas that address societal and community needs. It implies that there is too much money in areas where it is not needed but not enough on areas where it is needed. To address this imbalance, it is recommended to redesign the system so that the current extraction-based financial system is replaced by institutions which can support the real economy and thereby serve the ecology and community.

It should be noted that shifting of wealth is also observed in various geographical regions. When this occurs, geo-political instability can result. This goes to show that if we want lasting stability and security within the limits of our intrinsic reality, then we should not tolerate unabated greed. Otherwise, *polarity of wealth* will continue to persist and we cannot realize our full potential as individuals and collaborators for common good.

5. Way Forward...

Recognizing our intrinsic reality leads us to greater understanding of our very nature bounded by the constancy of mass and energy. It shed light on why growth stabilizes after addition of more inputs. Within this purview, trade-offs are also encountered which necessitates for R&D. R&D initiatives however should be underpinned on sustainability.

Looking at the practice of chemical engineering from the lens of intrinsic reality, the approach should be futuristic while considering relevant aspects related to factors of production and human beings.

The bigger picture shows that Engineers particularly Chemical Engineers are major players in driving the real economy. As Scharmer (2010) aptly puts it, "The essence of technology and knowledge creation, and the ultimate driving force of the economy, is the same: it is the human capacity to create." As such, if sustainability is to be achieved, Chemical Engineers have a greater responsibility to innovate for regenerative and social technologies. In this way, we can stem the polarity of wealth by providing affordable technology to enable disadvantaged individuals enhance their creativity, innovation and entrepreneurship. This can help them partake in the benefits of inclusive capitalism.

As demonstrated in this paper, the synergy emanating from the convergence of Chemical Engineering and Economics can powerfully tackle the world's woes. As such, there is a good reason for chemical engineers and economists alike to collaborate particularly on sustainability initiatives.

6. References

[1] 'China's economy overtakes Japan's in real terms' 2010, *The Economist*, 16 August, viewed 5 March 2016, http://www.economist.com/node/16834943

[2] Clark, G 2007, *A Farewell to Alms: A Brief Economic History of the World,* Princeton University Press, Princeton, NJ.

[3] Edgar, T, Himmelblau D & Lasdon, L 2001, *Optimization of Chemical Processes*, McGraw-Hill, New York.

[4] Huang, Y & Wang, B 2011, 'From the Asian Miracle to an Asian Century? Economic Transformation in the 2000s and Prospects for the 2010s', in H Gerard & J Kearns (eds), *Proceedings of the Reserve Bank of Australia annual conference*, Reserve Bank of Australia, Sydney, viewed 30 January 2016,

http://www.rba.gov.au/publications/confs/2011/huang-wang.html

[5] Kolko, J 2015, 'Design Thinking Comes of Age', *Harvard Business Review*, pp. 66-71, viewed 24 February 2016, https://hbr.org/2015/09/design-thinking-comes-of-age

[6] K.N.C. 2012, 'Mis-charting economic history: more 2,000 years in a single graphic', *The Economist blog*, web log post, 20 June, viewed 29 February 2016,

http://www.economist.com/blogs/graphicdetail/2012/06/mis-charting-economic-history

[7] Levitt, S & Dubner, S 2005, *Freakonomics: A Rogue Economist Explores the Hidden Side of Everything*, William Morrow, New York, NY.

[8] Malthus, T 1798, *An Essay on the Principle of Population*, J. Johnson in St. Paul's Church-yard, London.

[9] Maslow, A 1943, 'A Theory of Human Motivation', *Psychological Review*, vol. 50, pp. 370-396, viewed 24 February 2016, http://psychclassics.yorku.ca/Maslow/motivation.htm

[10] Monod, J 1949, 'The Growth of Bacterial Cultures', *Annual Review of Microbiology*, vol. 3, pp. 371-394, viewed 2 February 2016,

http://www.annualreviews.org/doi/abs/10.1146/annurev.mi.03.100149.002103

[11] Nye, J 2012, 'Fascinating new graph shows the 'economic history of the world since Jesus'', *MailOnline*, 23 June, viewed 15 November 2015,

http://www.dailymail.co.uk/news/article-2163610/Fascinating-new-graph-shows-economic-history-world-Jesus.html

[12] Oxfam International 2016, *An economy for the 1%, how privilege and power in the economy drive extreme inequality and how this can be stopped*, Oxford, viewed 5 March 2016,

https://www.oxfam.org/sites/www.oxfam.org/files/file_attachments/bp210-economy-one-percent-tax-havens-180116-en_0.pdf

[13] Samuelson, P & Nordhaus W 2009, *Economics*, McGraw-Hill Higher Education, New York, NY.

[14] Scharmer, O 2010, 'Seven Acupuncture Points for Shifting Capitalism to Create a Regenerative Ecosystem Economy', *Oxford Leadership Journal*, vol. 1, pp. 1-21, viewed 24 February 2016, http://www.oxfordleadership.com/journal/vol1_issue3/scharmer.pdf

[15] Senge, P 2006, *The Fifth Discipline: The Art & Practice of the Learning Organization*, Doubleday/Currency, New York.

[16] Smith, R 2005, *Chemical Process Design and Integration*, John Wiley & Sons Ltd., West Sussex, England.

[17] Thompson, D 2012, 'The economic history of the last 2,000 years in 1 little graph', *The Atlantic blog*, web log post, 19 June, viewed 29 February 2016, http://www.theatlantic.com/business/archive/2012/06/the-economic-history-of-the-last-2-000-years-in-1-little-graph/258676/

Acknowledgment

The author would like to thank Hyundai Engineering and Construction Co. Ltd. for the support in this endeavor. The author is also thankful to Ms. Sofia Ayu for meticulously editing this manuscript and Mr. Marcial N. Barroga for the poster design.

443521 Maximize Tray Deck Performances

Abstract:

Refineries, gas processing, petrochemical and chemical producers continually strive to maximize their production volume at minimum investment cost. Generally, this is done in the form of revamps where existing columns are retrofitted with high performance internals. This paper will focus on recent advances in tray technology to meet these needs.

Column design focuses on the ability of the internals to simultaneously process both the vapor and liquid streams. As a result, tray research and development centers mainly on two areas – development of tray deck design for higher vapor handling capacity while maintaining good tray efficiency, and optimization of downcomers for higher liquid loads while maximizing the tray active area.

One of the unique features of Sulzer trays is the rectangular or trapezoidal shape of their deck orifices. These orifices shapes, by definition, release vapor preferentially to the sides of the orifices above the tray deck, perpendicular to the main liquid flow. This release pattern creates a different interaction of the vapor and liquid streams between the valves. Research on these interactions has shown that the capacity of Sulzer's high performance trays can be further enhanced with better utilization of the vapor flow throughout the tray active area.

The paper discusses the development and testing of trays with enhanced tray design features, which improve vapor distribution with optimized valve layout and enhanced aeration. Pertinent test data will be shown to confirm the tray performance increases.

444867 Cyclic Operation of Distillation Columns

Abstract:

Cyclic Operation of Distillation Columns

Carsten Knösche, Jürgen Paschold, Steffen Bütehorn, BASF SE

Claudio Ribeiro, BASF Corporation

One of the earliest publications about the cyclic operation of distillation columns dates back to 1952 (Cannon, 1952). In the following decades numerous theoretical and experimental investigations were published. The scientific interest peaked in the mid 80-th. After that the number of publications decreased steadily presumably because of problems with the practical implementation of the cyclic operation modus. In 2011 Maleta et. al. (Maleta, 2011) published the design of a new cyclic separation tray which provides a promising way to put cyclic distillation into practice.

A theoretical analysis allows for predictions about the separation efficiency of cyclic distillation towers. Analytical solutions for binary mixtures and linear equilibrium lines are known (e. g. B. Horn, 1967). For arbitrary equilibria the corresponding set of describing equations can be solved numerically. It can be shown that depending on materials and operation conditions substantial improvements in separation efficiency are possible.

In the presentation a simple figurative explanation of the increase of separation efficiency in the cyclic mode of a distillation column is given. Theoretical predictions from numerical calculations are compared with experimental results obtained from cyclically operated columns. Moreover basic questions like minimum energy requirements and dimensioning of such columns are discussed.

Literature:

Cannon, M. R., Oil Gas J. 51, 1952

Maleta et. al., Chemical Engineering and Processing, 50, 2011

Horn, Ind. Eng. Chem. *Process* Des. Dev., 6, 1, 1967

444251 Prediction of Valve Tray Pressure Drop

Abstract:

Valve trays have been developed since the 1960's, and can perform with reduced entrainment and weeping, improved turndown and higher capacity, as compared to sieve trays.

Predicting the pressure drop of valve trays is an important design consideration. Numerous valve tray design parameters and shapes make the development of a generalized model particularly challenging. This presentation will focus on the status and application of existing valve tray pressure drop correlations from the literature.

445122 Sulzer UFM Tray - the Next Generation Valve Tray

Abstract:

The recently developed UFM™ tray, has received wide industry acceptance over the past 3 years. This paper will document several applications where the superior capacity and performance characteristics of this tray has enabled these towers to achieve better than expected performance. Operating data will be presented and tray efficiency will be discussed.

443734 A New Realization of Periodic Cycled Separation

Abstract:

A new realization of periodic cycled separation is presented which enables continuous vapor flow also during liquid drainage from the trays. With the proposed design a new operation form is thus introduced, where the trays are drained sequentially rather than simultaneously, such that the vapor flow is not interrupted during the liquid drainage.

Separation factors and column efficiencies are compared for a column, stripping ammonia from water with air, using three different sets of internals: Conventional sieve trays, Sulzer BX gauze packings and periodically cycled trays. The periodic trays are advantageous compared to their predecessors. For different ratios of countercurrent vapor/liquid flow rates, column efficiencies are two times greater for periodically cycled columns than for columns with sieve trays. Depending upon the holdup and gas flow, the tray efficiency of a single periodic tray can be up to 300 % of an ideal stage.

16AIChE
Spring Meeting, Houston, TX

Distillation Column Troubleshooting with Imaged Improved Gamma Scan Technique

Marcio Issamu Haraguchi
Tricom Tecnologia
Av. Conselheiro Rodrigues Alves, Centro, Piquete, SP, Brazil, 12620-000
marcio@tricomtecnologia.com.br

Hae Yong Kim
Sao Paulo University (USP)
Polytechnic School – Department of Electronic Systems Engineering (LPS)
Av. Prof. Luciano Gualberto, Tr. 3, 158, São Paulo, SP,Brazil, 05508-900
hae@lps.usp.br

Wilson Aparecido Parejo Calvo
National Nuclear Energy Commission
Institute for Nuclear and Energy Research -IPEN-CNEN/SP
Av. Prof. Lineu Prestes, 2242, São Paulo, SP, Brazil, 05508-000
wapcalvo@ipen.br

Patented INPI BR 10 2012 024416-0

Prepared for Presentation at
American Institute of Chemical Engineers
2016 Spring Meeting
11th Global Congress on Process Safety
Houston, Texas
April 10-14, 2016

UNPUBLISHED

Distillation Column Troubleshooting with Imaged Improved Gamma Scan Technique

Marcio Issamu Haraguchi
Tricom Tecnologia
Av. Conselheiro Rodrigues Alves, Centro, Piquete, SP, Brazil, 12620-000
marcio@tricomtecnologia.com.br

Hae Yong Kim
Sao Paulo University (USP)
Polytechnic School – Department of Electronic Systems Engineering (LPS)
Av. Prof. Luciano Gualberto, Tr. 3, 158, São Paulo, SP,Brazil, 05508-900
hae@lps.usp.br

Wilson Aparecido Parejo Calvo
National Nuclear Energy Commission
Institute for Nuclear and Energy Research -IPEN-CNEN/SP
Av. Prof. Lineu Prestes, 2242, São Paulo, SP, Brazil, 05508-000
wapcalvo@ipen.br

Keywords: Distillation Column, Troubleshooting, Gamma scanning,

Abstract

Gamma scanning is one of the most widely used nuclear inspection technique worldwide to troubleshoot industrial equipment in refineries and petrochemical plants. The technique simply involves a sealed radiation source (Nal(TI)) paired with a radiation detector positioned on opposite sides of the equipment to be tested. The pair moves simultaneously alongside the vertical axis of the equipment generating a one-dimensional density profile. Despite a few incremental improvements proposed over the years (including automated and remote scanning, wireless detectors and new field and data output programs), the technique remains essentially unchanged since first proposed decades ago and its usage still results in a one-dimensional median density profile of the target. The present work (developed as part of the requirement for a Master of Science degree) describes a new approach for gamma scanning that generates a two-dimensional density distribution plot using industrial tomography tools. This two-dimensional image reveals far more information on the equipment under analysis, information not previously available under conventional testing, and represents a significant advance in non destructive testing of this type of equipment. This innovative technology, patented, also granted the inventors with 2013's Petrobras Technology Prize, for Master Degree at Refining and Petrochemicals category.

1. Introduction

In continuous production plants like refineries and petrochemical sites, process equipment performance are analyzed with the help of a process model, based on the operational variables, quality of feed and products [2]. With increasing complexity and restricted boundaries of process, design and operations, the use of NDT (Non-Destructive Testing) has been widely used to on-line validate, check and troubleshoot these process models.

Among the available technologies, nuclear techniques stand out by not perturbing or affecting the process in analysis, allowing that on-line testing be performed. Modern equipments and methods permit that nuclear techniques found only in literature migrate to the field. Distillation column gamma scanning, neutron backscattering, chemical and radioactive tracers and industrial computerized tomography are common practices nowadays and represents one of the most powerful techniques to on-line analyze process equipments.

Within these technologies, the distillation column profiling consolidated as one of the best options to perform mechanical and operational troubleshooting. However, although it relies on simple principles, the practical use is complex due to the high specific knowledge required on all phases of a typical project. This barrier also turn out to be an extra difficulty for a wider divulgation and acceptance on the market, as the final customer need to have a good experience with the technique too.

To solve technical difficulties and in the search of a competitive edge, the industry developed several solutions. However they did not resulted on higher quality information. All technologies on the market are based on the same principle: register gamma ray attenuation to generate a median density profile. This paradigm would break if a new technology presents a more direct and comprehensive result, with much wider acceptance.

The computerized tomography is available for medical use for more than 30 years and, to this date, reaches a surprising level of complexity and precision. Imaging industrial equipment with the gamma scan profiling technique appears as a good alternative for a new approach. Although it users the same concepts, imaging industrial equipments faces a series of new challenges, difficulties and limitations. The initial achievements of this research are described on this article.

2. Process Equipment Analysis

2.1. Process equipment t

The performance of equipments can be measured, understood and designed trough process models, which input variables can be directly measured (temperature, pressure and flow rate) or determinate (composition, heat consume, mixture, reaction and feed stock). These models are built, considering some projects premises, and are robust enough to account the normal measuring errors and parameter variations.

Obviously the process model cannot account for characteristics of random, uncontrolled, unknown and even human nature:

- Physical – corrosion, mechanical damage and assembly faulty;
- Process – contamination, unexpected physical-chemical phenomena, fouling and saturation;

- Operational – operational disturb, instrument lecture error and coking;
- Human – project problems, in house solutions and human mistakes.

This is the context where some special Non-Destructive Testing (NDT) has their main focus, because it is possible to validate process and troubleshoot problems in equipments, while they are on-line [1]. Nowadays the use of these techniques had been spread beyond the process and operations engineering to other fields of industrial plant engineering [3]:

- Maintenance – on-line evaluation of equipments;
- Shutdown planning – opening equipments, supply parts purchasing and extent of field work;
- Projects and revamps – critical points verification, start up monitoring, baselines and performance study;
- Predictive practices – critical equipment monitoring.

2.2. Gamma ray profiling or gamma scan

Gamma ray column profiling or gamma scan is one of the most used NDT techniques to evaluate the on-line mechanical and operational behavior of process equipments. In this technique, a radioactive source and detector are positioned around the equipment and simultaneously moved along its length. The radiation attenuation values measured thought the vessel results on a density longitudinal profile. The profile or scan plot is then analyzed and the results are present on a report [4,11,13].

The Figure 1 shows an example of a column gamma scan plot. On the left side the density profile are shown and on the right side a sketch of the column is presented.

Figure 1 : Gamma scan plot example

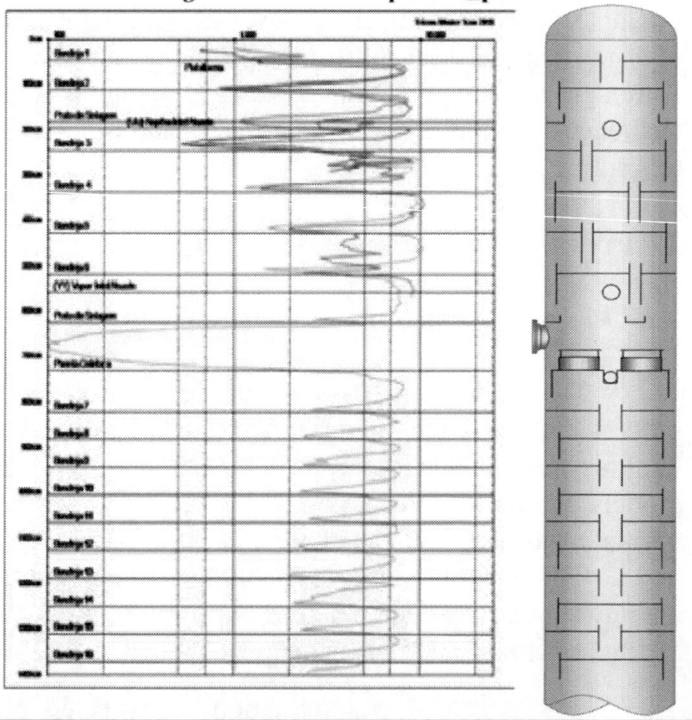

2.2.1. Limitations of the technique

Column scan gamma scanning has some disadvantages that restrict the use to some applications:

- Count rate – as the available time on field tests are finite, gamma scans are limited to those cases where a statistical count rate can be obtained. Source activity and energy, detector size and efficiency, column diameter and length, wall thickness and constitution of the analyzed equipment are fundamental variable on verifying the viability of a project. Radiation safety issues are equally relevant due to limitation of personal equivalent doses [12];

- Mean density – every single point on a gamma scan plot represents the mean density along the scan cord. Any analysis of a gamma scan data should take into account the mean behavior, not the punctual, of the density profile. Additionally, on bigger or trickier equipments the gamma ray profile tends to smooth problems and phenomena. This limitation has been partially contoured with baseline scanning, a common practice for the most critical equipments. Baselines can be performed with equipment off-line (dry-scan) or on-line (operational baseline) at optimum operational conditions;
- Positioning – since the gamma scan is a mean density profile, a data interpretation is possible only where the proprieties of the object are kept constant during the test and along the gamma ray path. That would considerably limit the positioning possibilities and the type of density profile that can be obtained. Many times, complex design equipments are hard, limited or impossible to be properly scanned;
- Dimensions – a density profile offers a one-dimensional view (elevation wise) of the mean density of equipment. Multiple scans or grid pattern scans are often utilized to obtain some space notion of the density distribution.

2.2.2. The art of gamma scanning

Although conceptually simple, the technique of gamma scanning requires a combination of knowledge, skills and talent of the crew, that directly interfere on the quality of the gathered information and trueness of the written report. In general, a gamma scanning specialist should master the following subjects:

- Preliminary evaluation – limitations and possibilities of the technique, possible results and relevant information estimation, preliminary evaluation of required source and count rate obtained, work and data collection strategy;
- Field work – mechanical, electrical, electronic, interference and noise problems, outdoor field work and weather, long and strenuous work journeys, self independence and manual skills;
- Radiation – absorbed doses care, radioactive source handling, radiation attenuation, detector saturation, collimation, scattering, back scattering, radiation physics, source energy and nuclear instrumentation;
- Previous knowledge – knowledge about the process and equipment under analysis, experience with similar cases and proposition of alternative testing;
- Data interpretation – experience to judge and filter all type of interference on the gathered data as a result either of a problem or an inherent characteristic of the equipment;

- Reporting – ability to explain and write down on a technical report the gamma scan, its limitations, the problems found and propositions of solutions; and
- Interpersonal – team work with the scanning crew and other parties involved on the project, relationship with industrial safety crew, radiation myth, communication and persuasion abilities, oratory skills, multi language and technical terms mastery, leadership and stillness.

The combination of these characteristics that compound a typical profile of gamma scan specialist are so restrict that historically it is one of the biggest difficulties that the industry faces. On these circumstances the industry of gamma scan has invested on some solutions:

- Field work – automatic movement systems, wireless detectors and pre-adjusted electronics;
- Data view – automatic report and plot generation, on-line web visualization of data gathering.

Even though there was a clear advance with some of these improvements, there are some remarks:

- User friendly equipments and procedures normally implies on less flexible systems and thus harder to adjust in the field;
- As a result of an easier training, less experience and more inexperienced crews reach the field;
- Distance monitoring do not eliminate the presence of an specialist on a gamma ray scan job, as many tests cannot be repeated;
- Not rare, some improvements resulted in loss of quality of the collected data.

2.3. Computed tomography

The processes of obtaining a space image of an object through some of its projections are called reconstruction. The mathematic process that permits an image reconstruction was published only in 1963/64 by the South African Allan Cornack and implemented on the first tomographic machine in 1972, at EMI laboratories, by Godfrey Hounsfield. This work awards then a Nobel Medicine Prize in 1979. Nowadays medicine, research and laboratory tomography has high degree of sophistication with good imaging quality and high reconstruction velocity. The first tomography used the radiation attenuation principles, but other physical principles can be used as electrical resistivity and sound propagation speed.

2.3.1. Physical principles

Vision is human's most important sense. As an example, it is easier to recognize a circle with a drawing than through its equation. Similarly, any object (a patient, a distillation column, a tree or a casted iron piece) can be easily analyzed with images where the eventual problems can be visualized (a cancer, an obstruction, a crack or a foundry fail). That´s is why a good image reconstruction is important.

There are several physical principles that can be used on a tomography:

- Radiation attenuation – conventional tomography [7,9,15-17];
- Velocity of sound – seismic, acoustic and ocean tomography [14];
- Electrical properties – resistivity, impedance and capacitance tomography [10];
- Light absorption – optical tomography;

Usually any physical property of an object that can be measured and varies only with its constitution can be used. The equation of radiation attenuation can be expressed as:

$$I = I_0 \cdot f(\mu, x) \quad \text{[Eq 1]}$$

Figure 2 : Radiation attenuation along a path

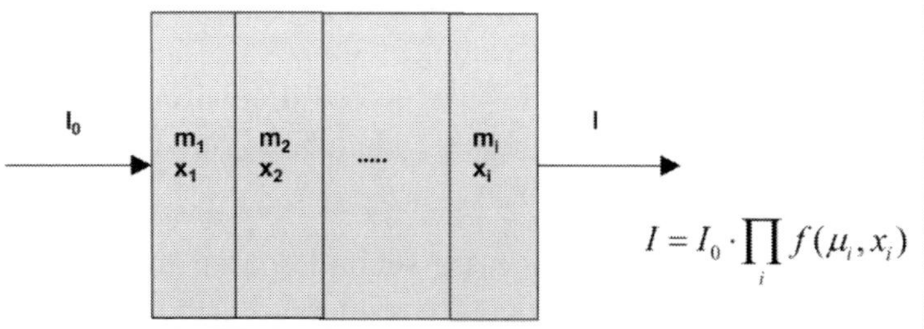

Were "I" and "Io" are respectively the initial and final radiation intensities, "μ" is linear attenuation coefficient for a given median and energy and "x" is the object thickness [6]. If the physical property varies along the path on the object, the function can be expressed as a product of functions as it is shown in Figure. 2. The process of image reconstruction calculates the property distribution within a region, with a limited number of measurements using a mathematical or iterative algorithm.

2.3.2. Conventional tomography basics

In principle, as long as the emission and detection coordinates are known, it is possible to obtain a tomographic image of any spatial arrangement. On most of the cases, the layout of the apparatus imposes a geometry or the design starts with a know geometry for construction convenience or due to the data treatment routine. Specifically for this work, the conventional tomography uses the radiation attenuation principle, expressed by the Beer-Lambert law.

Figure 3 : General tomography types as shown in Radioisotope Gauges for Industrial Process Measurements [8]

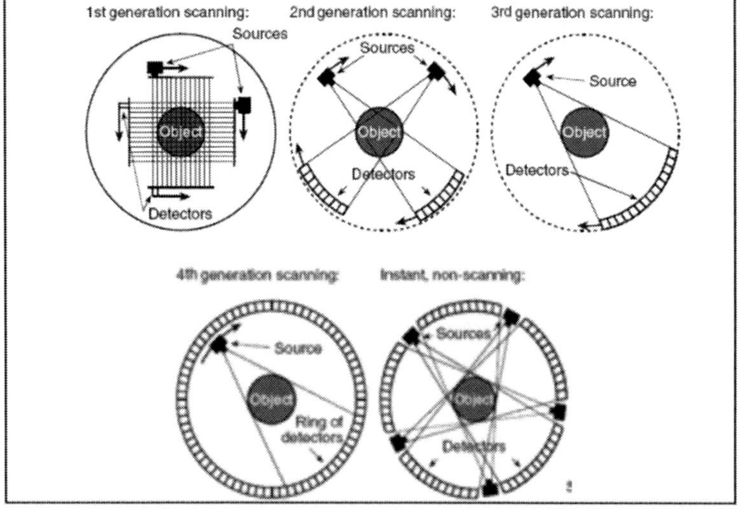

There are several tomographic equipment configurations. Fig. 3 shows the general arrangements of the five tomography generations, where the differences basically reside on increasing complexity with reducing data time acquisition and flexibility [8]. Thus a 1st generation apparatus is versatile, but it can take hours of sampling, on the other hand there are already 5th generation tomographic machines registering ten thousands frames per second [5,7,16].

3. Image improved gamma scans

The Image Improved Gamma Scans concept arises from the development of a conventional industrial tomography methodology back in 2008, carried out by Tricom Tecnologia and the Department of Electronic Systems Engineering (LPS), Polytechnic School of Sao Paulo University (USP). In 2010 the project was developed as part of the requirement for a Master of Science degree at Center of Radiation Technology (CTR) of Institute for Nuclear and Energy Research (IPEN). The inventors were granted with 2013´s Petrobras Technology Prize, for Master Degree at Refining and Petrochemicals category.

3.1 Objectives

In this work, we propose to obtain a two dimensional longitudinal density profile of industrial equipments representing an advance over the conventional gamma ray profiling techniques with some advantages:

- Dimensional – a two dimensional density distribution image would ease the visualization and identification of problems, process and phenomena on an equipment;
- Positioning – as a second spatial dimension is included, the number of longitudinal profiles needed are reduced. Furthermore, features hidden on a conventional (gamma scan) technology, due to the mean density values, would be revealed on an image technology. Similarly, density profiles or equipments that cannot be scanned with gamma ray profile have now possibilities of testing; and
- Interpretation – image results are much easier to analyze and present compared to the conventional technique. This might result on a better acceptance of the inspection by the customer and also easier the training of field personal.

3.2 Simulation studies

Several resources and tools from Tricom and LPS previous experience in industrial tomography systems were revised or adapted for this new technology. One direct application were the ability to simulate irradiations and reconstruct images which opened possibilities to:

- Compare several source-detectors positioning arrangements with resulting image quality;
- Fine tune reconstruction algorithm parameters;
- Reduce personal radiation exposition

Figures 4 and 5 show examples of two simulated column arrangements, the resulting gamma ray scanning conventional plot, and the reconstructed image of the column. On Figures 4 and 5 the images in the left represents two types of typical distillation columns cross sections, with tray and packing assembly. The simulations are numerically performed were each pixel color represents different
steel, steel packing and liquid distribution were considered. The software simulates the count rate obtained from a Cobalt 60 source with NaTl detectors.

3.2 Column model scanning

Two types of column layouts (one pass trayed and single packing bed - Figure 6) were build at CTR facilities in order to research several key aspects of data gathering:

- Source size and type
- Calibration and statistics
- Detector size and type
- Scanning pitch
- Radioprotection safety
- Acquisition software

These models were built to consider perpendicular to downcomer scanning on the trayed column and severe liquid mal distribution with dense pockets and packing voids at the packed column. The resulting images obtained through model irradiation and imaging reconstruction techniques are shown on Figures 7 and 8. Important features observed on analyzing the images are also noted.

Figure 4 : Computed simulated column arrangement (a), resulting gamma ray profile (b) and reconstructed tomographic image of a trayed column (c).

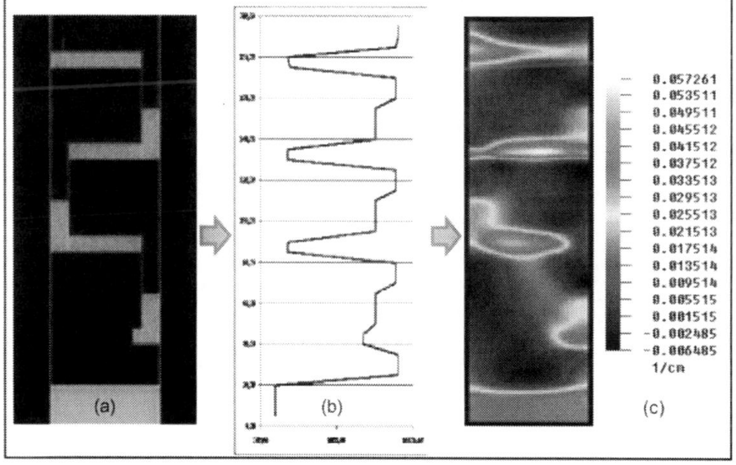

Figure 5 : Computed simulated column arrangement (a), resulting gamma ray profile (b) and reconstructed tomographic image of a packed column (c).

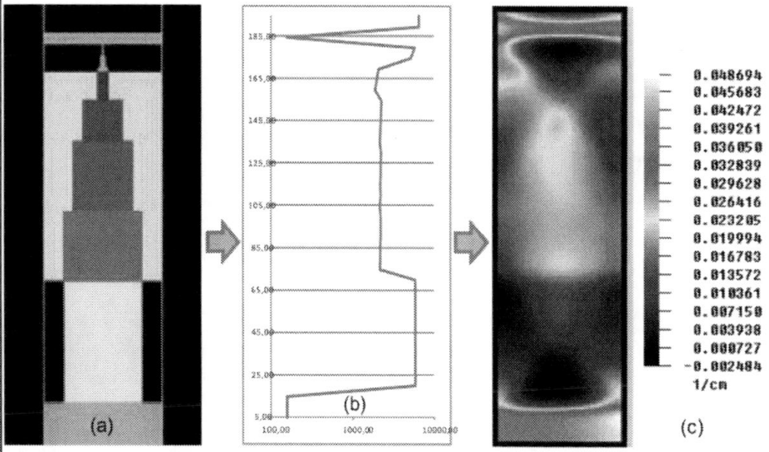

Figure 6 : Column models: trayed (a), packed (b)

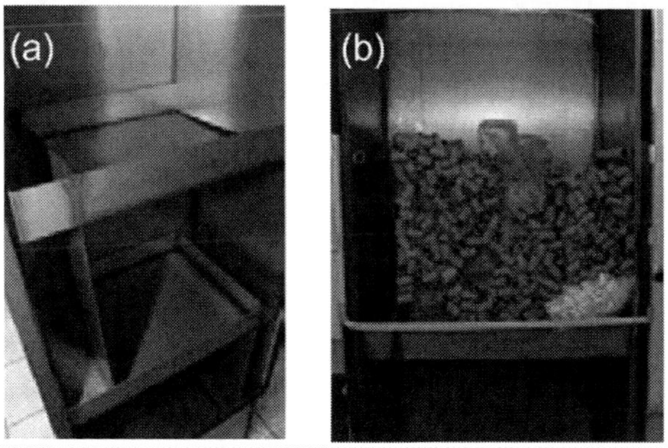

Experimental data were also used to fine tune image reconstruction and filtering routines. As a result, a reiteration process could be included in the methodology (Figure 9). The reiteration process uses the traditional gamma scanning troubleshooting experience to analyze the first image obtained, propose a possible output result and obtain a second improved image.

Figure 7 : Features observed on a tomographic image of a trayed column.

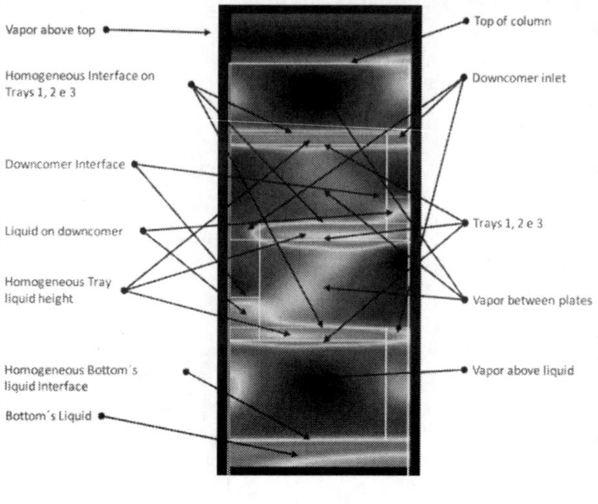

Figure 8 : Features observed on a tomographic image of a packed column.

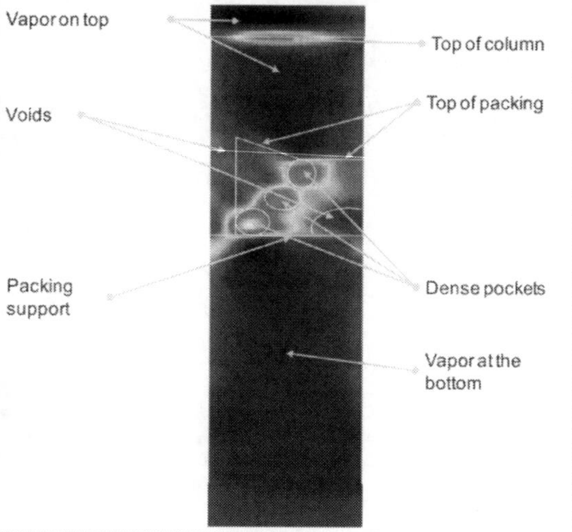

4. Conclusion

Figure 9 : reiteration process

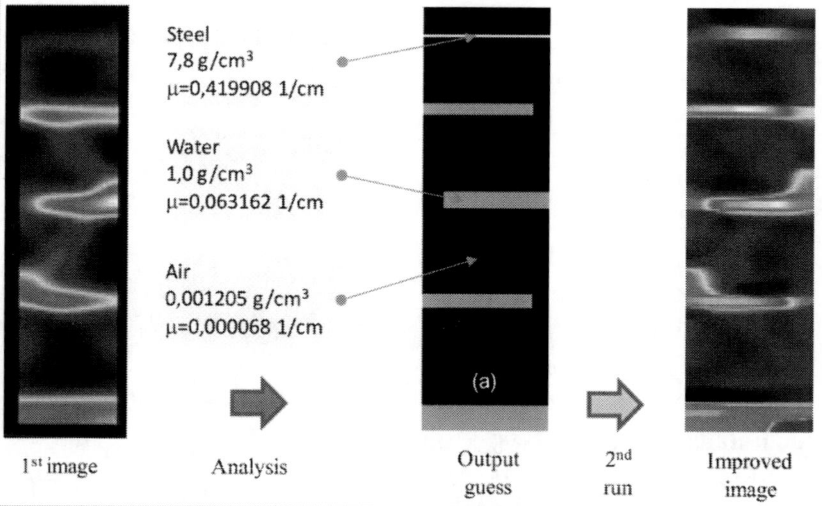

The initial objective of this work was truly achieved with tomographic images of process equipment revealing a two dimensional density distribution. Features hidden on a conventional gamma ray density profile were seem with this technique adding new perspective on troubleshooting process columns and other industrial equipments were a conventional gamma scanning would only offer limited information and possibilities.

Although promising there are some fields that need improvements and further development such as software, equipments, filters, data acquisition, calibration, resolution, accuracy, etc. The simulation tools developed proven to be very useful on fine tuning the results and also planning the irradiation process saving time and personal exposition. The experience gathered indicated also concepts to be employed when adapting the technology from lab scale to real field application, goal of a current PhD degree thesis.

5. References

[1] Charlton, J. S. (1986). Radioisotope techniques for problem solving in industrial process plants. Gulf Pub. Co. (ISBN 978-08-7-201778-8).

[2] Eldridge, R. B. (2006). Advanced hydraulic and mass transfer models for distillation column. Optimization and design. The University of Texas and Oak Ridge National Laboratory.

[3] Haraguchi, M. I. (1999). Online tests for distillation columns, reactors and industrial furnaces. 4th Congress of Equipment and Automation for Chemical Industry, Sao Paulo/SP, Brazil.

[4] Haraguchi, M. I.; Pless, L. T.; Strandmeier, D. (2001). Diagnosis of distillation columns. XVI Meeting of producers and consumers of industrial gases, Salvador/BA, Brazil.

[5] Haraguchi, M. I.; Kim, H. Y.; Machado, R. O. (2010). Industrial computered tomography. XXVIII – National congress of non-destructive testing and inspection, Santos/SP, Brazil.

[6] IAEA (2005). IAEA-TECDOC-1459 - Technical Data on Nucleonic Gauges.

[7] IAEA (2008). IAEA-TECDOC-1589 - Industrial Process Gamma Tomography. Final report of a coordinated research project, October 3, 2008. (ISBN 978-92-0-104508-9).

[8] Johansen, G. A.; Jackson, P. (2004). Radioisotope gauges for industrial process measurements. John Wiley& Sons, Ltd. (ISBN 0471-48999-9).

[9] Kim, J.; Jung, S. (2006). A study on industrial gamma ray CT with a single source-detector pair. KAERI. Nuclear Engineering and Technology, Vol.38, No.4.

[10] Peyton, A. J.; Beck, M. S.; Borges, A. R. et al. (1999). Development of Electromagnetic Tomography (EMT) for Industrial Applications. Part 1: Sensor Design and Instrumentation. 1st World Congress on Industrial Process Tomography, Buxton, Greater Manchester, April 14-17, 1999.

[11] Pless, L.; Xu, S. (2001). Flooding phenomenon in distillation columns and its diagnosis. PETROBRAS Distillation Seminar, Belo Horizonte, Brazil, August 10, 2001.

[12] Sanches, M. P.; Haraguchi, M. I.; Beckmann, F. S.; Calvo, W. A. P. (2007). Radiological safety in the gamma scans procedures. INAC 2007, Santos, Brazil.

[13] Urbanski, N. F.; Resetarits, M. R.; Shakur, M. S. M. et al. (1999). Gamma Scanning a Column Containing Closely Spaced Trays. Annual Meeting AIChE 1999 - Separations Topical Conference. Session #T1007 - Distillation Hardware and Applications II. Dallas, Texas, USA, November 4, 1999.

[14] Wiens, T.; Behrens, P. (2009). Turbulent Flow Sensing Using Acoustic Tomography. Inter Noise 2009, Otawa, Canada.

[15] Xu, S.X.; Kennedy, G.; Conforti, C. et al. (1999). Troubleshooting Industrial Packed Columns by Gamma-Ray Tomography. CE Expo'99, Houston, July 10, 1999.

[16] Xu, S. X. (1999). Gamma CAT-Scans for distillation columns - from theory to practice. 4th Topical Conference on Separation Science and Technology, AIChE Annual Meeting, Dallas, November 4, 1999.

[17] Xu, S. X. (2000). Quantitatively measure and assess misdistribution in industrial packed towers. CE Expo'2000, June 7-8, 2000.

445225 Integration of Divided Wall Columns in a Shale Oil Processing Unit

Abstract:

Use of Divided Wall Column (DWC) in petroleum processing has significantly increased in the past decade. Savings of 20-30% in capital and energy have been reported. This paper discusses the application of DWC in a Shale Oil Processing unit. Impact of utility constraints on the economics for application of DWC is discussed. The paper will also discuss how Pinch Technology and Process Integration techniques can be used to integrate the DWC with the rest of the process and the insights that can be developed to optimize the process configuration.

443937 Dividing-Wall Column Screening Guidelines and Applications

Abstract:

A dividing-wall column (DWC) is a distillation column with a vertical partition that can be implemented as a potentially attractive alternative to conventional multi-column or sidecut column arrangements; use of a DWC may enable capital cost and energy savings of around 30% or even higher to be achieved. Rules-of-thumb involving process conditions like feed composition and relative volatility are often cited for consideration when trying to determine if a DWC might make sense (or not) in a given flowscheme. The intent of this work is to present and review these guidelines in the context of refining and petrochemical processes and obtain an updated perspective that will allow for more detailed guidance to be provided for future screening evaluations.

444354 New Superstructure-Based Optimization for Process Intensification

Abstract:

Process intensification enables "multi-tasking" by combining reaction, separation and other operations in a single unit, and leads to a substantially smaller, cleaner, safer, and more energy-efficient technology [1-3]. However, there is a lack of a systematic method for the identification of intensification opportunities and the optimal synthesis of intensified processes at the flowsheet level. Traditional process synthesis and integration methods have focused on classical unit operations, where units are integrated but they are not merged to perform multiple operations *in situ*. Though research on process intensification has brought about many promising technologies including reactive distillation, divided wall columns, membrane reactors and task integrated columns, current methods are often suboptimal and consider only a few of all plausible alternatives. While recent phenomena-based design methods [4-6] incorporate process intensification, they are difficult to solve and are based on heuristics and/or decomposition approaches leading to suboptimal solution. Instead of considering classical unit-operation based superstructure representation of a flowsheet, we propose a new flowsheet based on building blocks in which each block represents a unit use of materials performing a specific function (reaction, separation, storage, etc.). An assembly of blocks of the same material would obtain a classical unit operation, while blocks with different materials or functionalities would result in intensified unit operation. This allows a systematic identification, representation and generation of intensification alternatives at the flowsheet level without exhaustive enumeration. We pose the optimal process intensification problem as a mixed-integer nonlinear optimization (MINLP) problem which aims to synthesize a process with intensified units by minimizing or maximizing a process metric given the feed and product specifications, feed and product prices, material properties and bounds on flow rates. In this talk, we will present the overall process intensification method with specific applications.

References

[1] Reay, D.; Ramshaw, C.; Harvey, A. Process intensification: engineering for efficiency, sustainability and flexibility. Waltham, MA: Butterworth-Heinemann, 2013.

[2] Stankiwicz, A.I. , Moulijn, J.A.,. Process intensification: transforming chemical engineering. Chem. Eng. Process, January, 2000, 22-34.

[3] Baldea, M. From process integration to process intensification. *Computers & Chemical Engineering* 2015, Doi.10.1016/j.compchemeng.2015.03.011.

[4] Arizmendi-Sanchez, J. A., and Sharratt, P. Phenomena-based modularisation of chemical process models to approach intensive options. *Chemical Engineering Journal* 2008,135, 83-94.

[5] Lutze P, Babi DK, Woodley JM, Gani R. Phenomena based methodology forprocess synthesis incorporating process intensification. *Ind Eng Chem Res* 2013, 52(22):7127–44.

[6] Babi, D. K., Holtbruegge, J., Lutze, P., G_orak, A., Woodley, J. M., and Gani, R. Sustainable process Synthesis–intensification. *Comput. Chem. Eng.* 2015, .doi:10.1016/j. compchemeng.2015.04.030.

443166 Novel Dividing Wall Columns for Ternary and Quaternary Distillation

Abstract:

Dividing wall columns for fully thermally coupled distillation can provide significant cost benefits. For many applications, they are known to offer savings on capital and operating costs by as much as 30% compared to conventional sequences using direct/indirect splits. However, during operation, these benefits may not necessarily be realized because the vapor flows in the different sections of the DWCs are not conveniently controlled by external means. To solve the issue of the unregulated vapor split at the bottom of a vertical partition, we propose new, alternate DWCs for any thermally coupled configuration. The vapor flow in each section of the new DWCs can be regulated by means external to the column, thereby making optimal/near-optimal operation, a practical possibility. We will illustrate our method by presenting novel DWCs for 3-component and 4-component distillations.

443932 Bridging the Know-How Gap - Hydraulic Design of Packed Dividing Wall Columns

Abstract:

By virtue of its design a Dividing Wall Column (DWC) minimizes energy and equipment as well as plot area requirements of three and more component separations by distillation, which makes it to the most sustainable among distillation technologies. Although proven and well established in practice, and highly appealing regarding energy and capital saving potential, the number of industrial applications of DWCs is still limited. Indeed, as elaborated in greater details elsewhere, e.g. Jansen et al., Chemical Engineering, August 2014, 40-48, a DWC is an atypical distillation column and there are certain process and mechanical design and operation-related concerns and potential constraints that may discourage inexperienced practitioners and make them reluctant to consider design and implementation of a DWC in prospective applications.

In the meantime, many process engineers have developed skills and are capable of using available tools within commercial software packages to simulate performance, i.e. to establish reflux and stage requirements of a DWC. However, what is not readily available in commercial software packages is a proven DWC dimensioning method. Indeed, a reliable hydraulic design method is needed to translate performance simulation results into diameter(s) and height of the shell of a new DWC, or to examine appropriateness of an existing column shell in a retrofit case, including detailed internal layout and proper dimensions of normal and partitioned sections, depending on chosen type and size of structured packing and auxiliary equipment. This must be done with certain rigor if one wishes to arrive at a reliable basis for evaluation of cost-effectiveness of both new DWC designs and retrofits.

To provide basis for filling the existing know-how gap, the present paper addresses all relevant aspects of packed DWC dimensioning, using engineering tools available in public domain. Emphasis is on equipment choice and a predictive model that enables accurate estimation and proper balancing of the pressure drop in parallel sections. The latter being the key to successful design and operation of DWCs. Capability of quantification in this respect opens the door for exploration of possibilities for expansion of the application window of DWCs, both as a new design to enhance potential benefits and as an energy saving retrofit option.

Dividing Wall Column – Design for a Retrofit

Steven G. Schon, P.E.

Engineering Fellow

Arkema Inc.

AIChE Spring 2016
Houston, Texas
April 10 – 14, 2016

Overview

- **Existing products produced for many years at existing plant**
  - Campaign operation (2 products)
  - Derived from impure olefins; key impurities are isomers and heavy homologs
- **Vacuum distillation train used trayed columns**
  - Plant built prior to introduction of high efficiency structured packing
  - HETP of 12" tray spacing < HETP of earlier-generation random packing
  - Small-diameter columns (4 foot)
- **Market opportunity for higher-purity products**
  - New market
  - Higher added-value
  - Need to reduce heavy homologs to < 80 ppm (vs. 2000 ppm for standard grade)
- **Good process simulation model of distillation train**
  - T profiles & purity vs. reflux agree well with plant data

Product Distillation Train

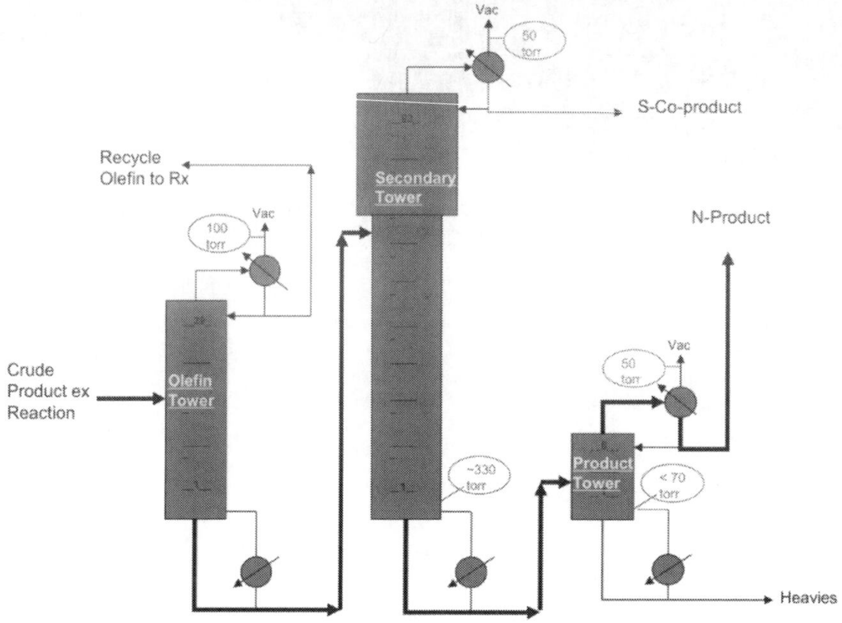

High-Purity Product Requirements

- **Existing 6-tray product tower insufficient for separation**
- **Proposed extending product tower shell and replacing trays with structured packing**
- **BUT, foundation, reboiler & condenser inadequate**
- **Revised proposal: New product tower**
  - Higher cost (new foundation, longer piping)
  - Longer lead time
  - Constructability concerns (congested area)

Conventional Train for High-Purity Product

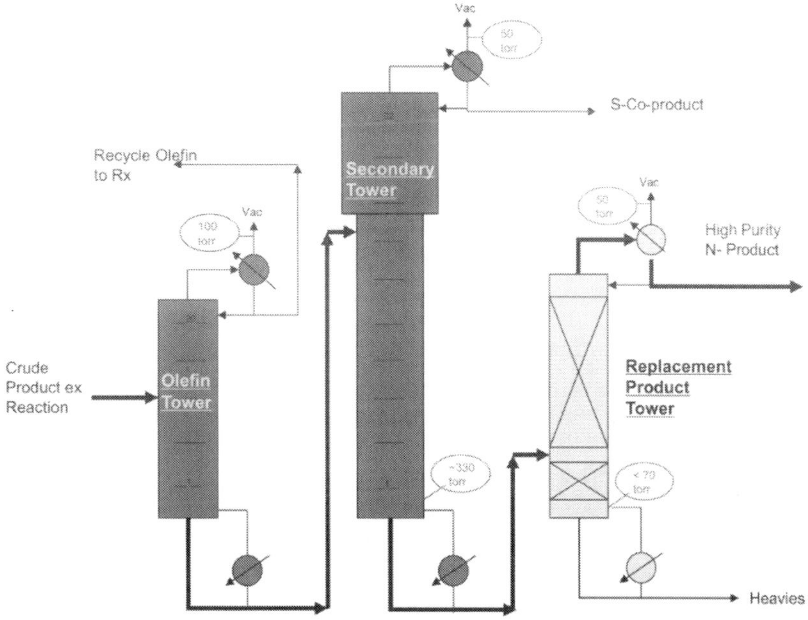

Other Considerations

- **Interest in debottlenecking entire production unit**
- **Olefin tower near flood limit, and reboiler near capacity**
- **Potential thermal degradation in olefin tower bottoms**
- **Consider revamping secondary tower with structured packing**
  - Lower HETP → more theoretical stages / easier separation
    - Sieve trays @ 70 % Efficiency, 12" spacing ~17 " HETP
    - 250 "Plus" style structured packing ~12" HETP
    - Well suited for high V/L → allows higher boil-up w/o vapor flood
  - Less dP → lower bottoms T → increase reboiler capacity; less degradation

Insight & Inspiration

- **Olefin tower has much higher capacity if packing vs. trays**
  - Vapor-flooding limited (vacuum service)
  - Low HETP / more theoretical stages with structured packing ➔ reduce reflux
  - ~ 60 % flood with Plus-style structured packing
- **New product tower vapor rates < ½ that of olefin tower**
  - Equivalent to ~ 30 % of olefin tower flooding capacity
- **Packed olefin tower P profile close to that of new product tower**
- **Dividing Wall Column (DWC) does work of 2 towers in series**
  - If both conventional towers at comparable operating pressure
- **Can olefin tower be converted to DWC ?**
  - Use "excess" hydraulic capacity (with structured packing) for purification-side hydraulics
  - Remove both lights and heavies simultaneously in one column
  - Potentially reduce energy requirement ➔ less boilup required
  - No new product tower ➔ faster, cheaper project

Dividing Wall Column (DWC)

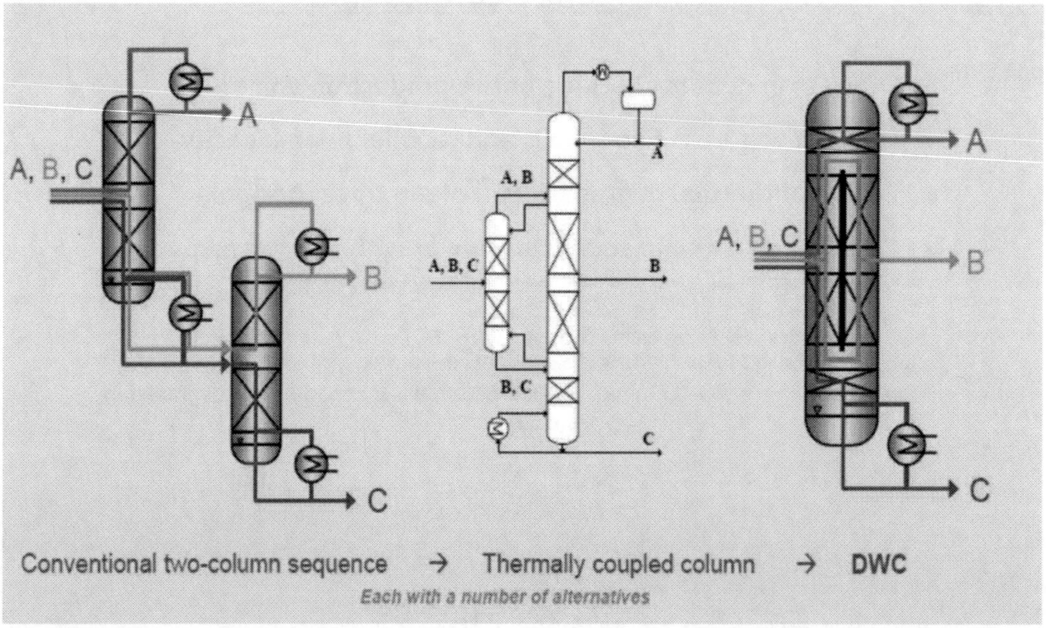

Courtesy of Julius Montz GmbH

DWC Retrofit Concept

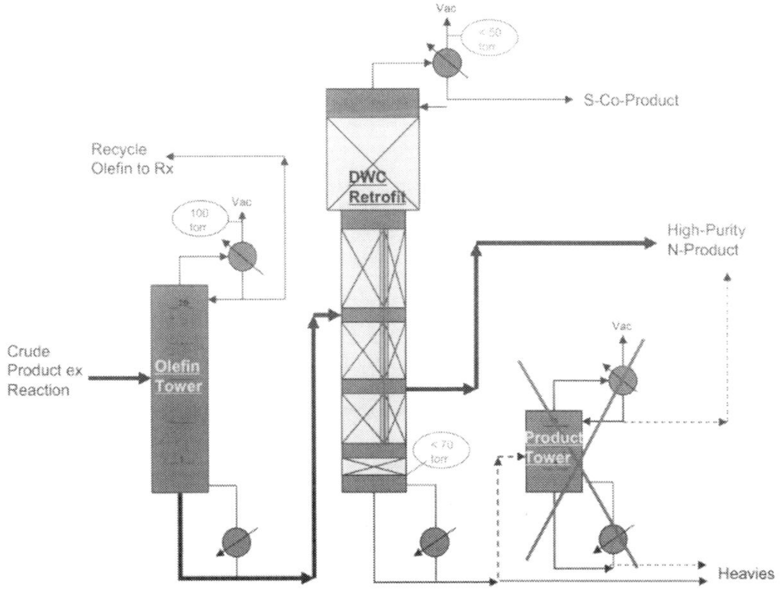

DWC Equivalency / Model

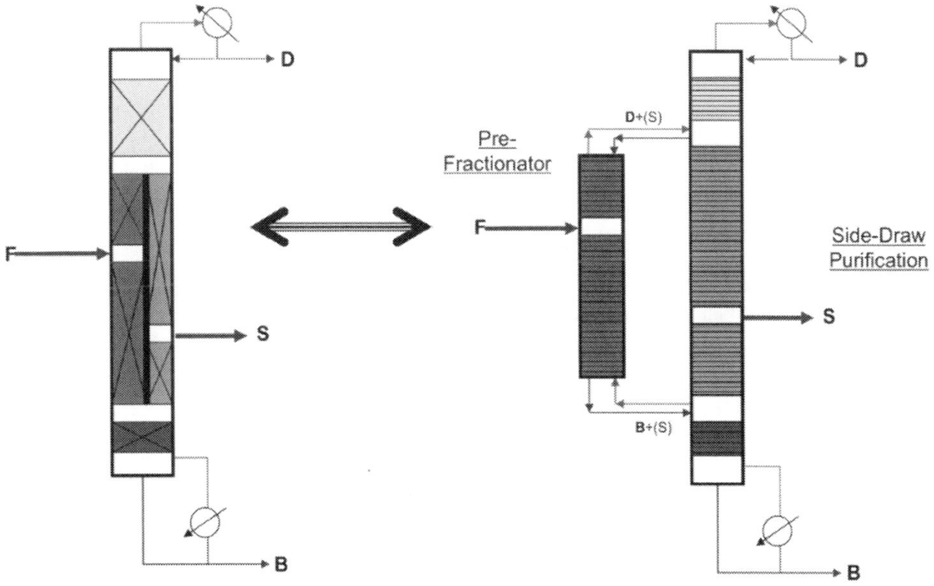

Aspen Simulation of DWC vs. Conventional

Conventional 2-Column Sequence

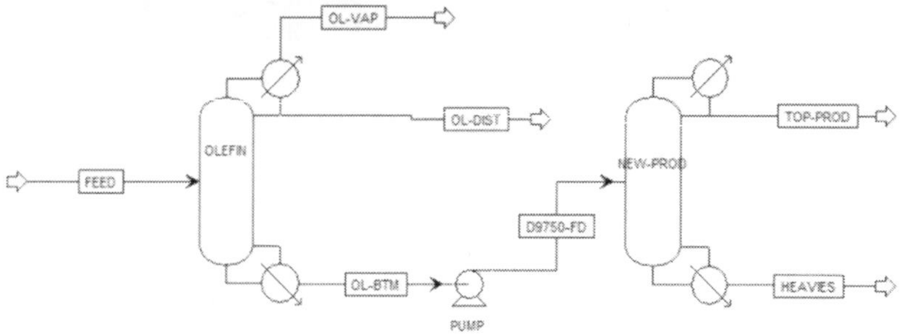

Dividing-Wall Sequence

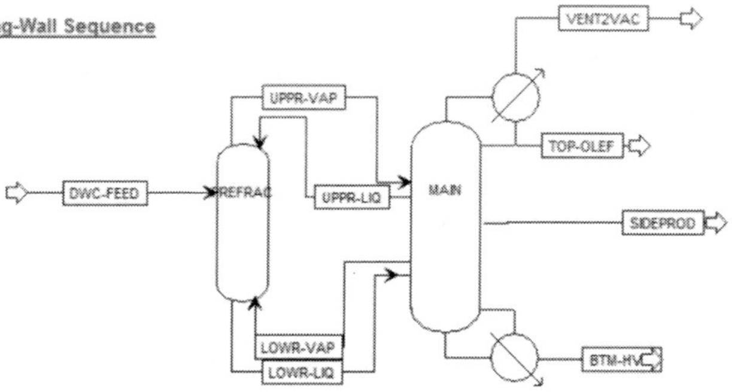

DWC Simulation Approach

- **Start with optimized conventional 2-column simulation**
- **Model DWC as prefractionator coupled to regular column**
  - Main column
    - Total stages = larger of 1st or 2nd conventional column stages
    - Upper connection stage = # stages below 1st conventional column rectifying stages
    - Lower connection stage = # stages above 2nd conventional column stripping stages
    - Side-draw = # 2nd conventional column rectifying below upper connection
  - Prefractionator
    - No condenser or reboiler
    - Same # stages as main column between upper & lower connections
    - Initialize top liquid & bottom vapor feeds based on corresponding conventional 2nd column stage flows & compositions
- **Activate dP calculations** (specify packing, diameter, HETP)
- **Adjust section diameters and prefractionator inlet flows until prefrac & corresponding main column section dPs match**
- **Optimize # stages & feed locations to minimize energy**

DWC Initialization

Conventional 2-Column Sequence

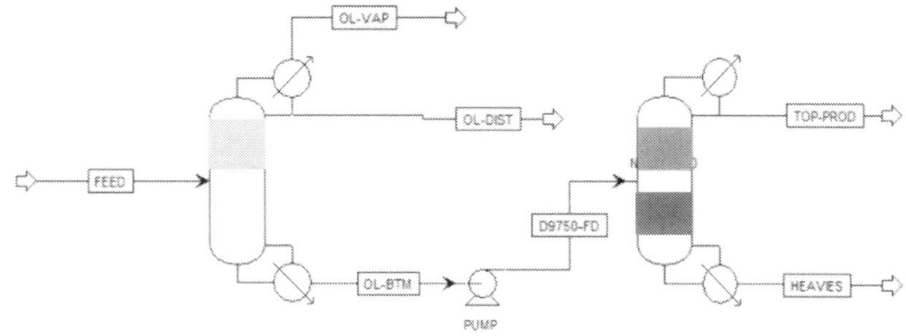

Dividing-Wall Sequence

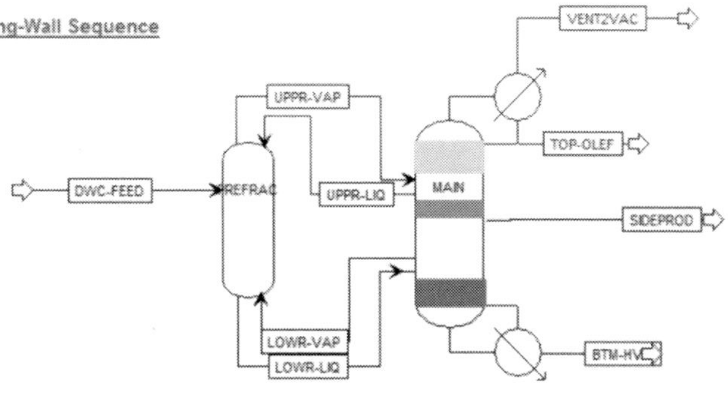

DWC Simulation Results

- **Revamping Olefin Tower as Packed DWC Appears Feasible**
  - Meets High-Purity Specs & Rates
- **73 Theoretical Stages**
  - Accounted for non-packed height "lost" for re-distributors
- **18 % Lower Reboiler Duty Than 2-Column Conventional**
- **5 Beds (3 Divided) of High-Efficiency Structured Packing**
  - Avg. HETP ~ 12" ; with distributors, fits in space of current 92 trays
  - Different packing to match section HETP & dP needs
  - < 80 % Flood in All Sections → potentially higher capacity

DWC Zone Distribution

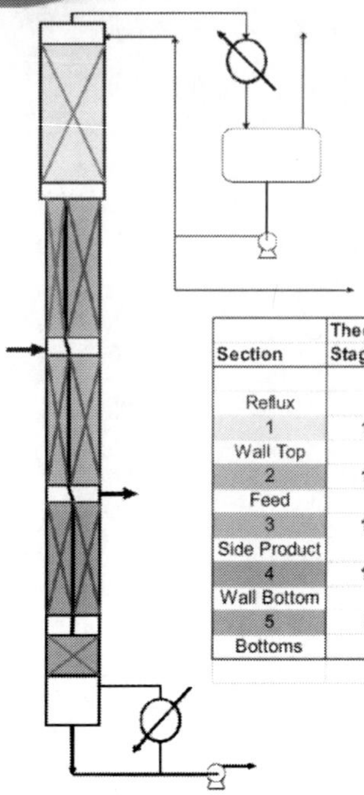

Section	Theor. Stages	Equivalent Diamter (feet) Pre-Frac	Common	Purif.	Overall	Packing (Mellapak Plus) Pre-Frac	Common	Purif.	Height (feet) Packing	Redistirb
Reflux										
1	18		4.5		4.5		452Y		18	
Wall Top										4
2	19	2.4		3.2	4.0	752Y		252Y	19	
Feed										4
3	16	2.6		3.0	4.0	252Y		252Y	16	
Side Product										4
4	15	2.8		2.8	4.0	252Y		252Y	15	
Wall Bottom										4
5	5		4.0		4.0		452Y		5	
Bottoms										
									89 total	

DWC Main Section Composition Profiles

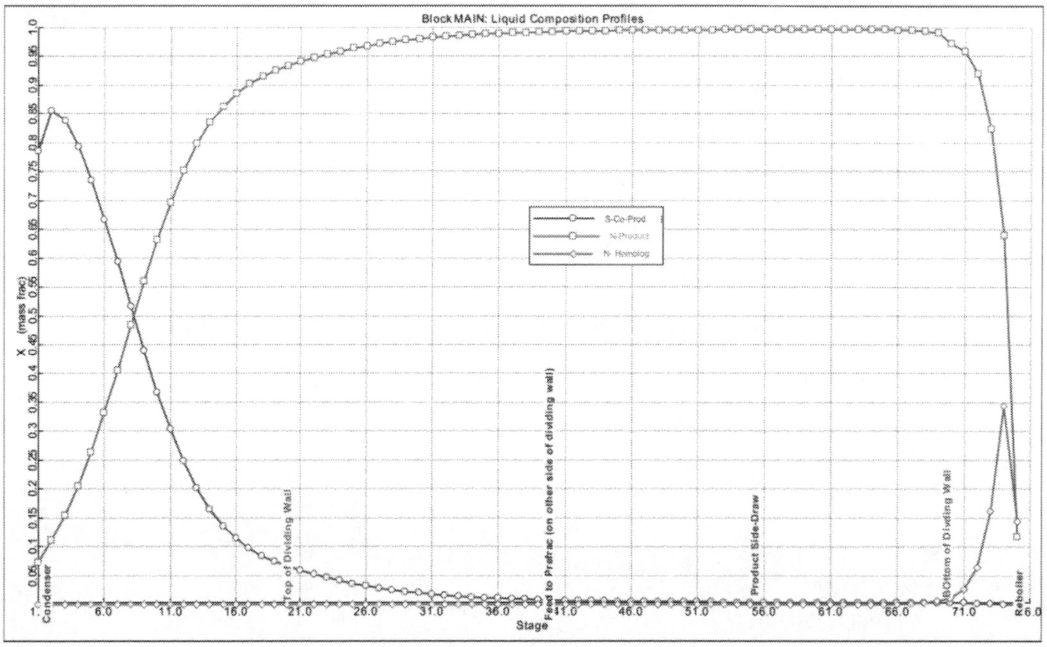

Predicted DWC Column Profiles

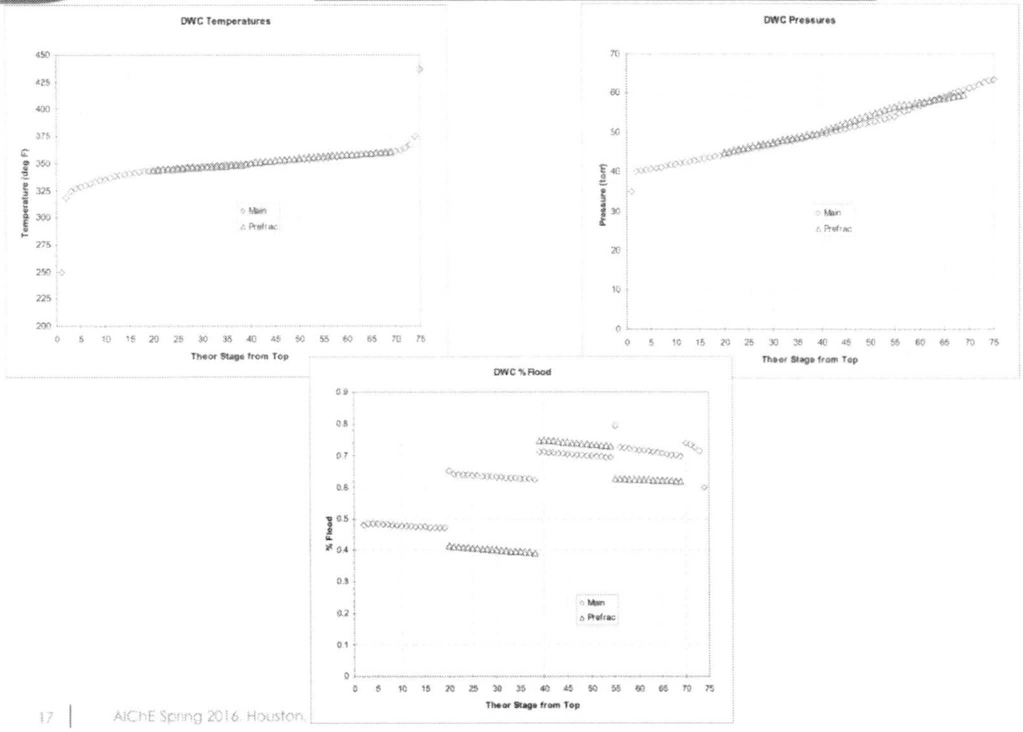

DWC Column Vendors' Assessments

- **Process excellent fit for DWC**
- **Un-equal sections / wall placement in small tower is manageable**
- **Options for DWC design**
  - Welded dividing wall
    - Small diameter requires multiple body flanges
    - Pre-install packing & distributors prior to setting column sections
    - Ensures no cross-wall leakage that can impact purity / efficiency
  - Non-fixed wall w/ wall wipers
    - No body flanges: Can do "replacement-in-kind" shell
    - Field-install wall, packing & distributors through manways after setting
    - Ensures good leveling of internals & liquid distribution
- **Other considerations**
  - New shell vs. retrofit
    - Removal of existing tray support rings labor-intensive
    - Cost
    - Down-time
  - More complicated control scheme
    - Recommend dynamic simulation & automated startup sequence

Preferred Option

- **Replacement shell with non-fixed wall**
  - "In-kind" replacement → easy permit approval
  - Cost comparable to grinding-off old tray support rings
  - Minimal down-time for column vessel replacement (setting & tie-ins)
  - Make use of existing infrastructure (foundations, piping, etc.)
  - Ensuring good liquid distribution outweighs cross-wall leakage concerns
  - No body flanges to leak (high T & vacuum service)
- **Additional studies to determine debottlenecking potential**
  - Column hydraulic limitations
  - Reboiler & Condenser capacities
  - Dynamic simulations by vendor / university partner
- **Keep old product column in place for other products; allow DWC to operate as (debottlenecked) conventional olefin tower if desired**
  - Option to feed both sides of wall & shut-off side-draw
  - Allay plant concerns (If it doesn't work, we won't be able to make current products)

Conclusions

- **Criteria when DWC retrofit may be appropriate**
  - Business driver for upgrading existing distillation train
  - At least 2 columns in conventional sequence at comparable pressure
  - Larger of existing columns has old / inefficient internals
    - Higher hydraulic capacity / more theoretical stages if modernized
  - Opportunities for energy savings
  - Reliable simulation of process
- **Practical considerations**
  - Mechanics / logistics of revamp; is a replacement shell justified?
  - Permit / regulatory implications
  - Additional instrumentation & controls needed: include in scope
  - Tricky hydraulics; leave final design to DWC experts

QUESTIONS ?

440511 Operating Characteristics of Sulzer's High Performance Nexring Random Packing

Abstract:

Sulzer's NeXRing^{TM} random packing is a new random packing that is characterized by a highly open and uniform structure. This low vapor side resistance of the packing structure significantly reduces the pressure drop of the packing, providing an ideal combination of very high capacity and efficiency. The design of this packing will be discussed and comparative results will be presented from Sulzer in-house data as well as from independent research and industrial sources.

443970 Distillation Performance Characteristics of RMP-Type Standard and High Capacity Structured Packings

Abstract:

A number of RVT's structured packings, including both conventional as well as high capacity packings, have recently been tested at the Separations Research Program (SRP). Among these packings include also two modifications of the current high capacity packings.

The packings were tested using the standard cyclohexane/n-heptane system at four operating pressures. Mass transfer efficiency, pressure drop and hydraulic capacity of the new modifications will be presented.

Highlighting specific results, one of these modifications resulted in a 30% lower pressure drop with 15% higher hydraulic capacity while maintaining the same mass transfer efficiency than state of the art high capacity structured packings.

443942 Olefins Column Troubleshooting and Process Improvement Project

Abstract:

An olefins column was having separation problems. A project team was assembled to first diagnose the source of the problem and then to recommend and implement a solution. First step taken by the project team was to use gamma scans to study the hydraulic performance of the column.

The column was configured with a top section of structured packing and a bottom section of one-pass trays. Initially it was not known what or where the potential problem was so both sections of the column were scanned at "normal" process conditions.

The gamma scan results, including tomography scans, indicated that the packed section was operating with some degree of liquid maldistribution while the trays appeared to be operating satisfactorily. Now the debate moved to whether the flow maldistribution through the packed bed was liquid (liquid distributor) related or vapor related.

The project team devised a novel approach to plant test whether the packing maldistribution was liquid related or vapor related. Gamma scans were used to ascertain the effects on the maldistribution from the plant test. The gamma scan results confirmed to the project team what then needed to be revamped in order to mitigate the column's separation problem. Some internal hardware was re-designed and later installed after which the column began to perform its intended separation.

This presentation will describe the plant test procedure and the thought processes based on the scan results, other process data and hardware design criteria. The process improvement plan that was designed and implemented will be discussed.

443175 Effect of Liquid Distribution on Structured Packing Performances

Abstract:

In the chemical processing industries (CPI), structured packings are widely used for distillation and absorption columns. To have the desired mass transfer performances, it is very crucial to distribute the liquid uniformly across the top of the packed bed with structured packings. The effect of liquid maldistribution on the separation performance of structured packing was studied. The tests were conducted in a 1.2 m diameter column using Mellapak 250.Y with the cyclohexane/n-heptane system at atmospheric pressure. Main test results from this study will be summarized and discussed.

It is also very important to have adequate drip points across the column cross section. A theoretical study on the required pour point density of a liquid distributor in a structured packing column was conducted. Based on the study, this paper presents a simple method of estimating the required pour point density for a packed bed with structured packings. The effect of structured packing parameters, such as specific surface area, crimple angle etc., on the required pour point densities will be discussed in this paper.

Key Words: Structured packing, packed column, liquid distributor, liquid distribution

*Corresponding author. E-mail: cai@fri.org , Phone: 1-405-385-0354, Fax: 1-405-385-0357

441009 A Potential Application of Ceramic Foam Material to Distillation: Structured Corrugation Foam Packing

Abstract:

This paper describes the application of SiC ceramic foam to distillation. SiC ceramic foam structured packing was developed according to the structured of corrugated packing. The hydrodynamic performance parameters including pressure drop for dry and wet packing, flooding F factor, and liquid holdup were determined in a plexiglass tower of 100 mm internal diameter. The mass transfer efficiency was measured in another glass tower of 100 mm internal diameter by total reflux experiments, using a mixture of n-heptane and cyclohexane at atmospheric pressure. The experimental results show SiC ceramic foam structured packing has low dry and wet pressure drop, high flooding F factor, liquid hold-up and mass transfer efficiency, when it compared with monolithic foam packing(same material but different structure) and BX gauze packing (same structure but different material). Besides, the influence of specific area, corrugated dip angle, foam diameter, corrugated thickness on SiC ceramic foam structured packing were also researched in order to optimize this new kind of packing and laid the foundation for industrial application.

444197 Distillation Tray Operation

Abstract:

Distillation trays come in different types. Examples include sieve trays, valve trays, bubble cap trays, baffle trays, and dual flow trays. Of these, sieve and valve trays are the most common.

The understanding of the hydraulic and mass transfer characteristics of trays is important for successful design, operation, and troubleshooting of trayed distillation columns.

This paper provides an overview of distillation tray design and operation with particular emphasis on the hydraulic phenomena observed in trayed columns. The influence of tray geometry on the performance of trays will be discussed.

Photos and videos from various FRI tray tests will be shown.

444200 Distillation Packing Operation

Abstract:

Distillation packings are widely used in industrial distillation columns, particularly for low-pressure-drop applications. Distillation packings are generally classified as random packings, structured packings, and grids.

This paper presents an overview of the operation of packed distillation columns, including the hydraulic and mass transfer characteristics of packed columns. The influence of packing geometry on the performance of packed distillation columns will be discussed.

Photos and videos from various FRI tests will be shown.

443684 Acid Gas Absorption Fundamentals

Abstract:

Absorption is a standard unit operation that has been studied by chemical engineering students for decades. One of its primary uses is for the removal of acid gases, like CO_2 and H_2S, from natural gas streams. Historically, aqueous amines and hot carbonate solutions have been used to perform these absorptions. Many different amines, solution strengths, activators and passivators have been tested and employed industrially. Approximately half of the industrial towers are trayed; half, packed. Foaming adversely affects the hydraulic performance in these systems.

Acid gas absorption is receiving appreciable new attention as concerns rise regarding CO_2 in the atmosphere. In addition, the United States has recently experienced significant growth opportunities in shale gas production which require improvements in gas treating technologies. This presentation reviews acid gas absorption and regeneration fundamentals. Several PFD's will be shown to illustrate important concepts. New absorbent solutions will be discussed, including ionic liquids. Flue gas applications will be discussed, including the difficulties associated with column pressure drops and very large column diameters. Rate-based modeling approaches will also be reviewed.

443442 Introduction to Rate Based Mass Transfer Models

Abstract:

This presentation is aimed at providing insights into rate based mass transfer simulations, and how they are different from an equilibrium (ideal stage) model simulations.

Rate based mass transfer models are built on five key elements –

- Mass and energy balances around individual phases on a tray or in a packed segment
- Thermodynamic phase equilibrium

- Equilibrium across interfaces
- Kinetic effects on mass transfer rates
- Mass and heat transfer rate models for transport across interfaces.

In addition, the five key elements have interwoven relationships with chemical, physical, phase and transport properties based on the chemical species being used and the actual mass transfer hardware. In sharp contrast, an equilibrium stage model uses only mass and energy balances around an entire ideal stage plus thermodynamic phase equilibrium and use empirical correction factors to account for efficiencies and mass transfer hardware. For applications such as absorption, where the mass transfer efficiencies are poor (5-25%) equilibrium stage based models prove to be grossly non-predictive, while rate based models are mechanistic, and predictive. This presentation will focus on the characteristics of true mass transfer rate based models, contrast them with the ideal stages and introduce briefly the ProTreat ® process simulator.

443535 Process Investigations Using Radioisotope Techniques

Abstract:

In all chemical plants and refineries there generally is not enough instrumentation in the right places in our processes. During the design phase, decisions were made between knowing more about the process, which means buying more instruments, and saving capital funds. When a problem appears with the operation of a vessel, there are always gaps in the knowledge of what is happening due to a lack of instruments in the places where knowledge is not routinely required. Process engineers and operations personnel need to know what is happening so they can take corrective action. When knowledge is missing, they often take actions that make matters worse.

For distillation and separation processes Chemical Engineers have historically relied upon plant process measurements such as flow rates, temperatures, and pressures and model or simulation results for data with which to solve troubleshooting or re-design problems. An array of on-line diagnostic services are available that can provide additional data offering real-time information on how pieces of process equipment are actually operating.

The most common application is Gamma Scanning. This test is primarily applied, but not necessarily limited to, distillation or separation columns. Gamma scans provide a density profile of the internal process of operating distillation columns and other process vessels. The density profile can be used to diagnose the hydraulic operating conditions of mass transfer devices such as damage to internals, flooding, degree of entrainment or weeping, liquid levels on trays and distributors, liquid distribution through packed beds, etc. The presenter will show case studies where scanning revealed vital process information that helped solve an operating problem, or helped make a revamp/re-design successful.

Additional derivatives from gamma scanning of detecting flow distribution patterns will be demonstrated. The ThruVision Scan technique is a specialized horizontal gamma scan used to generate a topographic profile of the internal cross-sectional density of process equipment. This profile is useful for the detailed study of liquid flow distribution through packed columns as well as having applications beyond distillation columns. A hybrid application involving tracers and scanning on fixed bed catalytic reactors will also be discussed.

Another technique involving radioactive materials is the Tracer Test, in which a radioactive tracer is introduced into the process for investigating the process flows. External radiation detectors are placed at strategically chosen locations outside vessels and piping to monitor the tracer flow through the process. Either liquid, gas or a solid tracer can be utilized depending on process conditions and the fluid to be studied. This tracer technique can identify liquid or gas maldistribution in fixed or fluidized bed reactors. The residence time through a tank or reactor can be measured and compared with the calculated residence time. Observed differences can relate directly to the degree of mixing or channeling inside the reactor. These tracer techniques have also been used in identifying leakage of heat exchangers and valves.

16AIChE
Spring Meeting, Houston, TX

Evaluation of the Axial Mixing and of the Mass Transfer Coefficients in the Distillation Column

František J. Rejl
University of Chemistry and Technology, Prague
Technická 3, Praha 6, 166 28, Czech Republic
rejlf@vscht.cz

Lukáš Valenz, Jan Haidl
University of Chemistry and Technology, Prague
Technická 3, Praha 6, 166 28, Czech Republic
valenzl@vscht.cz, haidl@vscht.cz

Prepared for Presentation at
American Institute of Chemical Engineers
2016 Spring Meeting
11th Global Congress on Process Safety
Houston, Texas
April 10-14, 2016

UNPUBLISHED

Evaluation of the Axial Mixing and of the Mass Transfer Coefficients in the Distillation Column

František J. Rejl
University of Chemistry and Technology, Prague
Technická 3, Praha 6, 166 28, Czech Republic
rejlf@vscht.cz

Lukáš Valenz, Jan Haidl
University of Chemistry and Technology, Prague
Technická 3, Praha 6, 166 28, Czech Republic

Keywords: enter a few keywords here to aid searching

Abstract

In our previous paper we assessed success of the recalculation of the mass-transfer data from absorption to distillation conditions when the procedure took into the account the axial mixing of the liquid phase or when the usual assumption of the plug flow was introduced. The absorption data in the form of the volumetric mass-transfer coefficients were measured in the absorption column of inner diameter of 0.29 m using proven aqueous test systems. The liquid phase Bodenstein number quantifying effect of the axial mixing has been measured in the same absorption column with the air – water system. These data were used for calculation of the distillation *HETP*s using BS model to transfer the data from absorption conditions to distillation ones. The comparison was made against the experimental distillation data in the form of *HETP*.

In this contribution we present results of the evaluation of the volumetric mass-transfer coefficients and of the liquid phase Bodenstein number directly from the distillation experiments performed under the total reflux in the 0.15m atmospheric column using methanol/ethanol system.

This procedure enables to compare the raw values of the mass-transfer coefficients and utilization of the *HETP* is avoided. Thus the comparison not only qualitatively assesses success of various recalculations, but provides direct comparison of the mass-transfer coefficients and the liquid phase Bodenstein number for both processes.

1. Introduction

Development of the common mass-transfer model for absorption as well as distillation on the packing is complicated by limited distillation data in the form distinguishing gas and liquid phase mass-transfer resistance and by contemporary inability to determine distillation effective mass-transfer area. The "profile method" [1] has been used for evaluation of k_La and k_Ga under distillation conditions and the results were in reasonable agreement with forecasts of the RBF and BS models [2,3] with exception of the dependences on the phase flows. It was assumed that this discrepancy could be caused by modeling of the distillation column with plug flow of both phases.

The axial mixing of the phases is rarely incorporated in the models of the distillation columns although significant mixing is known to occur in the dispersed (liquid) phase from the absorption experiments [4]. Its introduction revealed improvement in the

When only one phase is a subject of axial mixing than its degree can be evaluated directly from the composition profiles obtained during distillation experiments (for plug flow of the phases the molar fraction profiles of the liquid and of the gas phase are identical under total reflux; when axial mixing is present in any phase the molar fraction of the more volatile component is higher in the vapor phase). The axial mixing, presumed in the liquid phase, can be therefore deduced from the distillation data.

This contribution presents mass-transfer parameters evaluated from the experimental composition profiles when the axial dispersion in the liquid phase is assumed and is simultaneously evaluated.

2. Experimental

Volumetric mass-transfer coefficients of the packings under absorption conditions have been measured using standard aqueous systems (O_2/water – k_La, SO_2/aq. NaOH - k_Ga) for various flows of the gas and liquid phase. Experiments have been performed in the i.d. 0.29m column with four elements of the stainless steel Mellapak 250Y (Sulzer) and with the stainless steel Intalox-25 packing.

Distillation experiments have been performed with MeOH/EtOH systems in the atmospheric distillation column of i.d. 0.15m under total reflux. Composition profiles in the liquid and in the gas phase have been measured and were used for evaluation of k_La, k_Ga and liquid-phase axial mixing (in terms of Bo_L) using profile method.

2. Results

2.1 Evaluated dependences of the mass-transfer parameters on the phase flows

Table 1: Dependence of the Sh*a_{EFF} on the corresponding phase flow for M250Y (methanol/ethanol distillation)

Model/Experiments	$Sh_L{*}a_{EFF}{\sim}Re_L^A$; A=	$Sh_G{*}a_{EFF}{\sim}Re_G^B$; B=
Rocha-Bravo-Fair	0.597	1.20
Billet-Schultes	0.733	1.16
Olujic' (DELFT)	0.445	0.845
Absorption, plug flow	0.552	0.652

Distillation, plug flow	1.12	0.77
Distillation, axial mixing	1.06	0.70

The relation between mass-transfer parameter and a phase flow has been correlated as a Sherwood number as a power function of the Reynolds criterion. The evaluated dependences reveal rather significant differences among individual models, results of absorption experiments and of the distillation experiments.

For the gas phase this dependence seems to be overestimated for by all respected models. There was found quite good agreement between absorption and distillation, especially for the data evaluated with axial mixing in the liquid phase. It seems that the models underestimate the dependence for the liquid phase. Moreover there is a significant difference between absorption and distillation experimental data.

2.2 Contribution of individual phenomena to the resistance against separation

Resistance against separation is here assumed to be formed by mass-transfer resistance in each phase and by longitudinal mixing in each phase. The results for structured packing Mellapak 250Y and for random packing Intalox-25 are summarized in the tables Table 1, Table 2.

For the structured packing one can observe that – according to the expectations – the dominant contribution to the overall resistance is formed by the gas-phase mass transfer resistance, which rises with the load of the column from 40 % to about 60 %. The liquid-side mass-transfer resistance is not negligible as it forms 20 % - 28 % of the overall mass-transfer resistance and decreases slightly with the load. The axial mixing of the liquid phase is found to act with approximately the same effect, as it forms 23 % - 32 % of the overall resistance.

Table 1.

Distribution of the relative resistance against separation for M250Y (methanol/ethanol distillation)

Phenomenon	10kW	20kW	30kW	40kW
Liquid-side mass transfer resistance	26.4%	28.1%	22.6%	19.7%
Gas-side mass transfer resistance	41.5%	48.4%	53.2%	57.4%
Axial mixing of the liquid phase	32.1%	23.4%	24.2%	23.0%

For the random packing the mass-transfer resistance in the gas phase is found to be of even more importance (52 % - 65 %) without significant dependence on the load. The liquid-side mass-transfer resistance is found minor (9 % - 14 %) while the axial mixing of the liquid phase forms 26 % - 34% of the overall resistance and thus is clearly the second most important factor.

Table 2.

Distribution of the relative resistance against separation for Intalox-25 (methanol/ethanol distillation)

Phenomenon	10kW	20kW	30kW	40kW

Liquid-side mass transfer resistance	13.8%	10.2%	14.0%	8.7%
Gas-side mass transfer resistance	58.6%	57.1%	52.0%	65.2%
Axial mixing of the liquid phase	27.6%	32.7%	34.0%	26.1%

2.1 Transfer of the parameters between absorption and distillation

The volumetric mass-transfer coefficients evaluated from absorption and distillation experiments on the identical packing have been expressed in terms of Sherwood criterion and were mutually compared using dependences on Reynolds and Schmidt criteria in two different forms

$$Sh_G = C_1 * f_1(Re_G) * Sc_G^{0.333} \qquad \text{[Eq. 1]}$$

with the power of Schmidt number 0.333 taken from Chilton-Corbun analogy, or

$$Sh_G = C_2 * f_2(Re_G) * Sc_G^{0.61} \qquad \text{[Eq. 2]}$$

with the power of Schmidt number 0.61 taken from recent experiments on wetted-wall column [1]. Assuming the dependence holds for absorption as well as for distillation (and assuming the analogical behavior of the effective mass-transfer area), one should get identical curves

$$Sh_G * a_{EFF} / (Sc_G^{EXP}) = C * a_{EFF} * f(Re_G) \qquad \text{[Eq. 3]}$$

for absorption as well as for distillation if the recalculation according to Eq. 1 or Eq. 2 works.

As can be seen from the Figure 1, the distillation and absorption data in the form given by Eq. 3 agree fairly well for the structured packing. It seems that for higher loads the distillation data evaluated with axial mixing taken into account agree better than those for which the plug flow is assumed. The difference between the distillation data evaluated with power of the Schmidt number 0.333 or 0.61 is marginal and there can be hardly performed any decisive conclusion.

The distillation and absorption data for random packing do not agree well for the lower loads, but the agreement improves significantly for those evaluated with the axial mixing taken into account for higher loads.

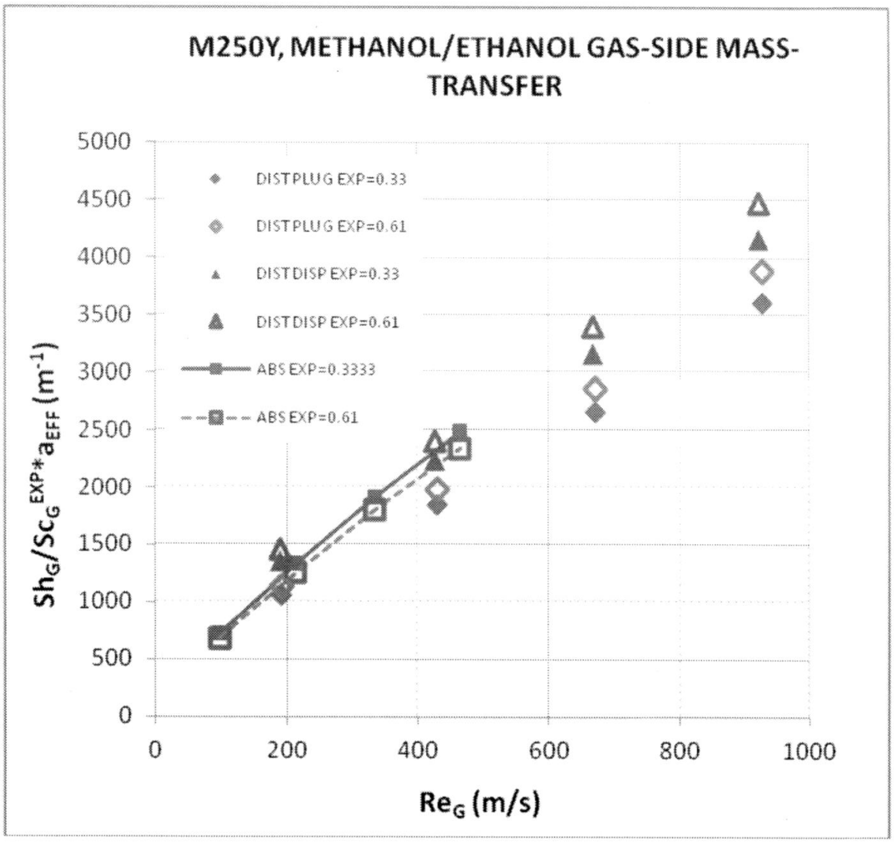
M250Y, METHANOL/ETHANOL GAS-SIDE MASS-TRANSFER
DISTPLUG EXP=0.33
DISTPLUG EXP=0.61
DISTDISP EXP=0.33
DISTDISP EXP=0.61
ABS EXP=0.3333
ABS EXP=0.61
Sh_G/Sc_G^EXP*a_EFF (m^-1)
Re_G (m/s)
0
500
1000
1500
2000
2500
3000
3500
4000
4500
5000
0
200
400
600
800
1000

Figure 1a

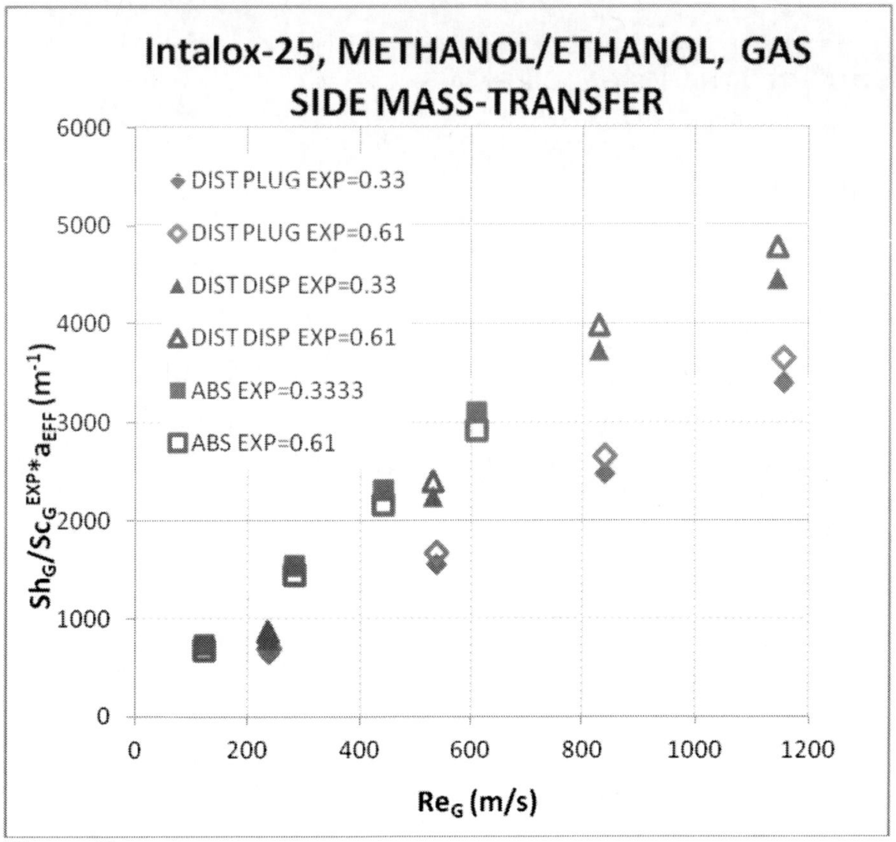

Figure 1b

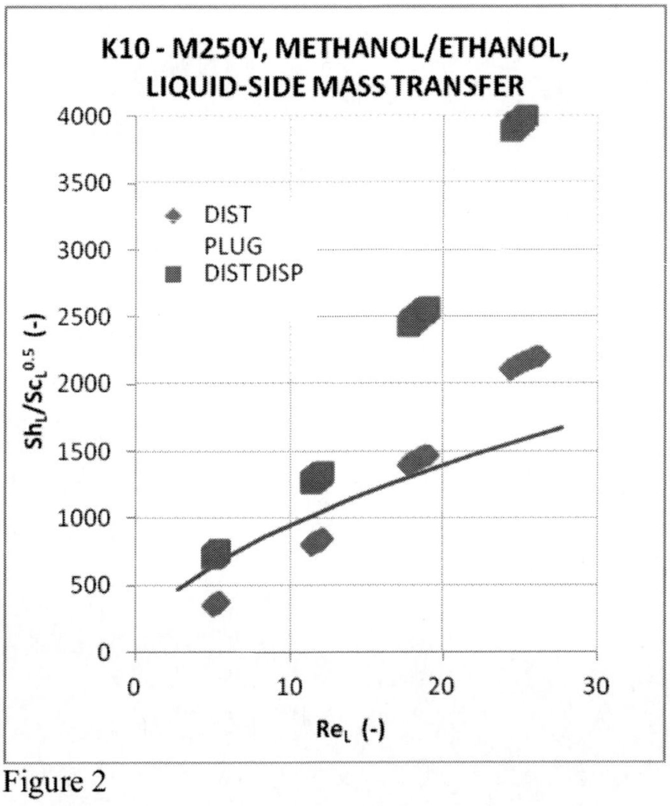

Figure 2

3. Conclusion

Evaluation of the distillation data from composition profiles along the packing reveals that gas-side mass-transfer resistance is, according to general expectations, the most important phenomenon acting against the separation in the column. There was also found, however, that axial mixing of the phases (assumed to occur mostly in the liquid phase) forms about 1/5 - 1/3 of the overall resistance against separation in the distillation column for random as well as for structured packing. The liquid-side mass-transfer resistance is of minor importance for random packing but is significant for structured packing.

The transfer of the mass-transfer data between absorption and distillation is complicated by almost non existing knowledge of effective mass-transfer area under distillation conditions. If we use strong assumption of its equivalence with that for absorption under the same conditions and we use dependence of mass-transfer coefficient on the hydrodynamics and composition through usual dimensionless equations, we get rather good agreement for gas-phase mass-transfer coefficient for structured as well as for random packing. The distillation data evaluated with axial mixing taken into account seem to provide better agreement.

However, the recalculation fails for liquid-side mass transfer, the dependence of Sh_L*a_{EFF} on the Re_L differs significantly for both processes and axial mixing taken into account does not help in this case. As the shapes of the dependences of the Bo_L on the Re_L also do not agree for absorption and distillation, it seems that the hydrodynamics of the liquid phase (and consequently mass-transfer in the liquid phase) differ substantially between both processes.

Acknowledgement

Authors are grateful to the Czech Science Foundation for supporting the project through Grant 13-01251S.

4. References

[1] V.Linek, T. Moucha, E. Prokopová and J.F.Rejl: „Simultaneous determination of vapour- and liquid-side volumetric mass transfer coefficients in distillation column", Chemical Engineering Research and Design, 2005, 83 (A8): 979-986

[2] Rocha, J. A.; Bravo, J. L.; Fair, J. R. Distillation columns containing structured packings: A comprehensive model for their performance. 2. Mass-transfer model. Ind. Eng. Chem. Res. 1996, 35, 1660.

[3] Billet, R. and Schultes, M. (1999), „Prediction of Mass Transfer Columns with Dumped and Arranged Packings. Updates Summary of the Calculation Method of Billet and Schultes.", in Trans IChemE, 77 (Part A), p. 498-504.

[4] L. Valenz, F.J.Rejl and V.Linek: "Gas and Liquid Axial Mixing in the Column Packed with Mellapak 250Y, Pall-Rings 25, and Intalox Saddles 25 under Flow Conditions Prevailing in Distillation Columns." *Ind. Eng. Chem. Res. 2010*: *Vol.* ***49****, Issue:****20****, pp 10016-10025*

443971 Predicting Pressure Drop of Structured Packings at Lowest Operating Pressures

Abstract:

Removal of impurities, consisting of free fatty acids (FFA) and other valuable components as well as contaminants from edible oils occurs on large scale by hot steam stripping under deep vacuum conditions, where, because of a strong deteriorating effect on separation, each Pascal of pressure drop counts. Similarly, subsequent isolation and purification of individual FFA-s and other valuable thermally sensitive components is carried out by rectification at as low as reasonable top of the column pressure. Therefore, modern continuously operating deodorization strippers and FFA distilling plants are mainly equipped with structured packings. Due to their formidable performance characteristics (good wetting and lowest pressure drop per equilibrium stage), high specific geometric area wire gauze structured packings are preferred choice, which, however, because of a rather small hydraulic diameter, may not be adequate when deep vacuum applications are considered. Another interesting aspect is that advanced geometric features of high performance corrugated sheet metal structured packings may not appear beneficial as generally believed in these applications. Due to a relatively large pressure drop (in the range of, or larger than the absolute pressure at the top of the column), operating conditions and consequently both hydraulic and mass transfer performance differ strongly at the top and the bottom of the bed. In order to arrive at an optimized design a tedious iterative procedure is required. Established, purely empirical predictive models, do not account properly for flow peculiarities of the prevailing flow regime. Namely, ascending gas moves under laminar flow conditions. Being based on common applications i.e. operating conditions where turbulent gas flow prevails, empirical models tend to underestimate the pressure drop in deep vacuum applications. One should note that a very low density gas ascends through a packed bed at a rather high effective velocity and tends to follow the shortest, vertically oriented sinusoidal flow path, which length exceeds the bed height, to the extent depending on the effective gas flow angle. The latter does not necessarily coincide with the corrugation inclination angle, and therefore needs to be determined for every type of structured packing from the pressure drop measured under operating conditions. As shown here, the effective gas flow angle may serve as single adjustable parameter in a simple phenomenological pressure drop model, which enables a theoretically founded approach to conceptual (re)design of packed columns for edible oil stripping, FFA fractionation, and similar deep vacuum distillation applications.

443158 Model for Methanol Partitioning in Fractionation Column

Abstract:

Abstract

Model for Methanol Partitioning in Fractionation Column

Rameshwar Hiwale* and Kathy Bigger

Linde Process Plants, Inc., Tulsa, USA

*Corresponding Author:rameshwar.hiwale@lppusa.com

Natural gas hydrates are formed in the presence of moisture, low temperatures and high pressure operations. Hydrate formation is a potentially serious problem in natural gas flow lines starting at the production well all the way through to the customer delivery system. Methanol is an effective hydrate inhibitor exhibiting low viscosity, high water affinity, and relatively low cost, thus making it a versatile solvent. The injection of methanol into pipelines has become common practice in the natural gas industry to prevent hydrate formation and prevent corrosion. The use of methanol has been increasing recently, specifically due to the enactment of environmental restrictions limiting the benzene, toluene, ethylbenzene and xylene (BTEX) emissions of Triethylene Glycol (TEG) Dehydration plants. For these reasons, methanol has been and will frequently be present in feed streams to natural gas liquid (NGL) fractionation plants. However, then total level of oxygenates in NGL streams is limited by specification. Therefore, the accurate prediction of methanol partitioning in a distillation column is critical to ensure that NGL products maintain their maximum value.

Apart from its desirable characteristics, methanol also exhibits an elevated vapor pressure and complex polarity. Thus methanol distribution can become a concern, particularly in fractionation columns such as a Deethanizer where product impurities are of high interest. The accuracy of simulation tools to predict the methanol partition coefficient is critical to maximize return on investment. Existing thermodynamic models are tailored toward systems of high methanol content and do not predict methanol distribution accurately when methanol is present as a low concentration contaminant.

In this study, in the first phase, the model is developed to predict methanol distribution along with column. The accuracy of a vapor-liquid equilibrium (VLE) model for methanol-hydrocarbon mixtures has been improved through modification of the simulation-embedded binary interaction parameters. The modified model accuracy to predict the methanol partition coefficient was confirmed by comparing model predictions for wide range of operating conditions with published VLE data from GPA Research Report 219.

In the second phase, the improved thermodynamic model results will be confirmed by comparing experimental results for methanol concentration in the overhead ethane product stream from Plant in 2016.

443777 Hydrodynamic Behavior and Residence Time Distribution of Winpak in a Concurrent Downflow Packed Column

Abstract:

The hydrodynamic behavior of Winpak was tested in a concurrent downflow packed column with air-water system. Models and empirical equations for predicting pressure drop and dynamic liquid holdup were proposed and compared with experimental results. The examination of Residence Time Distribution (RTD) relied on the pulse method and a conductivity meter which deduceprintd the axial Peclet number, axial dispersion coefficient and dynamic liquid holdup. The relations of dynamic liquid holdup obtained from gravimetric drainprinting experiments and RTD studies were discussed with static and total liquid holdup. Potential impacts of the liquid distributor and conductivity cell were also assessed. The results prove that models and empirical equations fit well with experimental data and are reliable for design and scale-up.

16AIChE
Spring Meeting, Houston, TX

Analogy of Mass Transfer in Absorption and Distillation Columns: Wetted Wall Column Study Distillation

Jan Haidl, MSc
UCT Prague
Prague, Czech Republic
jan.haidl@vscht.cz

Rejl, F.J., Valenz, L., Moucha T.
UCT Prague
Prague, Czech Republic

Schultes, M.
Raschig GmbH
Ludwigshafen, Germany

Prepared for Presentation at
American Institute of Chemical Engineers
2016 Spring Meeting
11th Global Congress on Process Safety
Houston, Texas
April 10-14, 2016

UNPUBLISHED

Analogy of Mass Transfer in Absorption and Distillation Columns: Wetted Wall Column Study Distillation

Jan Haidl, MSc
UCT Prague
Prague, Czech Republic
jan.haidl@vscht.cz

Rejl, F.J.[1], Valenz, L.[1], Moucha, T.[1], Schultes, M.[2]
[1]UCT Prague, Czech Republic, [2]Raschig GmbH, Germany

Keywords: enter a few keywords here to aid searching

Abstract

Rate-base modeling of packed distillation columns represents an advanced approach to design these devices. Compared to the traditional stage-wise approach the rate-based modeling offers more information about the conditions inside the column and enables to the column designer to simulate the behavior of the column if the operation conditions are changed. The possibilities of the rate-based approach are limited by the reliability of mass-transfer parameters, the mass-transfer coefficients. Unlike the absorption conditions, the measurement of the individual mass-transfer coefficients under distillation conditions in the industrial scale columns has never been mastered. Therefore, the values of those parameters for distillation conditions are usually recalculated out of the results of standardized absorption experiments. The possibility of such recalculation, i.e. the analogy of the mass-transfer in absorption and distillation columns, is tacitly assumed. This work is focused on the study of this analogy and the possibility of the recalculation of the mass-transfer coefficients between absorption and distillation conditions in the wetted wall column. The correlations for absorption gas- and liquid-side mass transfer coefficients (k_G and k_L, respectively), measured under conditions close to the distillation ones, were presented on this conference last year. In this paper we bring the comparison of the predicted and experimental values of distillation separation efficiency measured on the methanol-*n*-propanol system in terms of individual mass-transfer coefficients evaluated using the profile method. The distillation k_G values have been found to be 12 % and the k_L values 24 % lower than predicted out of the results of absorption experiments. The influences of the heat-transfer and the axial mixing of the phases on the results are discussed.

1. Introduction

Distillation and absorption in the packed column can be treated as the analogical processes of the mass transfer between the gas/vapor and the liquid flowing on the packing. The assumed analogy implies that the phenomena occurring in both the processes can be described with the same set of models dealing with the hydrodynamic of the phases, phase equilibria and the mass-transfer phenomena. The parameters of those individual models, usually the explicit functions of the process conditions, should not be dependent on the process which they are calculated for.

Numerous models for estimation of the fundamental mass-transfer characteristics (k_L, k_G, a) for given process conditions can be found in the literature. Hydraulic conditions (phase flow rates, liquid hold-up and pressure drop), physical properties and physical state of the phases are usually used as the input parameters for these models. However, the validation of the assumed dependencies of the mass-transfer parameters on the input variables is difficult in distillation because it: (1) requires extrapolation of the results of standard absorption experiments based on aqueous systems to the distillation conditions; (2) it is possible only through the *HETP* quantity as the contemporary results of the distillation mass-transfer data from industrial scale apparatus are provided in this form instead the values of k_L, k_G, and a. The inability of verification of the predictions of the mass-transfer models otherwise than by *HETP* has serious consequences. Predictions of the effective mass-transfer area a under distillation conditions vary substantially. In order to agree with the *HETP* the values of mass-transfer coefficients predicted by individual models must also differ. The present mass-transfer models can predict quite successfully *HETP* as they are developed to do so, but their results in terms of fundamental mass-transfer coefficients differ considerably.

An attempt to remedy this insufficiency of the distillation calibration experiments has been made recently by the introduction of the so called "profile method", which enables the determination of the volumetric mass-transfer coefficients under distillation conditions by the analysis of the concentration profiles shape. Its application on the atmospheric distillation of binary mixtures of primary alcohols provided reasonable results with exception of the substantially higher dependency of k_La on the liquid phase velocity than it is predicted by literature models as well as measured in absorption calibration experiments (Rejl et al., 2010). In the same time it provided somewhat lower dependence of k_Ga on the gas phase velocity. Due to the fact that the profile method provides only the values of volumetric mass-transfer coefficients, it is not possible to decide, whether the surprising dependences on the phase velocities are attributed to mass-transfer coefficients or to the interfacial area. Study of the basics of distillation in the packed column in this phase of the profile method implementation was found to be too complex.

The wetted wall column has been chosen as a suitable intermediate experimental device; in contrary to the packed column the interfacial area is known in this device and the volumetric mass-transfer coefficients can be easily recalculated to the ordinary ones. In the prior work the absorption mass-transfer coefficients of the wetted wall column have been measured. In this paper the distillation experiments were performed in the identical column and the results of those experiments have been interpreted using the profile method.

Such approach provided the same quantities – mass transfer coefficients – for both the processes enabling their direct comparison. It is assumed that under the same process conditions the obtained mass-transfer coefficients should be identical. The results of these comparisons should suggest, whether the absorption and distillation are, in the contemporary way of their description, indeed analogical or whether there exist differences, which are necessary to be taken into account.

2. Literature

2.1 Absorption characteristics of wetted wall column

The absorption characteristics of the wetted wall column used in this work have been measured in the prior work (Haidl et al., 2016). The distillation-like systems and flow rates have been utilized providing the absorption Sh_L and Sh_G data measured in the range of gas and liquid phase Schmidt and Reynolds numbers achieved in distillation conditions.

The absorption k_L values have been measured by desorption of oxygen from primary alcohols – methanol, ethanol and *n*-propanol – at ambient and elevated temperatures. The data measured on all systems and at all temperatures have been correlated with a dimensionless correlation (1). The method of Tyn and Calus, 1975 has been used for the estimation of the diffusion coefficients of oxygen in the liquid; the dependence of the correlation coefficients on the liquid diffusivity model is reported.

$$Sh_L = 0.33 \mathrm{Re}_L^{-0.12} Sc_L^{0.47} \tag{1}$$

The absorption k_G values have been measured with absorption of sulfur dioxide from air, helium and sulfur hexafluoride into the aqueous solution of sodium hydroxide. The data measured with all three carrier gases have been correlated with a dimensionless correlation

$$Sh_G = 0.012 Re_{LG}^{0.90} Sc_G^{0.61}. \tag{2}$$

2.2 Distillation characteristics of wetted wall column

Although the wetted wall column was intensively used for the mass-transfer research in the last century, only a few of papers are dealing with its mass-transfer characteristics under distillation conditions. Several attempts to evaluate the contribution of the liquid-side resistance to the overall mass-transfer resistance, i.e. the relative resistance of the liquid phase have been made. The influence of the liquid side resistance on the distillation performance reported by the individual authors differs considerably.

While Jackson and Ceaglske (1950) and Kayihan et al. (1975) found no measurable liquid-side resistance in the wetted-wall column distillation of alcohol-water systems, the findings of Surowiec and and Furnas (1942) show that the liquid phase mass-transfer resistance is responsive for 30 – 50 % of the overall mass-transfer resistance for ethanol-water distillation.

Johnstone and Pigford (1941) estimated that the relative mass-transfer resistance of the liquid film is lower than 10 % of the overall mass-transfer resistance.

3. Experimental

The glass column of ID 25 mm (geometrical area $a_G = 4/d = 160\ m^2/m^3$) and length of 2.1 m was used. The column was equipped with 7 pairs of holes to enable placing liquid and vapor sampling devices along the column. Each pair consisted of two conical shaped holes positioned on the opposite sides of the column and was rotated by 90 degrees with respect to the previous one. The vertical distance between the sampling points was 30 cm. The sampling points were distributed evenly along the column with the first sampling point placed 15 cm below the column edge. Liquid was fed to the column by means of 8 needles which are inserted in to the liquid distributor body. The needles of the inner diameter of 0.8 mm and the length of 25 mm were aligned tangentially with respect to the column perimeter and inclined by 30 degrees with respect to the column axis. With this installment the liquid flow entering the column has tangential component of the velocity which ensures connection of individual rivulets and a formation of the uniform liquid film at the top of the column. It was visually observed that the rotation of the liquid along the column perimeter is suppressed via the viscous forces before the liquid reaches the first sampling point. On the bottom of the column, there was placed a liquid collector/vapor distributor separating the liquid outlet from the vapor inlet. All the parts of the equipment were made from glass, PVDF, PTFE, Viton, stainless steel and silicone which enabled usage of the various test systems and temperatures up to 120 °C. The column setup is schematically depicted on Figure 1.

The distillation experiments were performed with methanol-n-propanol test mixtures under the atmospheric pressure and the total reflux conditions. The column was operated in the steady state which was determined by the constant reflux flow rate and composition and by the constant temperature of the vapor leaving the column.

The liquid and vapor (as condensate) samples were withdrawn from the column and analyzed using the Atago RX refractometer. The concentration profiles along the column were measured for several reflux flow rates/vapor loads, the minimal reflux flow rate was limited by the stability of the liquid film, the maximal flow rate by the capacity of the column. Three batches of methanol-n-propanol mixture were used. The concentration profiles measured in distillation experiments were evaluated by the profile method (Linek et al., 2005; Rejl et al., 2010).

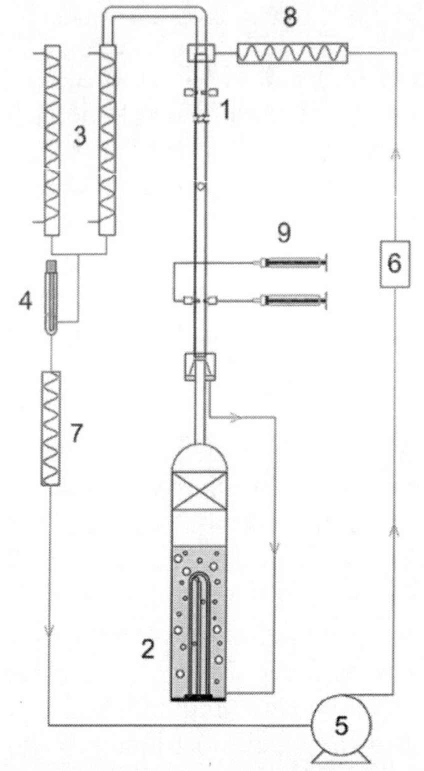

1 - glass column, 2 - electric reboiler with demister, 3 - total condender, 4 - liquid level meter, 5 - gear pump, 6 - liquid flow meter, 7 - reflux cooler, 8 - reflux pre-heater, 9 - syringe pump

Figure 1: The experimental setup for distillation experiments.

4. Results

The concentration profiles measured on the methanol-*n*-propanol system have been evaluated using the profile method. The Sh_L and Sh_G values required by the rate-based model of the distillation column incorporated in the profile method were calculated according to (1) and (2) and corrected by the optimized parameters b_L and b_G. The optimal values of those parameters found for individual reflux flow rates are depicted on Figure 2.

With respect to the standard deviation of individual b_L values no dependency of b_L on Re_L can be deduced and all the values can be described by the weighted mean value of $\bar{b}_L = 0.76$ with standard deviation of 0.17. On the other hand, the b_G values increase slightly with increasing value of Re_{LG}. Based on these results the dimensionless equations 3 and 4 were built for calculation of distillation Sh_L and Sh_G values.

$$Sh_L = 0.25\,Re_L^{-0.12}\,Sc_L^{0.47} \tag{3}$$

$$Sh_G = 2.1 \cdot 10^{-3}\,Re_{LG}^{1.07}\,Sc_G^{0.61} \tag{4}$$

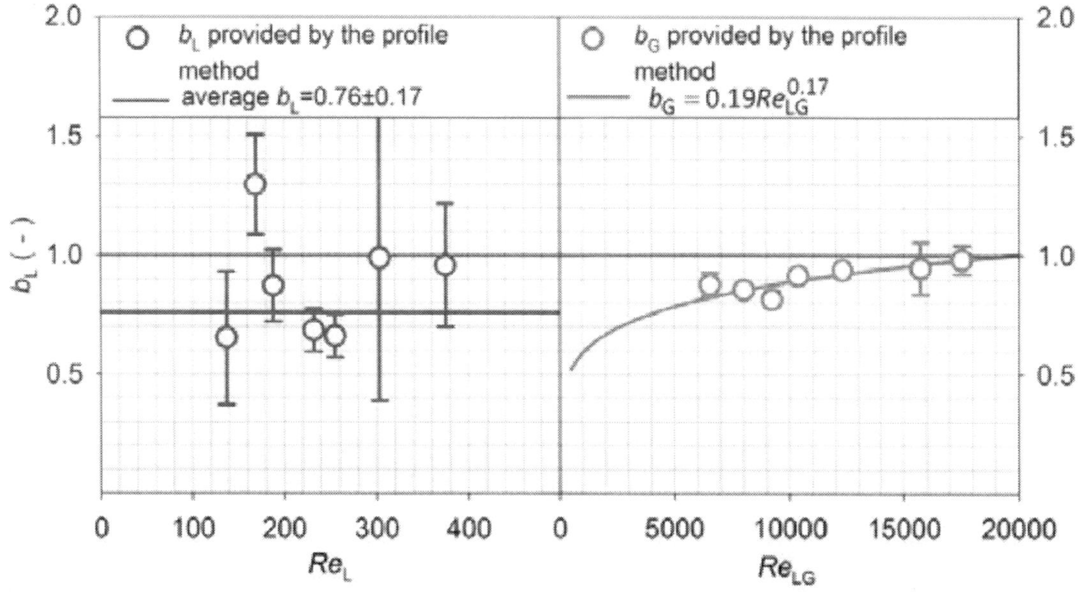

Figure 2: The b_L and b_G values of methanol-*n*-propanol system with their standard deviations.

While the lower values of the distillation Sh_L and Sh_G values are possibly explainable by the poor quality of the diffusivity models, simultaneous occurence of different dependences of absorption and distillation Sh_G on Re_{LG} suggests that there exists some additional difference between the absorption and distillation process not governed by the rate-based model.

The possible influences of two phenomena which could influence differently the mass transfer process in the absorption and the distillation conditions – the axial mixing of the phases and the effect of interfacial heat transfer on the mass-transfer characteristics – were examined.

1) Axial mixing is an undesirable phenomenon as it degrades efficiency of the separation equipment through a reduction of the mass-transfer driving force, and in the limit, a column behaves like a single equilibrium stage (Henley, 2011). In the distillation under the total reflux conditions the axial mixing of phases, if present, would demonstrate itself as the spacing between a liquid and vapor concentration profiles. On the left side of Figure 4 are shown the theoretical concentration profiles in the vapor and the liquid phase as calculated for different levels of the liquid phase axial mixing (characterized by the value of Bodenstein number, Bo_L). On the right side of the figure are depicted the experimental vapor and liquid concentration profiles measured with the methanol-*n*-propanol system. It can be seen, that the experimental profiles mutually overlap indicating no, or only low influence of the axial mixing on the distillation results.

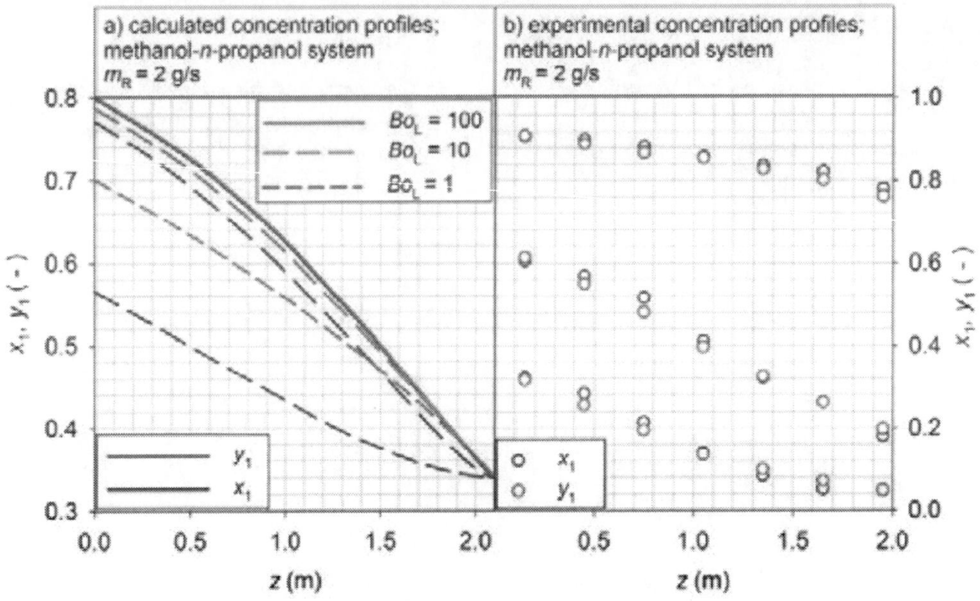

Figure 3: a) The vapor and liquid concentration profiles along the distillation column calculated for different levels of liquid axial mixing. b) The experimental vapor and liquid concentration profiles.

2) Both the phases, vapor and liquid, in the column are assumed to be saturated, i.e. liquid is on its bubble point and vapor on its dew point, respectively. As the temperatures of saturated phases differ considerably, the heat transfer between the vapor and the liquid occurs influencing the intensity of the interfacial mass transfer. To quantify the influence of the heat transfer on the mass-transfer characteristics a simultaneous heat- and mass-transfer model of distillation column (SHMT model) with the heat transfer controlled by the liquid and vapor phase heat-transfer coefficients, α_L, α_G, has been developed. In the model the sub-cooling of the liquid and over-heating of the vapor is enabled while the opposite behavior is suppressed by formation of bubbles in the over-heated liquid and/or mist in subcooled vapor. Thus, in this model three mass- and heat- transfer mechanism are assumed:

(1) the interfacial mass- and heat transfer controlled by k_L, k_G and α_L, α_G, respectively;
(2) the mass- and heat transfer via the bubbles of vapor generated in over-heated liquid;
(3) the mass- and heat transfer via drops of liquid (mist) formed in subcooled vapor.

The model has been applied for evaluation of the concentration profiles measured along the column[1] providing the identical b_G values as found with the model where the heat-transfer phenomena has been neglected. Also, the b_L values mutually agree within their standard deviations, although the values provided by the SHMT model are approx. 30 % higher than found earlier. This result indicates that the heat transfer has no or only marginal influence on the gas-phase mass-transfer characteristics in distillation.

[1] The heat-transfer coefficients have been calculated according to correlations of Dittus and Boelter (1930) for the turbulent vapor and Chun and Seban (1971) for wavy liquid film.

5. Conclusions

In this paper it is presented the unique study of the mass transfer analogy in packed absorption and distillation columns. The wetted wall column, a device with known interfacial area, was built for this purpose enabling to evaluate the mass-transfer coefficients instead the volumetric mass-transfer coefficients usually measured in the packed columns. The mass-transfer characteristics of the wetted wall column have been measured using both the absorption and distillation experiments.

The profile method has been applied to the experimental data measured on the methanol-*n*-propanol system. The values of both the mass-transfer coefficients provided by the profile method have been found to be lower than those predicted from the absorption results. Moreover, the slightly higher dependence of the distillation Sh_G values on Re_{LG} has been revealed, confirming the results obtained earlier on the structured packings. While the lower values of distillation Sh_L and Sh_G could be perhaps explained as the result of uncertainty of the diffusivity data, no such simple explanation can be found for the different phase flow dependence of Sh_G. As the observed difference was found not only for structured packings, but also for the wetted wall column, it can't be accounted to the different formation of the interfacial area. Apparently, there is at least one, phenomenon which affects differently the mass transfer performance of absorption and distillation packed columns. Two such phenomena – axial mixing of the phases and the heat transfer phenomenon – have been discussed showing neither of those phenomena is able to explain observed differences.

The results presented in this work show that the mass-transfer coefficients predicted on the basis of the absorption results agree within their standard error with values evaluated out of the distillation experiments. Based on this finding, it can be concluded that the transfer of the mass-transfer characteristics between the absorption and the distillation is, in general, possible. Certain differences between the absorption and distillation results have been found though. Their further study, however, is not possible until the reliable diffusivity data will be available. The improvement of the diffusivity models is necessary for further development, improvement and use of the rate-based modeling.

Acknowledgement

Authors are grateful to Raschig GmbH for financial support of the project, and to the Czech Science Foundation for supporting the project through Grant 13-01251S.

References

1) Chun, K.R.; Seban, R.A. Heat transfer to evaporating liquid films. *Journal of Heat Transfer* **1971**, 93 (4), 391-396.

2) Dittus, F.W.; Boelter, L.M.K. Heat transfer in automobile radiator of the tubular type. 2nd Ed. *University of California at Berkley Publ. Eng.* **1930**.

3) Haidl, J.; Rejl, F.J.; Valenz, L.; Kordač, M.; Moucha, T.; Labík, L.; Schultes, M. Absorption in wetted-wall column with phase properties close to distillation conditions. Chem. Eng. Sci. **2016**, 144, 126-134.

4) Henley, E.J.; Seader J.D.; Roper, D.K. Separation process principles, 3rd ed. *John Wiley & Sons, Asia* **2011**.

5) Jackson, M.L.; Ceaglske, N.H. Distillation, Vaporization, and Gas Absorption in a Wetted-Wall Column. *Industrial and engineering chemistry* **1950**, 42 (6), 1188-1198.

6) Johnstone, H.F.; Pigford, R.L. Distillation in a wetted-wall column. *Transactions of American Institute of Chemical Engineers* **1942**, 38 (38), 25-51.

7) Kayihan, F.; Sandall, O.C.; Mellichamp, A. Simultaneous Heat and Mass Transfer in Binary Distillation – 2. Experimental. *Chemical Engineering Science* **1976**, 32 , 747-754.

8) Linek, V.; Moucha, T.; Prokopova, E.; Rejl, J.F. Simultaneous determination of vapour- and liquid-side volumetric mass transfer coefficients in distillation column. *Chemical Engineering Research and Design* **2005**, 83 (A8), 979-986.

9) Rejl, F.J.; Valenz, L.; Linek, V. Profile Method for the Measurement of kLa and kVa in Distillation Columns. Validation of Rate-Based Distillation Models Using Concentration Profiles Measured along the Column. *Industrial & Engineering Chemistry Research* **2010**, 49 (9), 4383-4398.

List of symbols

a_G	geometrical area of the column, (m^2/m^3)
b_L, b_G	parameters optimized by the profile method, (-)
d	column diameter, (m)
D	diffusion coefficient, (m^2/s)
g	gravity acceleration, $g = 9.82$ m/s^2
k_G	gas phase mass-transfer coefficient, (m/s)
k_L	liquid phase mass-transfer coefficient, (m/s)
m	mass flow rate, (kg/s)
u_G	gas superficial velocity, (m/s)
x_1	liquid molar fraction of methanol, (-)
y_1	gas/vapor molar fraction of methanol, (-)

Greek symbols

α_L	liquid phase heat-transfer coefficient, ($W/m^2/K$)
α_G	gas phase heat-transfer coefficient, ($W/m^2/K$)
μ	viscosity, (Pa·s)
ρ	density, (kg/m^3)

Subscripts

G gas phase

L liquid phase

Dimensionless numbers

$$Re_L = \frac{4 \cdot \dot{m}_L}{\pi \cdot d \cdot \mu_L}$$ liquid phase (film) Reynolds number

$$Re_{LG} = \frac{d \cdot u_{LG} \cdot \rho_G}{\mu_G}$$ gas phase Reynolds number relative to the liquid surface

$$Sc = \frac{\mu}{\rho \cdot D_{12}}$$ Schmidt number

$$Sh_G = \frac{k_G d}{D_{12,G}}$$ gas phase Sherwood number

$$Sh_L = \frac{k_L}{D_{12,L}} \left(\frac{\mu_L^2}{\rho_L^2 \cdot g} \right)^{1/3}$$ liquid phase (film) Sherwood number

443677 Comparative Analysis of Phase Doppler Interferometry (PDI) Data Obtained in Forward-Scatter and Back-Scatter for Entrainment Applications

Abstract:

Entrainment characterization is critical for a variety of unit operations in the chemical process industry, particularly for distillation and absorption. Liquid entrainment is a key aspect of these applications, and it is usually poorly understood. Therefore, this work utilizes phase doppler interferometry to characterize spray morphology of entrained droplets with particular focus on the back-scatter technique. In addition, this work quantifies the effect of gas saturation on the resulting entrainment. Spray analysis was conducted within a system at controlled conditions using the PDI in forward-scatter and back-scatter configurations. These results were then compared in order to determine the agreement between the two measurement configurations and whether any meaningful correction factor, if necessary, could be developed using comparative data. These bench-scale data are critical because they guide the larger research effort that focuses on obtaining entrainment data at pilot scales using PDI in the back-scatter mode.

AUTHOR INDEX

AUTHOR INDEX

AIChE
120 Wall Street, FL 23
New York, NY 10005-4020

ISBN 978-1-5108-2482-9
9 781510 824829

CURRAN ASSOCIATES INC.
proceedings.com